Biological control of insect pests, plant pathogens and weeds is the only major alternative to the use of pesticides in agriculture and forestry. As in all technologies, there are benefits and risks associated with their utilization.

This book is the outcome of a unique gathering of international experts to discuss and debate the benefits and risks associated with biological control. After intensive discussion it was concluded that we must emphasize the benefits more, while not ignoring the potential risks. The authors address the various techniques and approaches used in biological control, including state-of-the-art reports and economic and risk analyses.

The book will be of interest to researchers and postgraduate students, in biotechnology, agriculture, forestry and environmental sciences.

T0212550

PLANT AND MICROBIAL BIOTECHNOLOGY RESEARCH SERIES: 4
Series Editor: James Lynch

Biological Control: Benefits and Risks

PLANT AND MICROBIAL BIOTECHNOLOGY RESEARCH SERIES
Series Editor: James Lynch

Biological Control: Benefits and Risks

Edited by

Heikki M. T. Hokkanen
University of Helsinki, Finland

and

James M. Lynch
University of Surrey, Guildford, UK

CAMBRIDGE
UNIVERSITY PRESS

PUBLISHED BY THE PRESS SYNDICATE OF THE UNIVERSITY OF CAMBRIDGE
The Pitt Building, Trumpington Street, Cambridge, United Kingdom

CAMBRIDGE UNIVERSITY PRESS
The Edinburgh Building, Cambridge CB2 2RU, UK
40 West 20th Street, New York NY 10011–4211, USA
477 Williamstown Road, Port Melbourne, VIC 3207, Australia
Ruiz de Alarcón 13, 28014 Madrid, Spain
Dock House, The Waterfront, Cape Town 8001, South Africa

http://www.cambridge.org

First published 1995
First paperback edition 2003

A catalogue record for this book is available from the British Library

Library of Congress cataloguing in publication data

Biological control: benefits and risks/edited by H. M. T. Hokkanen
and J.M. Lynch.
  p. cm. – (Plant and microbial biotechnology research series; 4)
Includes bibliographical references and index.
ISBN 0 521 47353 5 (hardback)
1. Agricultural pests – Biological control – Congresses.
2. Biological pest control agents – Congresses.   I. Hokkanen, H. M. T.
II. Lynch, J. M. (James Michael)   III. Series.
SB975.B54   1995
363.7′8—dc20   94-41489 CIP

ISBN 0 521 47353 5 hardback
ISBN 0 521 54405 X paperback

Contents

Part V   Economics and Registration

Contributors

Jean-Paul Aeschlimann
CSIRO/BCU
335, Avenue Abbé Parquel
F-34090 Montpellier
France

Claude Alabouvette
INRA Station de Recherches sur la Flore Pathogène dans le Sol
17, rue Sully
BP 1540
F-21034 Dijon
France

David A. Andow
Department of Entomology
University of Minnesota
1980 Folwell Avenue
219 Hodson Hall
St. Paul, MN 55108
USA

M. J. Bazin
Life Sciences Division
King's College, Campden Hill Road
Kensington, London W8 7AH
UK

Bernd Blossey
Department of Natural Resources
Fernow Hall
Cornell University
Ithaca, NY 14853
USA

Bert Boesten
Department of Microbiology
University College
Cork
Ireland

Raymond J. C. Cannon
Central Science Laboratory
Hatching Green
Harpenden, Herts. AL6 2BD
UK

J. Choi
Life Sciences Division
King's College, Campden Hill Road
Kensington, London W8 7AH
UK

R. James Cook
USDA-ARS
Root Disease and Biological Control
Research Unit, Pullman
Washington 99164-6430
USA

Raymond L. Correll
CSIRO Division of Soils and
CSIRO Biometrics Unit
Private Bag 2
Glen Osmond SA 5064
Australia

Norman E. Crook
Horticulture Research International
Worthing Road, Littlehampton
West Sussex BN17 6LP
UK

J. M. Cullen
CSIRO Division of Entomology
Canberra, ACT 2601
Australia

Geneviève Défago
ETH-Zürich
Phytomedizin, LFW
ETH Zentrum
CH-8092 Zürich
Switzerland

David N. Dowling
Department of Microbiology
University College
Cork
Ireland

Torgeir Edland
Norwegian Plant Protection Institute
Boks 70
N-1432 Ås-NLH
Norway

Ralf-Udo Ehlers
University of Kiel
Institute of Phytopathology
Biological Control Laboratory
Klausdorfer-Str. 28–36
D-2313 Raisdorf
Germany

Nyckle Fokkema
Department of Ecology and Soil Ecology
Research Institute for Plant Protection
IPO Box 9060
NL-6700 GW Wageningen
The Netherlands

Dan Funck Jensen
Royal Veterinary and Agricultural University
Thordvalsensvej 40, 83
DK-1871 Frederiksberg-C
Denmark

David Greathead
Imperial College at Silwood Park
Centre for Population Biology
Ascot, Berks. SL5 7PY
UK

Heikki M. T. Hokkanen
Department of Applied Zoology
PO Box 27
FIN-00014 University of Helsinki
Finland

Keith R. Hopper
European Biological Control Laboratory
USDA-ARS
BP 4168
Parc Scientifique Agropolis II
F-34092 Montpellier
France

Jürg Huber
Biologische Bundesanstalt
Institut für Biologische Schädlingsbekämpfung
Heinrichstr. 243
D-6100 Darmstadt
Germany

Christoph Keel
Swiss Federal Institute of Technology
ETH Zentrum
CH-8092 Zürich
Switzerland

Fred A. J. Klingauf
Biologische Bundesanstalt
Messeweg 11/12
D-3300 Braunschweig
Germany

C. P. Lane
Department of Entomology
University of Minnesota
1980 Folwell Avenue
219 Hodson Hall, St. Paul
MN 55108
USA

Robert D. Lumsden
Biocontrol of Plant Diseases Laboratory
Plant Sciences Institute
Beltsville, MD 20705
USA

James M. Lynch
School of Biological Sciences
University of Surrey
Guildford, Surrey GU2 5XH
UK

John Morris
Department of Microbiology
University College
Cork
Ireland

Fergal O'Gara
Department of Microbiology
University College
Cork
Ireland

Daniel J. O'Sullivan
Department of Microbiology
University College
Cork
Ireland

D. M. Olson
Department of Entomology
University of Minnesota
1980 Folwell Avenue
219 Hodson Hall
St Paul
MN 55108, USA

Arne Peters
University of Kiel
Institute of Phytopathology
Biological Control Laboratory
Klausdorfer-Str 28–36
D-2313 Raisdorf
Germany

David Pimentel
Department of Entomology
Cornell University
Ithaca, New York 14853
USA

Jonathan Robinson
Department of Applied Zoology
PO Box 27
FIN-00014
University of Helsinki
Finland

Maarten H. Ryder
CSIRO Division of Soils
Glen Osmond, SA 5064
Australia

C. Steinberg
INRA, Station de Recherches sur la Flore Pathogène dans le Sol
17, rue Sully, BP 1540
F-21034 Dijon
France

Peter Stephens
Department of Microbiology
University College
Cork
Ireland

Linda S. Thomashow
USDA-ARS Root Disease and Biological Control Research Unit
Pullman, Washington 99164-6430
USA

Timo Törmälä
Kemira Ltd
Box 330
SF-00101 Helsinki
Finland

Kenichi Tsuchiya
National Institute of Agrobiological Resources
2-1-1 Kannondai 305
Japan

Joop C. van Lenteren
Department of Entomology
Agricultural University
Box 8031, NL-6700
EH Wageningen
The Netherlands

Jeff Waage
International Institute of Biological Control
Silwood Park, Buckhurst Road
Ascot, Berks. SL5 7TA
UK

J. F. Walter
Process Development
W. R. Grace and Co.
Columbia, MD 20705-2350
USA

C. Howard Wearing
HortResearch
Clyde Research Orchard
Earnscleugh Road
RD1, Alexandra
New Zealand

David M. Weller
Root Disease and Biological Control
USDA-ARS-PWA
367 Johnson Hall
Washington State University
Pullman, WA 99164-6430
USA

Max J. Whitten
Chief, Division of Entomology
CSIRO
Canberra, ACT 2601
Australia

Doreen Winstanley
Horticulture Research International
Worthing Road, Littlehampton
West Sussex BN17 6LP
UK

Hanne Wolffhechel
Royal Veterinary and Agricultural University
Thordvalsensvej 40, 83
DK-1871 Frederiksberg-C
Denmark

Series Preface
Plant and Microbial Biotechnology

The primary concept of this Series of books is to produce volumes covering the integration of plant and microbial biology in modern biotechnological science. Illustrations abound, for example the development of plant molecular biology has been heavily dependent on the use of microbial vectors, and the growth of plant cells in culture has largely drawn on microbial fermentation technology. In both of these cases the understanding of microbial processes is now benefiting from the enormous investments made in plant biotechnology. It is interesting to note that many educational institutions are also beginning to see things in this way and integrating departments previously separated by artificial boundaries.

Having set the scope of the Series, the next objective was to produce books on subjects which had not been covered in the existing literature and, it was hoped, to set some new trends.

Two of the first books of this Series addressed the opportunities of protein engineering (Peter Shewry and Steve Gutteridge) and transformation (Kan Wang, Alfredo Herrera-Estrella and Marc van Montagu), while another concerned the release of genetically engineered and other microorganisms (John Fry and Martin Day). One of the major targets of agricultural technology is the biological control of pests and diseases. Some of the debates on genetic engineering have clouded the issue that biological products could reduce the chemical load on the environment while providing more effective products that are less prone to resistance. On the other hand, introduction of any exotic, whether it is genetically engineered or not, into the environment should only be sanctioned when the risks are fully evaluated.

In the Organisation for Economic Co-operation and Development (OECD) Programme, two of the three themes are relevant to biological control. Theme 1 concerns modification of plant/soil/microbial interactions to reduce inputs in farming systems (co-ordinator Jim Lynch), while Theme 3 concerns the risks of introducing new organisms in agricultural practice (co-ordinator Heikki Hokkanen). In March 1992, we organised a joint workshop of the two themes in Saariselka, Finnish Lapland. We were grateful for funding from the OECD, the Academy of Finland, the Finnish Ministry of Agriculture and Forestry, and the University of Helsinki Applied Zoology Department. Many stimulating discussions made it clear that it would be useful to produce a text on biological control where we would ask authors to consider benefits as well as risks; most other texts seemed to consider the subject from only one of these two dimensions. Inevitably, not everybody could attend the Saariselka workshop and we felt we needed to supplement our workshop participants with other authors, particularly to cover *Bacillus thuringiensis* in depth, as it is the most extensively used microbial biopesticide. The completed volume seemed ideal for the Cambridge University Press Plant and Microbial Biotechnology Series and I am delighted to offer it as my first personal volume in the Series. I must thank fellow editor, Heikki Hokkanen, who handled so much of the material for this volume in his office, ably assisted by Aana Vainio and Jonathan Robinson, and for the excellent co-operation of the contributors.

Our task in editing the volume became a little complicated in that we both moved to university chairs during the preparation, and we thank our families for their support and understanding. Finally, we thank OECD for their support in our attempts to generate an international forum for agricultural biotechnology.

Jim Lynch

Preface: Overview of benefits and risks of biological control introductions

Introductions of organisms in general

Introductions of organisms into new environments can be intentional, accidental, or migratory and may encompass practically any kind of living material. Intentional introductions fall into two broad categories: crop plants and domestic animals, and agents for the biological control of pests or other beneficial purposes, such as microbes to degrade toxic compounds. Benefits from the introduction of crop plants and domestic animals are obvious, but good economic assessments of these introductions are rare. Rigorous evaluations have been done for some classical biological control programmes (see Cullen and Whitten, Chapter 26; Greathead, Chapter 5; and Pimentel, Chapter 2), and overall, a return to investment of 30:1 has been estimated. An example of outstanding economic success is the biological control of Rhodes grass scale in Texas, which cost only US\$ 0.2 million, but from which *annual* economic benefit has been estimated at about US\$ 200 million. Other benefits of introducing biocontrol agents include the avoidance of unwanted side-effects associated with chemical pesticides, important from environmental, occupational health and social points of view.

Introductions of living material are always associated with a risk of unwanted side-effects (see Pimentel, Chapter 2; and van Lenteren, Chapter 3). Thus, of all intentional introductions of species some 25–68% result in permanent establishment, and 0–2% of these organisms become pests. In contrast, only about 5% of unintentional introductions result in species establishment, but 7% of those become pests in their new environment. In total, 128 intentionally introduced crop and ornamental plants have become weeds in the USA (2.2% of all introductions of plants), including Johnson grass, Missisippi chick corn, goatsrue and crotalaria (crop plants); and multiflora rose, water hyacinth and lantana (ornamentals). Also, 11 out of 18 of the world's worst weeds are crop plants in other regions of the world. Of the introduced domestic animals, 9 species out of 20 have become serious environmental problems in the USA. The ratios for game animals and fish are similar.

Thus, it is obvious that introductions of exotic living material into an ecosystem bear a risk. The risk has been remarkably low for introductions concerning the biological control of pests: of over 5500 introductions only a few have resulted in any known negative effects (see Greathead, Chapter 5). Practically all negative effects deal with early introductions of vertebrate predators (including mongoose and giant toad, but also the African snail). Insect introductions have not resulted in known, permanent negative side-effects. There may be one exception: possible extinction of the target host pest, coconut moth in Fiji.

Genetic engineering poses difficult questions about the environmental safety of such introductions. In a way, prediction of problems in this context is particularly difficult: if the problem is novel, or unique to transgenic organisms, it has not happened yet, and the prediction lacks supporting data. On the other hand, if there are data on the problem, then either the problem is not novel (implying that 'we can deal with it'), or it is not unique to transgenic organisms. Evidence accumulated from thousands of field tests with engineered organisms all over the world suggests that with proper precautions, these introductions may not pose a greater risk than introductions in general.

In light of previous experience it appears that we will have to accept that about 1–2% of all introductions will result in environmental or related problems. The magnitude and importance of these problems must be weighed against the expected benefits, to reach a decision on whether to carry out the introductions according to current practice, or whether certain procedures need to be modified.

Biological control introductions

The rationale behind the introduction of bio-control agents into the environment is to protect crops from pests and pathogens that will, if left unchecked, cause losses in yield and quality and thereby impact on the economics of crop production. All methods of crop protection have associated benefits and risks that are influenced by a whole range of temporal, climatic, biological and economic variables. A decision on whether to implement a particular strategy will be made taking account of the variables and assessing the balance between the benefits which might accrue and the risks likely to be faced. If the control strategy has not been attempted previously, or has not been attempted under similar conditions to those prevailing, the balance between benefits and risks, and thereby the outcome of implementation, cannot be predicted with certainty. In as much as biocontrol is concerned, predicting the outcome of implemention of a particular strategy is likely to be a much more difficult task than if it were, for instance, an agrochemical control strategy that were under consideration. The principal difficulties of working in biocontrol concern experimenting with the agents under controlled conditions, which are likely to be artificial, in terms of their representativeness of natural conditions, and subsequently extrapolating the results to the environment into which the biocontrol agent is to be introduced. Given the diversity of crop protection problems and the large numbers of prospective remedial measures that can be

attempted, assessment of the balance between prospective benefits and risks is far from easy. Some of the considerations in such comparisons are outlined in Table 1.

The papers included in this book arise from an OECD workshop on the benefits and risks following the introduction of biocontrol agents, which was held in Saariselkä, Finnish Lappland. They cover the use of insects, nematodes, fungi, bacteria and viruses in biocontrol programmes for crop pests, diseases and weeds. The papers range in scope from classical biocontrol to the use of genetically-modified organisms, and represent accounts of benefits and risks associated with particular biocontrol strategies from the viewpoints of scientists, administrators, economists and industry.

There currently exists considerable concern on the part of the public, as represented by national and international pressure groups, for maintenance of a clean and safe environment, particularly with respect to human health, in the face of increasing industrialization. The environmental lobby has been growing steadily in strength since the publication of Rachel Carson's *Silent Spring*, which dramatically brought to the attention of the public the possible consequences of uncontrolled technological advance in the absence of considerations for the environment. This concern naturally extends to the agrochemical and farming industries, which respectively produce and use vast amounts of compounds to promote crop growth directly, and also to control the pests and pathogens that are intimately associated with crops. As a consequence of this concern, alternative, non-chemical, strategies for combatting pests and diseases, which reduce the deleterious impact of the application of large amounts of agrochemicals, are sought. Biocontrol, the use of naturally occuring organisms and their products to combat pests and pathogens, represents one possible approach. Despite meeting with high degrees of success in the past, however, biocontrol does not represent a panacea as each crop protection strategy is to some extent unique, and the outcome is not always as predictable as could be hoped for.

To be able to compete successfully with agrochemicals in the opinions of the crop producers, manufacturers and end-users, biocontrol agents should match their performance to some degree. Complete control of a pest or pathogen with a

Table 1. Comparative features of chemical and various biological control approaches to control of pests (including insects, plant pathogens, and weeds) (after Lisansky, 1984; and Payne and Lynch, 1988)

Features	Chemical control	Microbial control agents: strategy of use			Parasites and predators: strategy of use		
		Biopesticide	Classical inoculative	Habitat management	Biopesticide	Classical inoculative	Habitat management
A. Method development							
Discovery	Screen 15 000 compounds, discover afterwards what targets they control	Often easy to find	Often difficult to find	Effective microbes may need to be introduced	Often easy to find	Often difficult to find effective agents	Often easy to discover methods
Cost of R&D	£12 000 000	£400 000	Variable	Variable	Variable	Variable	Variable
Market size required for profit	£30 000 000/year to recoup investment; limited to major crops	Markets less than £600 000/year can be profitable	Seldom profitable to private firms; should be done by public organizations or growers themselves	Variable	Variable, but even small markets can be profitable	Not profitable to private firms; should be done by public organizations	Usually not applicable; carried out by growers themselves
Patentability	Well established	Becoming established	Seldom necessary	Not necessary	Not possible except for delivery methods	Not applicable	Not applicable
B. Method use							
Reliability	Usually reliable	Variable, but usually reliable		Can be reliable, when applicable	Variable, but usually reliable		Variable, but sometimes very reliable when applicable
Kill	Up to 100%	Usually 90–95%	Variable, can be high	Can be high	Usually 90–95%		Variable; usually enough to keep pest populations under economic threshold level
Speed	Usually rapid	Sometimes slow	Sometimes slow	Sometimes slow	Sometimes slow	Sometimes slow, but aims at a permanent balance and control	Sometimes slow
Spectrum	Generally broad	Generally narrow	Generally narrow	Generally narrow	Generally narrow	Generally very narrow	Variable, but generally narrow
Resistance	Often develops	May develop	Seldom, if ever	Seldom, if ever	No evidence	No evidence	No evidence
C. Method safety							
Toxicology	Expensive to test, some toxic to user	Testing relatively inexpensive; usually non-toxic	Of no concern	Of no concern	Of no concern	Of no concern	Of no concern
Environmental hazard	Various well known examples; easy to discontinue use	None yet shown but some hazard possible; easy to discontinue use	As for biopesticides	No known hazards, practice easy to discontinue	None yet shown but some hazard possible; easy to discontinue use	Usually none, but great care must be practised	No known hazards, practice easy to discontinue
Residues	Interval before harvest usual	Crop may be harvested immediately	Of no concern on products	Of no concern	Not applicable	Of no concern on products, but populations will persist in environment	Not applicable

biocontrol agent is seldom possible, and in, for instance, the interests of reducing selection pressure on insect pests to overcome resistance to specifically deployed genes, it is not necessarily desirable either. It is usually possible for an equilibrium to become established between the target and control organisms, whereby the effect of the target organism on the crop is reduced to minimum. This minimum has to be acceptable in terms of the economic requirements and the requirements of industry and the consumer. In this respect, it may be necessary in the future to modify the standards that currently have to be met for the produce to allow for this. Decreases in absolute quality may be acceptable in the light of the benefits that derive from the reduction of agrochemical application in favour of biocontrol. Unfortunately, until a strategy becomes widely adopted the costs associated with it will remain high. Costs of production are generally high for biocontrol products, and there may be problems of timely supply and storage, which often makes them unattractive propositions in comparison with agrochemicals. This touches on an additional problem associated with biocontrol, namely that of the spectrum of action of prospective biocontrol agents.

Pesticides are mostly broad spectrum products, and are consequently able to be used successfully under a wide range of conditions against a variety of target organisms. Biocontrol agents are generally more specific in their action. For example, a hymenopterous parasitoid may have evolved to parasitize a single insect species or a narrow range of species. This may be true of other prospective biocontrol agents. The baculoviruses are very host specific, they infect arthropods only, and then mainly insects; specific viruses often being associated with particular insects. Biocontrol agents are also influenced to a much greater degree by the environment than agrochemicals. For instance, entomophagous nematodes are only able to survive below the soil surface, or in other sheltered microenvironments (e.g. under bark), where they are protected from the harmful effects of ultraviolet radiation and desiccation. The effectiveness of prospective fungal and bacterial biocontrol agents for the control of soil-borne diseases is very dependent on edaphic factors. Climate also plays a major role in the relative success and reliability of many biocontrol organisms.

The host specificity of biocontrol agents might suggest that they can be better targeted at pests and pathogens than broad spectrum agrochemicals. This specificity can work against their use, making them unattractive when mixtures of pests and diseases need to be controlled. It may be difficult to control all the pests and diseases on a crop with biocontrol agents, and it may at the same time be impossible to use a strategy that combines agrochemicals and biocontrol agents because of incompatibility. Moreover, the costs of the production of organisms with such a narrow range of action are likely to be prohibitive. These are not the only problems associated with narrow host ranges. It is possible that non-target organisms, which are neither pests nor pathogens, can be affected by biocontrol agents. Agrochemicals generally control organisms in the area where they are applied; although sprays may drift and affect non-target organisms, they do not drift anywhere near as far as is theoretically possible with a biocontrol agent, which has the potential to disturb the non-immediate environment to a far greater extent. Should the introduced biocontrol agent affect endangered species related to the primary target organism, or affect beneficial organisms, the consequence of its introduction will be serious. It must be borne in mind that although agrochemicals, or their breakdown products, may reside in the environment long after application, biocontrol agents can become new components in the biota and thereby reside indefinitely in the environment. There are ample records of natural invasions of pests and diseases, and if an invader was benign in its original habitat it does not indicate that it will be so in its new one. It is a mistake to believe that what is manufactured is inherently bad and what is natural is necessarily good. It may, furthermore, be difficult to remove an established but unwanted biocontrol agent from the environment. In this respect augmentative control, whereby levels of the biocontrol agent are periodically topped-up, as opposed to single introductions in space and time, may be more attractive ecologically, although it is unlikely to be economically advantageous. A thorough understanding of the ecology of the particular situation can help to reduce problems in this connection.

This naturally leads to the subject of monitoring

the spread and survival of introduced organisms in the environment. In addition to this being time consuming and expensive it may, depending on the nature of the introduced organism and the environment, be very difficult. An introduced insect biocontrol agent may be large and obvious enough to be successfully monitored. Following introduction its populations may decline, and therefore monitoring should not be stopped too soon if the biocontrol agent has been thought to have disappeared. Moreover, if it is capable of flight it should probably be monitored over a large area. However, considerable problems ensue if the insect is an introduced biotype of a species that already exists in that environment. Under such circumstances classical taxonomy may not assist in identification of the introduced biotype and resort may have to be made to molecular finger-printing or similar, expensive, specialized methods. Keeping track of introduced fungi, bacteria and viruses may prove even more problematical given their small size, cryptic habits and relatively high rates of mutation. There is of course the possibility that it is not an introduced organism that needs to be monitored but merely introduced genes. For instance, many crop plants have been transformed with *Bacillus thuringiensis* (*Bt*) genes, which code for compounds toxic to insects. The pollen and seeds of such transgenic plants may become widely dispersed and will be difficult to monitor.

In connection with biocontrol agents in general, and bacteria in particular, there is the ever present possibility that resistance will be overcome, just as with development of resistance to agrochemicals and host-plant resistance in pest and pathogen species. It has been suggested that inherent traits, including those possessed by biocontrol agents, can be enhanced with genetic manipulation; through insertion of new genes or modification of the existing genome. This is a particularly attractive option for use with bacteria. There is, however, a school of thought that promotes exploitation of existing genetic diversity and suggests that modification of genomes is not necessary given that adequate variation probably exists in nature. There already exist several cases of insect pest resistance to *B. thuringiensis* applied as a biopesticide. Moreover, with increasing deployment of the genes coding for *Bt* toxins in transformed crop plants, the likelihood of widespread

resistance developing in insect pests will increase. This will be especially problematical with poly-phagous insects, which might have the opportunity to feed on several different transgenic species and as a consequence develop a broad-based resistance to *Bt* genes. The management of biocontrol agents is, therefore, not necessarily a simple short-term prospect.

Biocontrol through biotechnology is a very complex subject and has attracted the attention of a wide range of disciplines as it impinges on ethics, has a high news profile and has important consequences for registration and patenting. Given the poor public perception of bacteria, and the general lack of understanding of genetics, molecular or otherwise, outside of the scientific community, it is little wonder that there is concern over the benefits and risks associated with the introduction of genetically-modified organisms into the environment. It is one thing to introduce a familiar, colourful ladybird into the environment to control aphids, but to introduce a bacterium, or its genes, is entirely another. This concern over genetic manipulation is amplified by the secrecy that generally shrouds 'cutting-edge' research and development, and which is necessary if private companies are to be able to develop profitable products. Although secrecy may slow the progress of research, and may result in a considerable lag between discovery of a technology and the public becoming aware of it, and that cooperation with competitors is hardly possible, it is necessary to protect technical and financial interests. There seems to be no easy solution to the problem, but it is essential that industry is not unnecessarily hampered in its efforts to produce new technologies, but that it is as open as possible about developments. Only in this manner will mutual trust become established between industry and the general public.

Biocontrol has been generally successful and safe. There have been no major perturbations to the environment or adverse effects on human health, and certainly nothing on the scale of recent industrial/nuclear accidents. Given the buffering capacity of the environment there are unlikely to be any significant deleterious effects on it that arise from the release of biocontrol agents, if adequate research is undertaken beforehand. While it is recognized that it is always better to address problems before control is required, this will not always be possible. Biocontrol represents one means of improving crop protection technology, which can be used alongside others. The inherent risks associated with biocontrol have to be faced, though many can no doubt be reduced through education of the people concerned with, and about, its use. It is easy to overstate the dangers of any controversial technology in the absence of adequate information and in the presence of very restrictive legislative frameworks that convey a sense of danger. It would, therefore, be beneficial in the long-run if the benefits and risks associated with biocontrol of crop pests and diseases could be communicated clearly. This book represents an attempt to do this.

Heikki M.T. Hokkanen
James M. Lynch
Jonathan Robinson

Note added in proof

This volume has had a longer gestation than originally intended, but it results from the need for interactive discussion between authors and addition of chapters not contracted at the outset. However, the main purpose of the book was to establish some general principles and concepts which would stand the test of time. I hope this has been achieved.

James M. Lynch

References

Lisansky, S. G. (1984). Biological alternatives to chemical pesticides. *World Biotechnology Report*, vol. 1, pp. 455–466. Online Publications, Pinner.

Payne, C. C. and Lynch, J. M. (1988). Biological control. In *Micro-organisms in Action: Concepts and Application in Microbial Ecology* (ed. J. M. Lynch and J. E. Hobbie), pp. 261–287. Blackwell Scientific Publications, Oxford.

Part I BIOLOGICAL INVASIONS

1

Suppressiveness of soils to invading micro-organisms

Claude Alabouvette and C. Steinberg

Introduction

Many attempts have been made to control soil-borne plant pathogens and to improve the growth of plants by inoculation of seed or soil with selected strains of micro-organisms; mainly bacteria. More recently, selected bacteria that are expected to degrade xenobiotics have been suggested for the bioremediation of polluted sites (Short *et al.*, 1990). Nowadays, risk assessment studies also need to address the fate of engineered telluric soil and non-telluric soil bacteria (Tiedje *et al.*, 1989; Doyle *et al.*, 1991).

Besides increased crop yields claimed by Russian workers (Schroth and Becker, 1990), most of the tentative applications of micro-organisms in agricultural soils have failed. The widely practised inoculation of legume seeds with *Rhizobium* spp. is one of the few examples of success of application of micro-organisms to improve crop yield (Stacey and Upchurch, 1984). Application of the strain K84, and more recently of the modified strain K1026, of *Agrobacterium radiobacter* to control crown gall of plants is one of the few examples of a biological control method commercially applied with some success in several countries (Ryder and Jones, 1990). In most cases, the beneficial effects expected from the microbial inoculation are not consistently reproduced under field conditions. The poor survival of the introduced micro-organisms in soil is the main explanation for these failures. In fact, the population density of the introduced micro-organisms decreases to the limit of the carrying capacity of the soil. Moreover, the soil represents very heterogeneous environments

in which the introduced micro-organisms must find suitable habitats, some of which are very strain specific (Hattori and Hattori, 1976). It is often admitted that exogeneous populations of micro-organisms are difficult to establish in the soil in such a way that they express the specific activity for which they have been selected. Therefore, many studies aim to improve the saprophytic or rhizospheric competence of beneficial micro-organisms in the soil environment (Ahmad and Baker, 1987).

Contrasting with this belief is the fact that it is almost impossible to eradicate diseases due to soil-borne plant pathogens, which indicates that these micro-organisms are able to survive in soil for long periods and under various climatic conditions.

Currently, there is a tendency to believe that genetically engineered micro-organisms will help to improve the efficacy of antagonists or plant-growth-promoting rhizobacteria (PGPR) and many scientists expect the use of such modified micro-organisms in agriculture. Even though insertion genes will probably cause micro-organisms to have reduced ecological competence (Schroth and Becker, 1990), there is great concern about the risk of losing control of these manipulated organisms and of the transmission of exogenous genes to wild micro-organisms in soil. Most of the risk assessment studies only consider transformed micro-organisms, but there is also a need to improve knowledge of the ecology of naturally occurring micro-organisms. Plant pathologists studying soil-borne plant pathogens are among the microbiologists with the longest experience in the field of microbial ecology. They have been accustomed to infesting soils with pathogens to reproduce disease symptoms and with antagonists to control diseases. Twenty-five years of studying soil that is naturally suppressive to diseases induced by soil-borne pathogens has produced a large amount of data related to the interactions between the pathogens and the antagonistic micro-organisms, and between the microbiota and the abiotic characteristics of the soils (Cook, 1990). The existence of soils that do not enable disease expression even after infestation with the pathogen demonstrates that the soil environment can suppress either the establishment or the activity of an introduced population of micro-organisms (Alabouvette, 1990). If soil suppressiveness to diseases is mainly due to microbial interactions between the pathogen and the antagonistic microflora, then it is not independent from the abiotic characteristics of the soils. Indeed, soils suppressive to fusarium wilts provide an example where the suppression of the disease is not necessarily due to the suppression of the pathogen; however, pathogen survival and disease suppressiveness are under the indirect control of the abiotic soil properties. Therefore, it is interesting to review the subject of suppressiveness of soil to the invasion of micro-organisms in relation to the concepts linked to soil suppressiveness of fungal diseases.

Survival of microbial populations introduced into soils

Studying the survival kinetics of a microbial population introduced into soil requires accurate methods to assess the population density. Many techniques based on direct or indirect assessments have been developed; all of them have some inconveniences that makes it difficult to compare data obtained by different workers, and different techniques. However, it appears that most of the micro-organisms studied are able to survive for some time in soil. The following examples illustrate different kinetics of survival after experimental introduction of bacteria or fungi into soil; but this is not intended to be an exhaustive review of the literature.

Bacteria

Assessing the fate or predicting the activity of a bacterial population introduced into soil appears to be suited to a case by case study, and is difficult to generalize. However, some general principles can be mentioned as will be shown by inspection of the kinetics of different bacterial species introduced into soil.

A comparison of the population dynamics of various telluric or non-telluric strains of bacteria in a clay loam soil (clay, 32%; silt, 36%; sand, 32%; pH 6.5) in microcosms (Richaume *et al.*, 1990) resulted in a primary discrimination of two groups of bacteria : the surviving one and the non-surviving one (Fig. 1.1). When inoculated at a high density, close to 10^8 bacteria/g soil, the various telluric strains first rapidly declined, then established at a lower level of 10^5 bacteria/g soil (*Azospirillum lipoferum, Agrobacterium radiobacter* or 10^6 bacteria/g soil (*Pseudomonas aeruginosa*) depending on the bacterial species. The slight increase of the *Bradyrhizobium japonicum* and *Agrobacterium radiobacter* densities observed on the days following the inoculation was interpreted by the authors as being growth at the expense of internal reserves such as poly-β-hydroxybutyrate because they had been grown on a rich culture medium. On the other hand, the population of

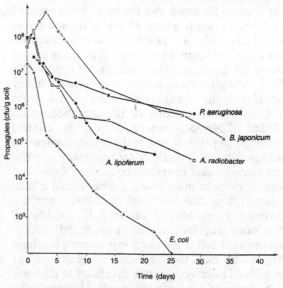

Fig. 1.1 Population dynamics of various bacterial populations introduced into a clay loam.

Escherichia coli declined rapidly and was no longer detectable after 25 days, as has often been reported (Devannas *et al.*, 1986; Recorbet *et al.*, 1992*a*).

The secondary discrimination obtained from these results concerned the telluric species. After one month of incubation, each of them persisted at a specific level. Moreover, Crozat *et al.* (1987) showed that strain G2 sp. of *Bradyrhizobium japonicum* established itself at a level 3 times higher (significant: $P > 0.05$) than the level reached by G49, a strain of the same species, inoculated in the same clay loam soil. The results indicated a soil–bacterium interaction which was strain dependent.

Nevertheless, the general feature of these kinetics was the rapid decline of the bacterial densities, which did not occur in sterile soil (Steinberg *et al.*, 1987) or with fungi (Couteaudier and Alabouvette, 1990). The question that arose was: why did the bacteria not maintain a high density following the inoculation? Two explanations can be provided depending on, once again, the type of bacteria.

1. The ability of the strain to compete successfully when introduced into the soil plays a major role in the establishment of non-telluric bacteria (Recorbet *et al.*, 1992*b*), as well as of engineered bacteria, compared with the parental strain (Van Elsas *et al.*, 1991).
2. With regard to telluric non-manipulated bacteria, it has been widely demonstrated that protozoan predation is the most important factor in the regulation of introduced populations (Habte and Alexander, 1977; Steinberg *et al.*, 1987). This regulating factor could explain why the equilibrium level reached by the bacteria was independent of the inoculum densities at which they were introduced. However, because this equilibrium level differed between strains, it must be stated that very specific interactions between the bacteria and the environment occurred.

Both biotic and abiotic factors acted rapidly because the decline generally occurred in less than one month. Nevertheless, field experiments carried out with *B. japonicum* strains showed that the population density remained stable at around 10^3 or 10^4 bacteria/g soil after 10 years in the presence or absence of soybeans. It is noteworthy that not only was the bacterial density stable during these years but some genetic and physiological charac-

teristics were also. Isolates of one of these strains were recovered by plant or immunological trapping and no genetic or physiological modifications were detected as compared with the parental strains retained in the laboratory (Brunel *et al.*, 1988).

Fungi

Most of the studies are related to the survival of soil-borne plant pathogens and their antagonists. Only two examples are given to show that fungi are able to survive for long periods in soil not only as resting structures but also as active saprophytes.

Studying the population dynamics of a strain of *Trichoderma harzianum* introduced in a field soil, Davet (1983) demonstrated that the fungus was able to survive saprophytically for 3 years (Fig. 1.2). Grown on straw, the inoculum was applied at a rate of 2 kg/m²; the population of *Trichoderma* reached a density greater than 1×10^5 colony forming units (cfu)/g soil immediately after inoculation, then decreased slowly. At the end of the experiment, the population density of *Trichoderma harzianum* was still 10 times greater in the treated plot than in the control, where the wild population of *Trichoderma* spp. established at 3×10^3 cfu/g soil. Davet (1983) was also able to demonstrate that the decline of the population during summer was due to drought; the population increased again in the autumn after a period of rain.

In contrast to these experimental conditions in

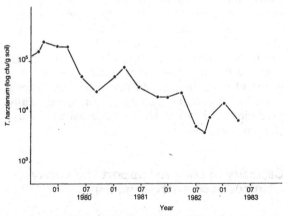

Fig. 1.2 Population dynamics of *Trichoderma harzianum* introduced into a sandy loam.

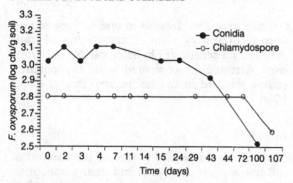

Fig. 1.3 Population dynamics of *Fusarium oxysporum* f.sp. *lini* introduced either as chlamydospores or microconidia into a clay loam.

an open field, Couteaudier and Alabouvette (1990) studied the population dynamics of a strain of *Fusarium oxysporum* f.sp. *lini* introduced into raw soil under well standardized conditions. Not only the chlamydospores but also the microconidia were able to survive for at least 100 days. Introduced at the initial concentration of 1.4×10^3 cfu/g soil, the density of the chlamydospore population remained stable for 72 days but fell significantly between 72 and 107 days (Fig. 1.3). The population dynamics of the conidial inoculum introduced at 3×10^3 cfu/g soil were similar to those described with chlamydospores. However, the decline of the population was greater and after 100 days of incubation only 21% of the conidia had survived compared with 70% of the chlamydospore inoculum. When *F. oxysporum* f.sp. *lini* was introduced at higher inoculum concentrations, i.e. 1.5×10^6 and 3×10^6 cfu/g soil respectively for chlamydospores and conidia, the survival kinetics were similar to those described for the lower inoculum concentration. Therefore, after 100 days of incubation in soil, the higher and the lower inoculum concentrations were significantly different; this indicates that the mechanisms of regulation of the microbial populations described for other micro-organisms, especially predation, did not apply to *Fusarium oxysporum* (Levrat *et al.*, 1991).

Capacity of soils to support the survival of introduced micro-organisms

The previous examples indicated that most bacteria and fungi are able to survive for long periods

in soils. However, soils may differ greatly with respect to their abiotic (texture, pH, etc.) and biotic characteristics. Therefore, the survival kinetics of a given strain of micro-organism may be different depending on the soil type.

Survival of one bacterial strain in various soils

As previously mentioned, the establishment of a bacterial population depended on a soil–bacteria interaction that seemed to be strain dependent. In fact, this highly specific interaction could also be qualified as a soil dependent interaction.

The survival of the *Pseudomonas fluorescens* WCS374ln5 strain in a silt loam was significantly ($P = 0.05$) better than in a loamy sand. On the other hand, an introduced population of the *Bacillus subtilis* F6 strain established in very few days at a similar level both in the loamy sand and in the silt loam (Van Elsas *et al.*, 1986). The *Bacillus* population consisted mainly (60–100%) of spores when established at the stabilized level of 5×10^3 cfu/g soil. Because of their different cell wall properties, Gram$^+$ bacteria and Gram$^-$ bacteria must react differently with respect to biotic and abiotic soil components. Moreover, the ability of Gram$^+$ bacteria to form spores renders them less sensitive to regulating factors. These two types of bacteria should be considered separately.

Moffett *et al.* (1983) also found that the survival of two strains of *Pseudomonas solanacearum* were greater in a clay loam than in a sandy loam. All of these findings suggest that bacteria are better able to survive in clay soils than in sandy soils. Indeed, Heijnen and Van Veen (1991) demonstrated a protective effect of bentonite clay and to a lesser extent kaolinite, against predation of *Rhizobium*. This effect was attributed to modifications in the pore-size distribution. Previously, Stotzky and Rem (1966) suggested the possible relationship between clay minerals and activity and population dynamics of bacteria in soils. Even though Stotzky and Rem, were referring to naturally occurring micro-organisms, their hypothesis could be extended to introduced micro-organisms.

Corman *et al.* (1987) modelled the survival kinetics of three strains of *B. japonicum*. Each of them was inoculated into three soils of different textures (a sandy soil, a silty sand and a silt loam). The highest rate of decrease was always observed

Table 1.1. Estimation of rates of decline (days^{-1}) obtained for each kinetic by the sum of the least squares technique (*B. japonicum*)

	Soils		
Strain	Sandy soil	Silty sand	Silt loam
G2sp	1.2	0.16	1.1
G49	0.64	0.1	0.37
GMB1Ka	0.41	0.34	0.19

in the sandy soil while only for strain GMB1Ka the lowest rate was obtained in the silt loam. Strain G49 and strain G2sp declined significantly more rapidly in the silt loam than in the silty sand (Table 1.1). These results suggested that clays were not the only key factor involved in the survival and activity of bacteria.

The aim of inoculation of soil or seed with bacteria is to obtain some beneficial effect (increased yield, biological control, bioremediation), therefore, not only the ability to survive but mainly the ability of the introduced population to express its activity in soil must be assessed. In other words, the level at which the population density stabilizes should result in the expected beneficial effect. This is the case for seed inoculation of legumes with *Rhizobium* or *Bradyrhizobium*; N$_2$ symbiotic fixation is efficient after the introduced bacterial population has stabilized in soil.

Van Elsas and Heijnen (1990) gave examples of applications of selected or engineered inoculated bacteria, but without any reference to the real agronomic benefit. Nevertheless, because of the potential applications of inoculation of soil with bacteria, the influence of such an inoculation on the indigenous populations has to be assessed. Most authors conclude their papers with the assertion that this should be checked but this is done very rarely. Doyle *et al.* (1991) demonstrated that a genetically manipulated strain of *Pseudomonas putida* (strain pp.0301 (pR0103)), in the presence of the substrate on which its novel genes can function, was capable of inducing measurable variations in the composition of the indigenous fungal and bacterial populations.

The expression of the biological activity at a significant level from an agronomical point of view depends on both the survival density and the physiological state of the introduced micro-organisms after they have established. The results presented here show that this establishment is influenced by a very specific soil–bacteria interaction. This soil–bacteria interaction, in turn, depends on complex processes that involve bacterial cell surface properties, soil, biological and physical components and the interaction between abiotic and biotic factors.

Fungi

Davet (1983) compared the survival kinetics of a strain of *Trichoderma aureoviride* in two field soils. The soil from Dijon was a clay loam of pH 6.3; the soil from Perpignan was a sandy loam of pH 5.9. The fungus was grown on straw and introduced into soil at the rate of 450 g/m^2. After soil infestation, the first estimation of the population density gave similar results for both soils. However, the population rapidly evolved differently in the two soils and the population density was higher in soil from Dijon than in soil from Perpignan (Fig. 1.4). This difference was evident until the end of the experiment and demonstrated that the carrying capacity of the soil from Dijon was greater than that of the soil from Perpignan; this could be attributed to the greater clay content of the soil from Dijon.

Amir and Alabouvette (1993) compared the population dynamics of a strain of *F. oxysporum* f.sp. *lini* introduced at the same inoculum density

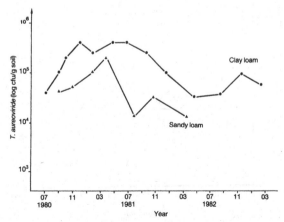

Fig. 1.4 Population dynamics of *Trichoderma aureoviride* introduced into a sandy loam and a clay loam.

into a sandy soil (sand 96%, silt 1.5%, clay 2.5%) and a clay loam soil (sand 19%, silt 44%, clay 37%). They observed that the population density dropped in both soils, but the decline of the population was greater in the sandy than in the clay soil. After 60 days of incubation, the residual populations were significantly different and represented 12% and 44% of the initial populations in the sandy and the clay soil respectively (Fig. 1.5). However, these two soils were not only different in texture; the sandy soil was conducive to fusarium wilt of flax compared with the clay soil, which was suppressive (Fig. 1.6). Although the clay soil supported larger populations of the pathogen, it had a lower inoculum potential than the sandy soil and therefore showed a lower disease incidence.

These results are not in accordance with previous data published by Alabouvette *et al.* (1985) indicating that the population dynamics of *F. oxysporum* f.sp. *melonis* were similar in a conducive and in a suppressive soil from another origin. The density of the introduced species remained almost stable for more than 1 year and was not significantly different between the soils. However, it must be noted that these two soils contained 18.4

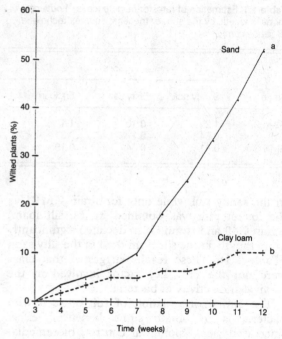

Fig. 1.6 Percentage of wilted plants after infestation of a sandy soil and a clay loam with *F.o.* f.sp. *lini* (4 × 10⁴ cfu/g soil).

and 15.5% clay respectively; their texture was closer to that of the clay soil than to that of the sandy soil described in the previous example.

Collectively, these results demonstrated that suppressiveness of soils to fusarium wilts is not correlated with the suppression of the pathogen in the suppressive soils. In suppressive soils, the introduced pathogen is able to survive at a high density without being able to induce the disease. On the contrary, in conducive soils, the density of the introduced pathogen may decrease significantly, but the remaining propagules are still able to induce the disease. This interpretation of our data leads to the hypothesis that the soil may exert an influence on the survival and on the activity of an introduced population of micro-organisms which is contrary to that expected. This ability of soils to control both the survival and the activity of a microbial population is not independent of their abiotic characteristics. Indeed, it is well established that clay soils support a larger biomass than sandy soils (Chaussod *et al.*, 1986) and clays have also been involved in soil suppressiveness of fusarium wilts (Stotzky and Martin, 1963).

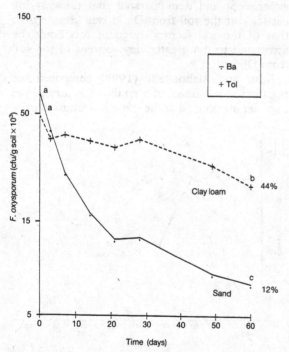

Fig. 1.5 Population dynamics of *Fusarium oxysporum* f.sp. *lini* introduced into a clay loam and a sandy loam.

Role of abiotic properties in mechanisms of soil suppressiveness

Amir and Alabouvette (1993) modified the texture of a sandy soil by the addition of fine particles of montmorillonite or talc (25% w/w). They observed that the survival kinetics of an introduced population of *F. oxysporum* f.sp. *lini* was modified; after 60 days of incubation, the inoculum density was not reduced to the same extent in the amended soil as in the control. The residual inoculum density represented 43% and 41% of the initial density compared with 2% in the control (Fig. 1.7).

The kinetics of CO_2 release after glucose amendment (1 mg/g soil) were significantly different (Fig. 1.8): after the addition of talc the amount of CO_2 released was lower than in the control, but it was enhanced after the addition of montmorillonite. These results suggest that montmorillonite enhanced the microbial activity of the soil, as already described by Stotzky and Rem (1966), and therefore increased the intensity of competition for nutrients between micro-organisms, leading to an increased level of fungistasis (Lockwood, 1988). This increased level of fungistasis could explain the better survival of the introduced population of *F. oxysporum* f.sp. *lini* in soil amended with montmorillonite. The beneficial effect of montmorillonite could be related to specific properties such as its large cation exchange capacity and its large surface area in comparison with talc.

Finally, the addition of talc or montmorillonite

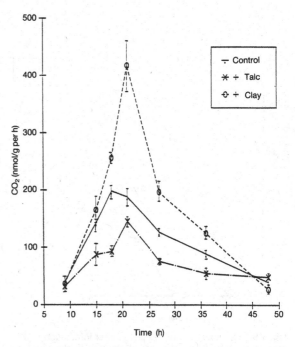

Fig. 1.8 Kinetics of CO_2 release from a sandy soil amended with glucose (1 mg/g) following the addition of talc or montmorillonite (25% w/w).

also modified the level of soil receptivity to fusarium wilt of flax. After soil infestation with 4×10^4 cfu/g soil of *F. oxysporum* f.sp. *lini*, disease incidence was significantly decreased after the addition of montmorillonite and increased after the addition of talc in comparison with the non-amended control. Talc made the soil more conducive to the disease and montmorillonite made it more suppressive (Fig. 1.9).

These results demonstrated that modification of the texture of the soil induced modification of the ability of the soil to support the survival of an introduced population of *Fusarium* and to control the infection activity of this inoculum.

Discussion

The experimental results reviewed in this paper demonstrate that most of the micro-organisms introduced into natural soil survive for weeks, or months, at levels high enough to be easily detected by the standard methods of enumeration.

The only exception concerns the non-telluric

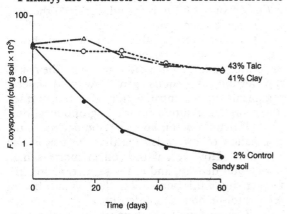

Fig. 1.7 Population dynamics of *F.o.* f.sp. *lini* introduced into a sandy soil following the addition of talc or montmorillonite (25% w/w).

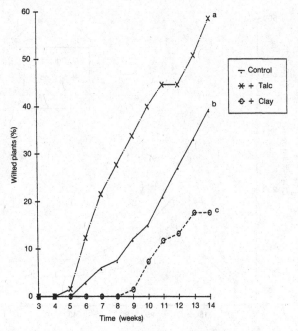

Fig. 1.9 Percentage of wilted plants after infestation of a sandy soil with *F.o.* f.sp. *lini* (4 × 10⁴ cfu/g soil) following the addition of talc or montmorillonite (25% w/w).

bacteria such as *E. coli*, which are not adapted to the soil environment and cannot be detected after a few days or weeks. Generally, the decline of bacterial populations occurs faster than that of fungi. However, it is necessary to distinguish between bacterial species unable to produce spores and spore-forming bacteria, the latter being able to survive at higher densities. Their kinetics of survival can be compared with those of fungi, especially of fungi that form specialized survival structures such as chlamydospores and sclerotia. However, most of the soil-borne fungi are also able to survive as active saprophytes growing on organic matter. Some bacteria such as the symbiotic *Bradyrhizobium japonicum* are also able to survive for long periods in soils without forming resting spores.

The density of the introduced population of either bacteria or fungi usually reaches a plateau that corresponds to the carrying capacity of the soil. This carrying capacity is both strain and soil dependent. Little is known about the factors that determine this carrying capacity. One possible mechanism is that the bacterial populations are related to the nutrient status of the soil. The

results from experiments reported in this paper show that the texture of the soil is also a factor that controls the capacity of soil to support the growth and activity of a microbial population. In general, the greater the clay and organic-matter contents of the soil, the better is the survival of the introduced population. However, most of the studies that deal with the behaviour of microbial populations in soil have been conducted after massive introduction of a given population. Only a few studies have considered the dynamics of a population introduced at a low inoculum density comparable to the levels that occur in natural soils. Therefore, it is not surprising to observe a quick decline of the population until it reaches the carrying capacity of the soil.

It must be stressed that most of the studies have only considered the kinetics of survival, i.e. the quantitative evolution of the population. However, studies of soils suppressive to diseases provide examples of soils that affect the activity of the population without necessarily affecting the density of the population. A soil can be conducive to the population of the pathogen and at the same time be suppressive to the disease induced by the pathogen. The soil affects the ability of the pathogen to survive and its ability to infect the host plant in various ways. This difference between suppressiveness towards a microbial population and its activity raises the question of what is the most important phenomenon to be considered. Obviously for the farmer, as well as for the plant pathologist, the suppression of the disease is the most important issue. If an antagonistic micro-organism or a PGPR is introduced into the soil, one expects its activity to result in a beneficial effect on the crop. In this case, the soil should support both the survival and the activity of the introduced population. And what about the risk of transferring exogenous genes to wild micro-organisms by the introduction of a transformed micro-organism into soil for an agronomical purpose? When the soil induces a rapid decline of the population one may assume that the risk is also declining. However, if the soil supports a high population density, and at the same time inhibits the activity of this population, the risk may be not related to the population density.

It is our feeling that most of the micro-organisms that could be introduced into soil will be able to survive for long periods, at least at low population densities. However, the large diversity of bacteria, actinomycetes and fungi that exists in every soil indicates that in the absence of a specific selection pressure no population is able to become dominant. The soil biotic environment is very well buffered and only drastic manipulations, such as soil disinfection, can modify the microbial balance over the long-term.

Conclusion

The suppressiveness of soils to invading micro-organisms must be considered from two points of view. The first concerns the population dynamics, the second the population activity, both of them being under the control of biotic and abiotic factors.

Most of the micro-organisms are able to survive for a long time when introduced into soils. The examples, used in this paper clearly show that telluric micro-organisms, both bacterial and fungal, have specific abilities to overcome adverse environmental conditions successfully. Moreover, they maintain their genetic and physiological characteristics, e.g. for *Rhizobium* the ability to nodulate soybean even after surviving for several years in soil.

Bringing a new property to a soil by the introduction of a novel genotype, as for example the introduction of *B. japonicum* into French soil to promote the nodulation of soybean, is easier than suppressing an established property of a soil. The introduction of an antagonist into a conducive soil will never completely eradicate fusarium wilt. However, the introduction of a pathogenic strain into a suppressive soil does not transform the soil into a soil conducive to the disease, thus indicating the role of abiotic factors on the expression of the activity of the population.

In fact, the level of the population at equilibrium and the expression of the activity of a microbial population are both strain and soil dependent. The soil components, especially the clay content, play an important role in the regulation of the microbial balance, but it is difficult to analyse their role precisely.

Regarding the risk, it is concluded that even if the quantitative diversity of the indigenous microbial populations may be temporarily affected by the introduction of an exogenous population, the soil is usually sufficiently well buffered to preserve the microbial balance.

References

Ahmad, J. S. and Baker, R. (1987). Rhizosphere competence of *Trichoderma harzianum*. *Phytopathology* 77, 182–189.

Alabouvette, C. (1990). Biological control of fusarium wilt pathogens in suppressive soils. In *Biological Control of Soil-borne Plant Pathogens* (ed. D. Hornby), pp. 27–43. CAB International, Wallingford.

Alabouvette, C., Couteaudier, Y. and Louvet, J. (1985). Soils suppressive to fusarium wilt: mechanisms and management of suppressiveness. In *Ecology and Management of Soil Borne Plant Pathogens* (ed. C. A. Parker, A. S. Rovira, K. J. Moore, P. T. W. Wong and J. F. Kollmorgen), pp. 101–106. American Phytopathological Society, St Paul.

Amir, H. and Alabouvette, C. (1993). Involvement of soil abiotic factors in the mechanisms of soil suppressiveness to fusarium wilt. *Soil Biology and Biochemistry* 25, 157–164.

Brunel, B., Cleyet-Marel, J. C., Normand, P. and Bardin, R. (1988). Stability of *Bradyrhizobium japonicum* inoculants after introduction into soil. *Applied and Environmental Microbiology* 54, 2636–2642.

Chaussod, R., Nicolardot, B., Catroux, G. and Chrétien, J. (1986). Relations entre les charactéristiques physicochimiques et microbiologiques de quelques sols cultivés. *Science du Sol* 24, 213–226.

Cook, R. J. (1990). Twenty-five years of progress towards biological control. In Hornby, D., *Biological Control of Soil-borne Plant Pathogens* (ed. D. Hornby), pp. 1–14. CAB International, Wallingford.

Corman, A., Crozat, Y. and Cleyet-Marel, J. C. (1987). Modelling of survival kinetics of some *Bradyrhizobium* strains in soils. *Biology and Fertility of Soils* 4, 79–84.

Couteaudier, Y. and Alabouvette, C. (1990). Survival and inoculum potential of conidia and chlamydospores of *Fusarium oxysporum* f.sp. *lini* in soil. *Canadian Journal of Microbiology* 36, 551–556.

Crozat, Y., Cleyet-Marel, J. C. and Corman, A. (1987). Use of the fluorescent antibody technique to characterize equilibrium survival concentrations of *Bradyrhizobium japonicum* strains in soil. *Biology and Fertility of Soils* 4, 85–90.

Davet, P. (1983). Introduction et conservation des *Trichoderma* dans le sol. In *Les Antagonismes Microbiens*, 24ème Colloque SFP, Bordeaux, 26–28 May 1983, Ed. INRA.

Devannas, M. A., Rafaeli-Eshbol, D. and Stotzky, G.

(1986). Survival of plasmid containing strains of *Escherichia coli* in soil: effect of plasmid size and nutrients on survival of hosts and maintenance of plasmids. *Current Microbiology* **13**, 269–277.

Doyle, J. D., Short, K. A., Stotzky, G., King, R. J., Seidler, R. J. and Olsen, R. H. (1991). Ecologically significant effects of *Pseudomonas putida* PPO301(pRO103), genetically engineered to degrade 2,4-dichlorophenoxyacetate, on microbial populations and processes in soil. *Canadian Journal of Microbiology* **37**, 682–691.

Habte, M. & Alexander, M. (1977). Further evidence for the regulation of bacterial populations in soil by protozoa. *Archiv für Mikrobiologie* **113**, 181–183.

Hattori, T. and Hattori, R. (1976). The physical environment in soil microbiology. An attempt to extend principles of microbiology to soil microorganisms. *CRC Critical Review in Microbiology* **4**, 423–460.

Heijnen, C. E. and Van Veen, J. A. (1991). A determination of protective microhabitats for bacteria introduced into soil. *FEMS Microbiology Ecology* **85**, 73–80.

Levrat, P., Pussard, M., Steinberg, C. and Alabouvette, C. (1991). Regulation of *Fusarium oxysporum* populations introduced into soil: the amoebal predation hypothesis. *FEMS Microbiology Ecology* **86**, 123–130.

Lockwood, J. L. (1988). Evolution of concepts associated with soilborne plant pathogens. *Annual Review of Phytopathology* **26**, 93–121.

Moffett, M. L., Giles, J. E. and Wood, B. A. (1983). Survival of *Pseudomonas solanacearum* biovars 2 and 3 in soil: effect of moisture and soil type. *Soil Biology and Biochemistry* **15**, 587–591.

Recorbet, G., Givaudan, A., Steinberg, C., Bally, R., Normand, P. and Faurie, G. (1992*a*). Tn5 to assess soil fate of genetically marked bacteria: screening for aminoglycoside-resistance advantage and labelling specificity. *FEMS Microbiology Ecology* **86**, 187–194.

Recorbet, G., Steinberg, C. and Faurie, G. (1992*b*). Soil survival of genetically-engineered *Escherichia coli* as related to inoculum density, predation and competition. *FEMS Microbiology Ecology* **101**, 251–260.

Richaume, A., Steinberg, C., Recorbet, G. and Faurie, G. (1990). Population dynamics of introduced bacteria in soil and effect on gene transfer. In *Ingegneria Genetica e Rischio Ambientale* (ed. S. Dumontet and E. Landi), pp. 67–75. Giardini Naxos (ME).

Ryder, M. H. and Jones, D. A. (1990). Biological control of crown gall. In *Biological Control of Soil-borne Plant Pathogens* (ed. D. Hornby), pp. 45–63. CAB International, Wallingford.

Schroth, M. N. and Becker, J. O. (1990). Concepts of ecological and physiological activities of *Rhizobacteria* related to biological control and plant growth promotion. In *Biological Control of Soil-borne Plant Pathogens* (ed. D. Hornby), pp. 389–414. CAB International, Wallingford.

Short, K. E., Seidler, R. J. and Olsen, R. H. (1990). Survival and degradative capacity of *Pseudomonas putida* induced or constitutively expressing plasmid mediated degradation of 2,4-dichlorophenoxyacetate (TFD) in soil. *Canadian Journal of Microbiology* **36**, 821–826.

Stacey, G. and Upchurch, R. G. (1984). Rhizobium inoculation of legumes. *Trends in Biotechnology* **2**, 65–69.

Steinberg, C., Faurie, G., Zegerman, M. and Pave, A. (1987). Régulation par les protozoaires d'une population bactérienne introduite dans le sol: modélisation mathématique de la relation prédateur-proie. *Revue d'Écologie et de Biologie du Sol* **24**, 49–62.

Stotzky, G. and Martin, T. (1963). Soil mineralogy in relation to the spread of fusarium wilt of banana in Central America. *Plant and Soil* **18**, 317–337.

Stotzky, G. and Rem, L. T. (1966). Influence of clay minerals on microorganisms. I. Montmorillonite and kaolinite on bacteria. *Canadian Journal of Microbiology* **12**, 547–563.

Tiedje, J. M., Colwell, R. K., Grossman, Y. L., Hodson, R. E., Lenski, R. E., Mack, R. N. and Regal, P. J. (1989). The planned introduction of genetically engineered microorganisms; ecological considerations and recommendations. *Ecology* **70**, 298–315.

Van Elsas, J. D. and Heijnen, C. E. (1990). Methods for the introduction of bacteria into soil: a review. *Biology and Fertility of Soils* **10**, 127–133.

Van Elsas, J. D., Dijkstra, A. F., Govaert, J. M. and Van Veen, J. A. (1986). Survival of *Pseudomonas fluorescens* and *Bacillus subtilis* introduced into two soils of different texture in field microplots. *FEMS Microbiology Ecology* **38**, 151–160.

Van Elsas, J. D., Van Overbeek, L. S., Feldmann, A. M., Dullemans, A. M. and De Leeuw, O. (1991). Survival of genetically engineered *Pseudomonas fluorescens* in soil in competition with the parent strain. *FEMS Microbiology Ecology* **85**, 53–64.

2

Biotechnology: environmental impacts of introducing crops and biocontrol agents in North American agriculture

David Pimentel

Introduction

Biotechnology (genetic engineering) and biological control offer many opportunities to improve agricultural production. Potential benefits include: achieving higher crop and livestock yields; improving nutritional make-up of crops and livestock; substituting biocontrols for pesticide use; as well as controlling some soil and water pollutants. At the same time, the use of biotechnology and release of genetically engineered organisms (for biocontrol and other purposes) into the environment could create serious ecological, social, and economic hazards (Pimentel et al., 1989).

One objective in the implementation of any new technology is to achieve maximum benefits with minimal risks to the environment, economy, and society. The other major objective is to be able to use the technology so that all future benefits are realized. If a serious problem results from the application of just one project, the future development of the entire technology is jeopardized. Note what has happened to nuclear energy in North America because of mismanagement by industry and the government.

In this article, the potential risks of the use of biotechnology and release of genetically engineered crops and biocontrol organisms into the environment are assessed. Various approaches are suggested that could be employed to minimize and reduce the risks of biotechnology to the environment and society.

Ecological issues

Crop resistance to pests

The engineering of crop resistance to insect and plant pathogen pests offers advantages to reduce the use of insecticides and fungicides in crop protection. This will generally reduce problems from pesticides (NAS, 1989; Pimentel et al., 1993) and improve the economics of pest control (Pimentel, 1986). Although crop resistance to pests generally offers environmental benefits from the reduction of pesticide use, care must be exercised to avoid the breeding of toxic chemical-like alkaloids into the crop and/or reducing the nutrient make-up of the crop (Pimentel et al., 1984).

Potential genetic transfer

Naturally occurring bacteria and some viruses are capable of transferring novel DNA sequences to bacteria of other species and genera. Although these transfers are rare in nature (Trevors et al., 1987), the potential for such transfer from engineered organisms to other organisms needs to be evaluated prior to release (Strauss et al., 1986). For instance, trans-conjugant bacteria that received plasmids via biotechnology have subsequently demonstrated 'unexpected alterations' in chemical make-up, virulence and antibiotic resistance (Stotzky and Babich, 1984).

In nature, genetic exchange may take place between closely related plants as well as in microorganisms. For example, major weed species have originated through the hybridization of two intrageneric species, such as the crosses of *Raphanus raphanistrum* × *R. sativus* (radish) and *Sorghum halepense* (Johnson grass) × *S. bicolor* (Sorghum corn) (Colwell *et al.*, 1986).

Genetic transfer and/or relatively small genetic change in an organism may convert an organism from being relatively well integrated in its ecosystem to becoming a major pest. This, in part, appears to be what happened with a strain of the sweet-potato whitefly (*Bemisia tabaci*) in California (Science, 1991). This insect (now termed the 'superbug') has developed into a serious pest, and in 1991 it caused about $200 million damage to California crops.

Single-gene changes and biocontrol pathogenicity

In nature most single-gene changes do not increase the pathogenicity and virulence of an organism in the environment, yet there are a few examples of such gene changes that have had detrimental results. Certain single-gene alterations in animal and plant pathogens have, for instance led to enhanced virulence and increased resistance to pesticides and antibiotics (Alexander, 1985; Georghiou, 1990). For example, Leonard (1987) reports that a single-gene change has transformed a specific microbe from being commensal with its plant or animal host to being pathogenic.

In another instance, two naturally occurring avirulent *Herpes simplex* viruses produced recombinants that were lethal to their host (Javier *et al.*, 1986). Also, a pathogen of grape with a limited host range was converted to a strain with a wide host range when only a single gene was transferred to it (Kerr, 1987). Following a single-gene change, some oat rusts, initially non-pest genotypes for a particular oat variety, became serious pests and were able to overcome resistance in the oat (Browning, 1974).

Similarly, with some insects a single-gene change has enabled an insect to overcome resistant wheat hosts, and other insects, like the housefly and anopheline mosquitoes, have evolved resistance to certain insecticides (Gallun, 1977; Roush and McKenzie, 1987).

These are all clear examples of how a single-gene change can convert various organisms from being non-pathogenic to being pathogenic biocontrol agents and/or pests. Thus, care must be exercised with single, as well as multiple, gene changes in genetically engineered organisms that are to be released into the environment.

Risks from modified native organisms

Some genetic engineers have suggested that there is little or no danger from the recombinant iceminus strain of *Pseudomonas syringae* (*Ps*), which could be used to prevent frost damage of crops, because *Ps* is a native organism that produces related phenotypes in nature (Lindow, 1983). However, there is no proof that just because some native organisms have the ability to alter their interactions within an ecosystem, that the genetic modification and release of the altered native species into the environment may not always be safe for the environment.

For instance, approximately 60% of the major insect pests of US crops once were harmless native organisms (Pimentel, 1987). Many of these organisms moved from benign feeding on natural vegetation to destructive feeding on introduced crops. Thus, the Colorado potato beetle moved from feeding on wild sandbur to feeding on the introduced potato (Casagrande, 1987). This beetle has become a serious pest throughout North America as well as the rest of the world.

Dispersal and control of released organisms

Once a biocontrol or other organism is released into the environment its dispersal depends on many factors (Andow, 1986; see Andow *et al.*, Chapter 10). For instance, the westward movement of the Dutch elm disease fungus from the east coast of the US has been relatively slow because of the prevailing westerly winds. These winds have also impeded the movement of both the European and American bark beetles, the carriers of the fungus (Sinclair and Campana, 1978). Conversely, southern winds facilitate the migrations of the potato leafhopper and the true armyworm to the north. Each spring these insect pests are known to travel from the southern United States to crops in the Northeast and Midwest – a distance of approximately 2500 km (McNeil, 1985).

Movements of natural organisms emphasize the fact that once a genetically modified organism is released, its dispersal will be difficult to monitor and, more importantly, to control. In fact, the only two micro-organisms that have been brought under control after their introduction into North America are: the human disease organism, smallpox; and the plant pathogen, citrus canker (reintroduced in 1984) (E. Civerolo, personal communication). The only macro-organism that has been exterminated from the United States after its introduction has been the Mediterranean fruit fly, the so-called medfly (Hagen *et al.*, 1981). So far, the medfly has been exterminated several times from Florida, Texas, and California.

Probability of environmental risks

The probability of an environmental problem occurring after a single release of any biocontrol or any other engineered organism cannot be accurately predicted at this time. Some ecologists and genetic engineers suggest that the risk is low (Alexander, 1985; NAS, 1987; OTA, 1988). Successful early releases may lead to public confidence that R-DNA organisms are risk-free. Nevertheless, as the number of releases grows, the probability of a problem occurring will increase if public confidence leads to relaxation of protocols. Also, widespread releases of genetically engineered organisms into diverse habitats will very likely increase the chances of a problem occurring.

It might also be expected that the potential environmental problems caused by the release of a genetically engineered organism will vary widely in severity. Experience has shown that exotic organisms have a wide range of impacts on native ecosystems. Consider, for example, the difference between: (i) the displacement of the native Hawaiian talitrid sandhoppers by an exotic amphipod sandhopper species, *Talitroides topitotum* (Howarth, 1985); and (ii) the devastating impact on trees and shrubs in the eastern US from the introduced European gypsy moth (Cameron, 1986).

Although the US public appears to be willing to accept some of the environmental risk associated with genetic engineering (OTA, 1987), one release resulting in an environmental disaster could rapidly change the public's attitude towards genetic engineering and the release of biocontrol organisms (Halvorson *et al.*, 1985; Panem, 1985).

In spite of the rigorous US Government Plant and Animal Quarantine programme and the low survival rate of introduced foreign organisms, some pest organisms have become established in the US (Pimentel, 1987). Recent examples are the reintroductions of the Mediterranean fruit fly into California during the summers of 1987 and 1988 (Holmes, 1987). The last medfly eradication effort in California required the application of massive amounts of insecticides, which cost the government and farmers a total of $174 million (Jackson and Lee, 1985). Major pests also cost the US $64 billion annually in lost crops and livestock, despite the annual application of more than $4 billion in pesticides (Pimentel *et al.*, 1993).

The potential costs of damage resulting from a new pest introduced via genetic engineering, or other causes, can be estimated from data on some current US pests. For example, corn rootworms cost the US about $2 billion annually (Pimentel *et al.*, 1989). Similarly, the European gypsy moth causes an estimated $100 million in damage to ornamental trees and shrubs, and commercial forests, and costs the US an additional $10 million in control each year (Pimentel *et al.*, 1989).

Intentional releases of crops and livestock

Some proponents of biotechnology suggest that the intentional introduction and release of foreign plants and animals for crops and livestock into the US is a reliable model for the prediction of potential problems from this technology (NAS, 1987). However, experience has demonstrated that several serious problems have resulted from the intentional introduction of what were believed at the time to be beneficial organisms. In the US, for instance, of the crop plants introduced, 128 species of agricultural and ornamental plants have become pest weeds (Table 2.1). Some of these are the most serious weed pests that occur in the US. For example, in agriculture these include, Johnson grass, Mississippi chick corn, goatsrue and crotalaria. The weed pests that originated

Table 2.1. Agricultural and ornamental plant introductions that became pest weeds in the United States

Plant introduced	Purpose of introduction	Weed problem	Sources
Johnson grass *Sorghum halepense* (L.) Pers.	Forage	Crops, pastures	McWhorter (1971); Williams and Hayes (1984)
Mississippi chicken corn (shatter cane) *Sorghum bicolor* (L.) Moench	Grain	Grain crops	DeWet and Harlan (1975)
Goatsrue *Galega officinalis* L.	Forage	Crops, pastures	Williams (1980)
Crotalaria *Crotalaria spectabilis* Roth *Crotalaria retusa* L.	Forage and green manure	Soybeans, rangelands	Patterson (1982)
Sicklepod milkvetch *Astragalus falcatus* Lam.	Forage	Rangelands	Williams (1980)
Reed canarygrass *Phalaris arundinacea* L.	Forage	Irrigation ditches, canals	USDA (1971)
Bermudagrass *Cynodon dactylon* (L.) Pers.	Forage	Pastures	Brown *et al.* (1985); Muenscher (1980)
Cogongrass *Imperata cylindrica* (L.) Beauv.	Forage	Crops, pastures	Patterson *et al.* (1983)
Bush beardgrass *Andropogon glomeratus* (Walt.) BSP	Forage	Displaces native species	Smith (1985)
Broomsedge *Andropogon virginicus* L.	Forage	Displaces native plants	Smith (1985)
Kikuyugrass *Pennisetum clandestinum* Hochst. ex. Chiov.	Forage	Displaces native species	Smith (1985)
Meadow ricegrass *Microlaena stipioides* (Labill.)	Forage	Displaces native grasses	Smith (1985)
California grass, tall panicum *Brachiaria mutica* (Forsk.)	Unknown[a]	Cultivated crops, native habitats and roadsides	Smith (1985)
Haola koe *Leucaena leucocephala* (Lam.) de Wit	Forage	Cultivated crops	Smith (1985); Haselwood and Motter (1983)
Guineagrass *Panicum maximum* Jacq.	Forage	Cultivated crops	Smith (1985); Haselwood and Motter (1983)
Molasses grass *Melinis minutiflora* Beauv.	Forage	Wastelands	Smith (1985); Haselwood and Motter (1983)
Kochia *Kochia scoparia* (L.) Schrad.	Forage, ornamental	Crops, pastures	Muenscher (1980); Wiley *et al.* (1985)
Kudzu *Pueraria lobata* (Willd.) Ohwi	Forage and erosion control	Forests, roadways	Hopson (1981); Laycock (1983)
Cordgrass *Spartina anglica*	Forage and bank stabilization	Wetland bird refuges	Johnson (1985)
Yellow Himalayan blackberry *Rubus ellipticus* Sm.	Fruit	Displaces native species	Smith (1985); Neal (1965)
Raspberry *Rubus glaucus* Bth.	Fruit	Displaces native species	Smith (1985)
Rubus nivalis Doug.	Fruit	Displaces native species	Smith (1985)

Table 2.1. (*cont.*)

Plant introduced	Purpose of introduction	Weed problem	Sources
Florida prickly blackberry (high bush blackberry) *Rubus argutus* Link	Fruit	Displaces native species	Smith (1985)
Guava *Psidium guajava* L.	Fruit	Wastelands and roadsides	Smith (1985); Neal (1965)
Buffelgrass *Cenchrus cillarus* L.	Cover for erosion control	Displaces native species	Smith (1985)
Chinese violet *Asystasia gangetica* (L.) T. Anders	Cover crop	Displaces herbaceous cover	Smith (1985); Neal (1965)
Klu (huisache) *Acacia farnesiana* Willd.	Perfume from flowers	Pastures	Smith (1985); Haselwood and Motter (1983)
Hemp or marijuana *Cannabis sativa* L.	Fibre, medicine	Pastures	Muenscher (1980)
New Zealand flax *Phormium tenax* J.R. and G. Forst	Fibre production	Displaces native species	Smith (1985); Neal (1965)
Multiflora rose *Rosa multiflora* Thunb. ex Murr.	Windbreaks, cover plantings	Pastures	USDA (1971)
Gorse *Ulex europaeus* L.	Hedge for sheep	Pastures and rangelands	Smith (1985); Haselwood and Motter (1983)
Opiuma *Pithecelobium dulce* (Roxb.) Benth.	Hedge planting	Displaces native species	Smith (1985); Neal (1965)
Cats claw *Caesalpinia sepiaria* Roxb.	Hedge ornamental	Pastures and rangelands	Smith (1985); Haselwood and Motter (1983)
Molucca albizia *Albizia moluccana* Miq	Forestation	Displaces native	Smith (1985); Neal (1965)
Mexican ash *Fraxinus uhdei* (Wenzig)	Forestation	Displaces native trees	Smith (1985); Neal (1965)
Melochia *Melochia umbellata* (Houtt.)	Forestation	Displaces native trees	Smith (1985); Neal (1965)
Slash pine *Pinus caribaea* Morelet	Forestation	Displaces native species	Smith (1985); Neal (1965)
Mexican weeping pine *Pinus patula* Schlecht. and Cham.	Forestation	Displaces native species	Smith (1985); Neal (1965)
Juniper berry *Citharexylum caudatum* L.	Arboretum	Displaces native species	Smith (1985); Neal (1965)
Koster's curse *Clidemia hirta* (L.)	Brought in by marijuana growers	Pastures and rangelands	Smith (1985); Haselwood and Motter (1983)
Tansy, common *Tanacetum vulgare* L.	Herb	Gardens, roadsides	Pammel (1911); Muenscher (1980)
Dyers woad *Isatis tinctoria* L.	Dyes	Rangelands, crops	Aspevig *et al.* (1985)
Henbane, black *Hyoscyamus niger* L.	Medicine	Roadsides	Williams (1980)
Belladonna *Atropa belladonna* L.	Medicine	Roadsides	Williams (1980)

Table 2.1. (*cont.*)

Plant introduced	Purpose of introduction	Weed problem	Sources
Melaleuca *Melaleuca quinquenervia* (Cau.) Blake	Afforestation	Wetland habitats	Williams (1980); Vietmeyer (1986)
Waterhyacinth *Eichhornia crassipes* (Mart.) Solms	Ornamental for pools	Lakes, waterways	Penfound and Earle (1948); Pierce (1983); Gopal (1987)
Hydrilla *Hydrilla verticillata* (L.f.) Royle	Ornamental for aquaria	Lakes, waterways	Anon. (1984*a*, *b*)
Brazilian peppertree *Schinus terebinthifolius* Raddi	Ornamental	Forests, parks, gardens	Williams (1980); Vietmeyer (1986)
Macartney rose *Rosa bracteata* J.C. Wendl.	Ornamental	Pastures	Williams (1980)
Slender speedwell *Veronica filiformis* Sm.	Ornamental	Turf, lawns	Cisar (1981)
Lantana, large leaf *Lantana camara* L.	Ornamental	Fence rows, pastures	Elton (1958); Haselwood and Motter (1983)
Dalmatian toadflax *Linaria genistifolia* subsp. *dalmatica* (L.) Maire & Petitmengin	Ornamental	Rangelands	Robocker (1974); USDA (1971)
Japanese honeysuckle *Lonicera japonica* Thunb.	Ornamental	Pastures, woodlands	USDA (1971); Muenscher (1980)
Japanese knotweed *Polygonum cuspidatum* Sieb. & Zucc.	Ornamental	Lowlands, homesites	Muenscher (1980)
French tamarisk *Tamarix gallica* L.	Ornamental	Pastures, floodplains	Williams (1980)
Stranglervine *Morrenia odorata* (Hook. & Arn.) Lindl.	Ornamental	Citrus orchards	Ridings (1985)
Yellow toadflax *Linaria vulgaris* Mill.	Ornamental	Rangelands	USDA (1971)
Bouncingbet *Saponaria officinalis* L.	Ornamental	Pastures	Pammel (1911), Muenscher (1980)
European buckhorn *Rhamnus cathartica* L.	Ornamental	Pastures, rangelands	Williams (1980)
Chinaberry *Melia azedarach* L.	Ornamental	Pastures, rangelands	Williams (1980)
Corn cockle *Agrostemma githago* L.	Ornamental	Wheat, grasslands	USDA (1971); Muenscher (1980)
Foxglove *Digitalis purpurea* L.	Ornamental	Pastures	Muenscher (1980)
Jimsonweed *Datura stramonium* L.	Ornamental	Pastures, croplands	Muenscher (1980); USDA (1971)
Precatory bean *Abrus precatorius* L.	Ornamental, medicine	Fence rows, roadsides	Williams (1980)
Purple loosestrife *Lythrum salicaria*	Ornamental	Wetland bird refuges, waterways	USDI (1987)
Black wattle, green wattle *Acacia mearnsii* Willd.	Ornamental	Displacement of native species	Smith (1985); Neal (1965)

Table 2.1. *(cont.)*

Plant introduced	Purpose of introduction	Weed problem	Sources
Java plum *Eugenia cumini* (L.) Druce	Ornamental and forestation	Displaces native trees	Smith (1985); Neal (1965)
Roseapple *Eugenia jambos* (L.)	Ornamental and forestation	Displaces native trees	Smith (1985); Neal (1965)
Chinese banyan *Ficus retusa* L.	Ornamental and forestation	Displaces native trees	Smith (1985); Neal (1965)
Kiawe *Prosopis pallida* (Humb. and Bonpl.)	Ornamental and forestation	Displaces native trees	Smith (1985); Neal (1965)
Silk oak *Grevillea robusta* A. Cunn.	Ornamental and forestation	Allelopathy inhibits native species	Smith (1985); Neal (1965)
Fountaingrass, Crimison *Pennisetum setaceum* (Forsk.) Chiov	Garden	Displaces native species	Smith (1985); Neal (1965)
German ivy *Senecio mikanioides* Otto ex Walp	Ornamental	Rangelands	Smith (1985); Haselwood and Motter (1983)
Palmgrass *Setaria palmifolia* (Koen.) Staph	Ornamental	Wastelands and roadsides	Smith (1985); Haselwood and Motter (1983)
African tuliptree *Spathodea campanulata* Beauv.	Ornamental	Displaces native trees	Smith (1985); Haselwood and Motter (1983)
Wild Marigold, stinkweed *Tagetes minuta* L.	Ornamental	Pastures and rangelands	Smith (1985); Haselwood and Motter (1983)
False kamani, Tropical almond *Terminalia catappa* L.	Ornamental	Displaces native species	Smith (1985)
Glorybush *Tibouchina urvilleana* (DC.) Cogn. Difam	Ornamental	Displaces native species	Smith (1985)
Common mullein, Velvet plant *Verbascum thapsus* L.	Ornamental	Displaces native species	Smith (1985)
Bamboo *Bambusa* sp.	Ornamental	Impenetrable thickets	Neal (1965)
Common ironwood, Australian-pine *Casuarina equisetifolia* L. ex J. R. and G. Forst.	Ornamental	Displaces native species	Smith (1985)
Swamp oak, Scaly-bark beefwood *Casuarina glauca* Sieb.	Ornamental	Displaces native species	Smith (1985)
Trumpet tree *Cecropia peltata* Sandmark	Ornamental	Displaces native species	Smith (1985)
Bocconia frutescens L.	Ornmanetal	Displaces native species	Smith (1985); Neal (1965)
Fayatree, firebush *Myrica faya* Ait.	Ornamental	Pastures and rangelands	Smith (1985); Haselwood and Motter (1983)
Octopus tree Brassaia actinophylla Endl.	Ornamental	Displaces native trees	Smith (1985); Neal (1965)
Fiddlewood *Citharexylum spinosum* L.	Ornamental	Displaces native species	Smith (1985); Neal (1965)
Glorybower *Clerodendron japonicum* (Thunb.) Sweet	Ornamental	Pastures and rangelands	Smith (1985); Haselwood and Motter (1983)
Trailing velvet plant *Rubus moluccanus* L.	Ornamental	Displaces native species	Smith (1985)

Table 2.1. (*cont.*)

Plant introduced	Purpose of introduction	Weed problem	Sources
Formosan koa *Acacia confusa* Merr.	Ornamental	Displaces other plants because of allelopathy	Smith (1985)
Sour bush *Pluchea odorata* (L.)	Ornamental	Displaces native species	Smith (1985); Neal (1965)
Strawberry guava *Psidium cattleianum* Sabine	Ornamental	Displaces native species	Smith (1985); Neal (1965)
Indian rhododendron *Melastoma malabathricum* L.	Ornamental	Displaces native species	Smith (1985); Neal (1965)
New Zealand tea, tea tree *Leptospermum scoparium* J. R. & G. Forst	Ornamental	Displaces native trees	Smith (1985)
Hairy cats-ear (Spotted cats-ear) *Hypochoeris radicata* L.	Ornamental	Cultivated crops	Smith (1985); Neal (1965)
Downy rosemyrtle *Rhodomyrtus tomentosa* (Ait.)	Ornamental	Displaces native species	Smith (1985); Neal (1965)
Castorbean *Ricinus communis* L.	Ornamental	Pastures and rangelands	Smith (1985); Neal (1965)
Sweet granadilla, Lemona *Passiflora ligularis* Juss.	Ornamental	Displaces native species	Smith (1985); Neal (1965)
Banana poka *Passiflora mollissima* (HBK.) Bailey	Ornamental	Displaces native species	Smith (1985); Neal (1965)
Huehue-haole (Corky stemmed passion flower) *Passiflora suberosa* L.	Unknown[a]	Displaces native species	Smith (1985); Neal (1965)
Hilo grass (sour paspalum) *Paspalum conjugatum* Berg.	Unknown[a]	Displaces native species	Smith (1985); Neal (1965)
Red mangrove, American mangrove *Rhizophora mangle* L.	Unknown[a]	Displaces native species	Smith (1985); Neal (1965)
Tree manuka *Leptospermum ericoides* A. Rich	Ornamental	Displaces native species	Smith (1985)
White ginger *Hedychium coronarium* Koenig	Ornamental	Displaces native species	Smith (1985); Neal (1965)
Yellow ginger *Hedychium flavescens* Carey	Ornamental	Displaces native species	Smith (1985); Neal (1965)
Kahili ginger *Hedychium gardnerianum* Roscoe	Ornamental	Displaces native species	Smith (1985); Neal (1965)
White moho *Heliocarpus popayaensis* HBK.	Reforestation	Displaces native species	Smith (1985); Neal (1965)
Velvetgrass, common *Holcus lanatus* L.	Ornamental	Displaces native species	Smith (1985)
Mauritius hemp *Furcraea foetida* (L.) Haw	Unknown[a]	Dense thickets	Smith (1985); Neal (1965)
Kahili flower, Haiku *Grevillea banksii* R. Br	Unknown[a]	Allelopathy inhibits native species	Smith (1985); Haselwood and Motter (1983)
Linociera *Linociera intermedia* Wight	Unknown[a]	Displaces native species	Smith (1985); Neal (1965)

Table 2.1. *(cont.)*

Plant introduced	Purpose of introduction	Weed problem	Sources
Beggar's tick, Spanish needle *Bidens pilosa* L.	Unknown[a]	Cultivated crops and roadsides	Smith (1985); Neal (1965)
Cluster pine *Pinus pinaster* Ait.	Reforestation	Displaces native species	Smith (1985); Neal (1965)
Indian fleabane *Pluchea indica* (L.) Less.	Unknown[a]	Pasture and rangelands	Smith (1985); Neal (1965); Haselwood and Motter (1983)
Glenwood grass *Sacciolepis indica* (L.) Chase	Unknown[a]	Displaces native species	Smith (1985)
Sweet vernalgrass *Anthoxanthum odoratum* L.	Unknown[a]	Invades disturbed areas	Smith (1985)
Shoebutton ardisia *Ardisia humilis* Vahl	Unknown[a]	Dense stands crowd out other species	Smith (1985)
New Zealand laurel *Corynocarpus laevigatus* J. E. & G. Forst.	Unknown[a]	Displaces native species	Smith (1985)
Woodrose *Merremia tuberosa* (L.) Rendle	Ornamental	Displaces native species	Smith (1985); Neal (1965); Haselwood and Motter (1983)
Triana *Miconia magnifica*	Ornamental	Displaces native species	Smith (1985)

[a]Some of the plants listed here from Smith (1985) may have been accidental introductions; however, the author indicates that most of these were introduced intentionally.

from ornamental crop introductions include the multiflora rose, waterhyacinth, and lantana (Table 2.1).

Also genetic similarities between many crops and weeds are evident from the fact that 11 of 18 of the most serious weeds of the world are crops in other regions of the world (Colwell *et al.*, 1985). The introduced crop plants escaped and became serious pests in their new ecosystems.

In addition, 9 out of a total of 20 introduced domestic animal species in the US have displaced or destroyed native species, and in general, have become serious environmental pests (Table 2.2). For example, these include feral goats, pigs, horses, and cats (Table 2.3). This does not mean that all plant and animal introductions should be halted.

Intentional release of biological control agents

Specialist predators and parasites of pest insects and weeds introduced for the purposes of biological control generally have had minimal environmental impacts and have reduced the need for ecologically disruptive pesticides (Pimentel *et al.*, 1984). However, there have been some environmental problems associated with the introduction and release of biocontrol agents in the US. For example, two fish species, the grass carp and blue tilapia, were introduced in order to control aquatic weeds in canals, streams, and ponds (Table 2.3). Although these fish reduced aquatic weeds, they also reduced the number of native sport fish and generally degraded the aquatic ecosystem (Avault, 1965; Cross, 1969; Courtenay and Robins, 1975; Pierce, 1983).

Table 2.2. Agricultural and ornamental plant introductions and agricultural, sport, and pet animal introductions that became pests[a] in the United States

Introductions	No. of species of intentional introductions	No. of species that became pests
Agricultural and ornamental plants	5800[b]	125
Domestic mammals and birds	20[c]	10
Sport mammals and birds	20[d]	9
Biological control vertebrates	6	6
Aquatic and sport fish	2000[e]	5

[a] A pest can be narrowly defined as an organism with direct impacts on human welfare, or more broadly defined to include negative impacts on indigenous organisms and habitats. Many of the authors cited used the narrow definition.
[b] Kresovich (1987).
[c] Estimated number of introduced mammals and birds.
[d] Estimated; however, this value does not include 51 exotic species introduced into Texas for game purposes – none of which to date have been classed as a problem (Armstrong and Wardroup, 1980).
[e] Estimated (McCann, 1984).

In addition, the English sparrow was introduced into the US for the biological control of caterpillars that fed on crops and trees (Southern, 1945; Robbins, 1973). But the sparrows consumed only a few caterpillars and instead preferred to feed on grain and fruit crops. They also have interfered with and displaced some native bird species, e.g. the bluebird (Zeleny, 1976). Later, the common myna bird was introduced from India into Hawaii for the biological control of armyworms, but the bird has proved to be relatively ineffective (Laycock, 1966). Also the myna bird unfortunately is having a negative impact on various native birds in Hawaii, including the wedge-tailed shearwater (Byrd, 1972).

One of the worst examples of an introduced biological control agent becoming a pest is the Indian mongoose (Pimentel, 1955). The mongoose was introduced into Puerto Rico, St Johns, St Croix, and several of the Hawaiian Islands for the control of rat pests in sugarcane. After the mongoose reduced the numbers of the Norway rat, the numbers of the tree rat increased and took the place of the Norway rat. Because the mongoose cannot climb trees, the tree-nesting tree rat was protected from predation from the diurnal mongoose (Pimentel, 1955). Thus, these tropical islands still have their original rat problem plus several other environmental problems.

For example, the introduced Indian mongoose eliminated most of the ground-nesting native birds and lizards (Pimentel, 1955). Some of the lizards were beneficial in controlling certain insect pests in sugarcane and other crops. With the reduction in numbers of certain ground lizards, it has been suggested that some crop insect pests increased in number. The mongoose also preys on poultry, and, in addition it is the major reservoir and vector of rabies in Puerto Rico (Pimentel, 1955).

Another vertebrate, the giant toad (*Bufo marinus*) was introduced to many tropical islands in the US for the biological control of pest insects that inhabit the soil and ground around sugarcane and other crops (Table 2.3) (Mooney and Drake, 1986). The toad was relatively ineffective and became a problem itself by poisoning vertebrates that preyed on it and also by displacing other amphibian species on the islands where it was introduced.

Note that most of the problems with introduced biological agents have been from the use of vertebrate biocontrol agents (Pimentel et al., 1984). The use of invertebrates for biological control has produced relatively few environmental problems, this includes the situations where insects have been employed for weed control (Pimentel et al., 1984; Howarth, 1991).

The use of 'new association biocontrol agents' (Pimentel, 1963; Hokkanen and Pimentel, 1989) has been proposed to be more effective than classical biocontrol. However, Goeden and Kok (1986) have suggested that there is a greater

Table 2.3. Agricultural, sport and pet animal introductions that became pests in the United States

Animal introduced	Purpose of introduction	Pest problem	Sources
Feral goat *Capra prisca*	Milk, meat	Forests, shrubs in Hawaii	Roots (1976); Baker and Reeser (1972)
Feral sheep *Ovis aries*	Meat, wool	Vegetation destruction in Hawaii and Channel Islands	Roots (1976); Van Vuren and Coblentz (1987)
Feral pig *Sus scrofa* (domestic)	Meat, leather	Vegetation destruction in Hawaii	Pimm and Pimm (1982)
Feral burro *Equus asinus*	Draught animal	Vegetation destruction	Laycock (1966); Presnall (1985)
Feral horse *Equus caballus*	Draught animal	Rangeland vegetation	McKnight (1964); Presnall (1985)
Feral cat *Felis catus*	Pet	Prey on wildlife	DeVos *et al.* (1956)
Feral dog *Canis familiaris*	Pet	Prey on wildlife	DeVos *et al.* (1956)
Reindeer *Rangifer tarandus*	Meat, leather	Rangeland vegetation	Roots (1976)
European rabbit *Oryctolagus cuniculus*	Meat	Vegetation destruction on Smith Islands	Roots (1976)
European gypsy moth *Lymantria dispar*	Silk production	Forests and ornamentals	Cameron (1986)
European hare *Lepus europaeus*	Sport hunting	Crops, orchards	Silver (1924)
Nutria *Myocastor coypus*	Fur	Crops, orchards, drainage canals	DeVos *et al.* (1956); Schitoskey *et al.* (1972); Kuhn and Peloquin (1974)
European wild boar *Sus scrofa*	Sport hunting	Crops, forests, wildlife	Stegman (1938); Pines and Gerdes (1973)
Sika deer *Cervus nippon*	Sport hunting	Displaced whitetailed deer in some locations	Flyger (1960); Feldhamer *et al.* (1978); Armstrong and Harmel (1981)
European red fox *Vulpes vulpes*	Sport hunting	Prey on wildlife	DeVos *et al.* (1956)
Ring-necked pheasant *Phasianus colchicus*	Sport hunting	Displaced prairie chickens in some locations	Sharp (1957); Laycock (1966); Vance and Westemeirer (1979); Robinson and Bolen (1984)
Muscovy duck *Cairina moschata*	Sport hunting	Might displace wood ducks or black-bellied whistling ducks	Weller (1969); Bolen (1971)
German carp *Cyprinus carpio*	Sport fishing	Muddied lake water, native sport fish	Cole (1904); Threinem and Helm (1954); Robel (1961)
Black tilapia *Tilapia melanpleura*	Food	Native sport fish	Courtenay *et al.* (1974)
Starling *Sturnus vulgaris*	Aesthetics, Shakespeare	Faeces and disease	Laycock (1966)

Table 2.3. (*cont.*)

Animal introduced	Purpose of introduction	Pest problem	Sources
Guppy *Lebistes reticuletus*	Aquarium fish	Displace native fish	Courtenay and Robins (1975)
Spotted tilapia *Tilapia mariae*	Aquarium fish	Displace native fish	Courtenay and Robins (1975)
Walking catfish *Clarias betrachus*	Aquarium fish	Native sport fish	Courtenay *et al.* (1974)
Black acara *Cichlasoma bimaculatum*	Fish trade, aquarium fish	Native sport fish	Courtenay *et al.* (1984)
Grass carp *Ctenopharyn godon idella*	Biological control of aquatic weeds	Native sport fish	Avault (1965); Cross (1969); Pierce (1983)
Blue tilapia *Tilapia aurea*	Biological control, aquatic plants	Degraded aquatic ecosystem	Courtenay and Robins (1975)
Indian mongoose *Herpestes auropunctatus*	Biological control of rats	Poultry, native birds and lizards, human diseases	Tierkel *et al.* (1952); Baldwin *et al.* (1952); Pimentel (1955); Nellis and Everard (1983)
English sparrow *Passer domesticus*	Biological control of caterpillars	Grain and fruit crops, native birds	Southern (1945); Robbins (1973); Zeleny (1976)
Common myna *Acridotheres tristis*	Biological control of armyworms	Native birds	Laycock (1966); Byrd (1972)
Giant toad *Bufo marinus*	Biological control of ground insects	Poisonous to vertebrates displaced native amphibians	Mooney and Drake (1986)

environmental threat from new associations than from the use of classical biocontrol agents particularly in the control of weeds. Actually the specificity of new association biocontrol agents has been shown to be equal to that of classical control agents (Hokkanen and Pimentel, 1989). In fact, the only case of an introduced biological agent for weed control moving to feed on crops and thus becoming a pest was the introduction of the lace bug (*Teleonemia scrupulosa*), a classical control agent, to control *Lantana camara* in East Africa (Greathead, 1973; Pimentel *et al.*, 1984). The lace bug began to feed on the African sesame crop. However, there has not been an instance of an environmental problem with new association biocontrol agents introduced for weed or other pest control (Pimentel, 1986).

Ecological niches

Some biotechnology specialists feel, that in general, ecological niches are filled, and therefore there is little or no threat from the release of genetically engineered organisms (NAS, 1987). However, this theory is not supported by fact. For example, over time 1500 exotic insect species have become established in the US and of these about 17% have become pests or have had an adverse impact on resident native species (Sailer, 1978, see Fig. 2.1). These data confirm that few of the niches in natural ecosystems are filled (Colwell *et al.*, 1985; Herbold and Moyle, 1986), and therefore there is ample opportunity for new species to become established in the US and elsewhere in the world. Thus, the argument that genetically engineered organisms will not become established because of competition for occupied niches is not substantiated.

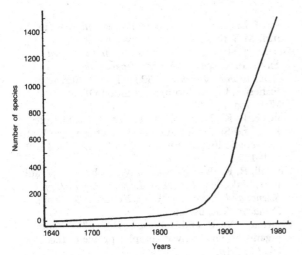

Fig. 2.1 Number of species of insects and mites introduced into the 48 contiguous states between 1640 and 1980 (after Sailer, 1983).

Conclusion

Although the financial rewards for successful research in biotechnology and releases of genetically engineered organisms into the environment are enormous, the incentives are unlikely to encourage innovation aimed at providing the greatest ecological and humanitarian benefits (Buttel *et al.*, 1985; Krimsky, 1991). In addition, the highly competitive and secretive climate that currently surrounds most biotechnology research in North America may slow down the total research process and its success. For these reasons, a stronger broad scientific and public role are needed for the definition and support of key research objectives, and for formulating standards to regulate the biotechnology industry. Such attention is needed to temper the substantial influence of private enterprise on the development of biotechnology.

Thus, despite the great potential benefits of biotechnology to crops and biological control of pests, the inherent risks of this new technology must be faced and its use must be effectively regulated. Under the current regulatory structure to ensure the safety of biotech products, the approval responsibility in the US is divided among five federal agencies (US Department of Agriculture, Food and Drug Administration, National Institutes of Health, Occupational Safety and Health Administration, and Environmental Protection Agency (EPA). This fragmented structure guarantees eventual disaster. The protocol for approval and all regulation should be placed in one agency, like the EPA, in order to ensure scientifically based and responsible regulation of biotechnology.

Failure to implement an effective regulatory structure and to exercise caution could lead to serious environmental, economic, and social problems in the US and elsewhere in the world. Although the pressure for immediate results is great, unexpected environmental problems could delay or jeopardize the realization of the potential benefits from biotechnology and biocontrol.

Acknowledgments

We thank the following people for reading an earlier draft of this paper and for their helpful suggestions: David Andow, University of Minnesota; Nyle Brady, USAID; Winston J. Brill, Agracetus, Wisconsin; Liebe Cavalieri, Memorial Sloan-Kettering Cancer Center; Joel Cohen, USAID; Robert K. Colwell, University of California at Berkeley; Jack R. Coulson, ARS-USDA; Charles Edquist, Linköping University, Sweden; Daniel J. Goldstein, Universidad de Buenos Aires; Fred Gould, North Carolina State University; Arthur Kelman, University of Wisconsin; Stephen Kresovich, USDA-ARS Germplasm Resource Unit, Geneva; Sheldon Krimsky, Tufts University; Joseph J. Molnar, Auburn University, AL; Norman Myers, Oxford, England; Zev Naveh, Israel Institute of Technology, Haifa; Gil Omenn, University of Washington, Seattle; A. Sasson, UNESCO, Paris; Guenther Stotzky, New York University; Trevor Suslow, Advanced Genetic Sciences, Oakland, CA; C.D. Upper, University of Wisconsin; and M. Coburn Williams, Utah State University.

At Cornell University, Ithaca, New York we thank: Fred Buttel, June Fessenden-MacDonald, Robert Kalter, Richard T. Roush, Mark E. Sorrells and Bruce Wilkins.

References

Alexander, M. (1985). Ecological consequences: reducing the uncertainties. *Issues in Science and Technology* **1**, 57–68.

Andow, D. A. (1986). Fate and movement of organisms in the environment. Part II. Dispersal of microorganisms with emphasis on bacteria. *Environment Management* **10**, 470–487.

Anonymous (1984*a*). The aquatic weed hydrilla has now invaded Washington, DC. *BioScience* **34**, 463.

Anonymous (1984*b*). Monster plant is choking U.S. waters. *International Wildlife* **14**, 32.

Armstrong, W. G. and Harmel, D. E. (1981). Exotic mammals competing with the natives. *Texas Parks and Wildlife* **39**, 6–7.

Armstrong, W. G. and Wardroup, S. (1980). *Statewide Census of Exotic Big Game Animals*. Federal Aid Project W-109-R-3. Texas Parks and Wildlife Department, Austin, TX.

Aspevig, K., Fay, P. and Lacey, J. (1985). *Dyers Woad: A Threat to Rangeland in Montana*. Montguide – MT – Agriculture – Montana State University Cooperative External Service August 1985 (8523), Bozeman, Mont.

Avault, M. N. (1965). Biological weed control with herbivorous fish. *Proceedings of the 5th Weed Control Conference* **18**, 590–591.

Baker, J. K. and Reeser, D. W. (1972). *Goat Management Problems in Hawaii Volcanoes National Park*. National Park Service National Resources, Report 2, Honolulu.

Baldwin, P. H., Schwartz, C. W. and Schwartz, E. R. (1952). Life history and economic status of the mongoose in Hawaii. *Journal of Mammology* **36**, 62–68.

Bolen, E. G. (1971). Some views on exotic water fowl. *Wilson Bulletin* **83**, 430–434.

Brown, S. M., Whitwell, T. and Street, J. (1985). Common Bermudagrass (*Cynodon dactylon*) (*Gossypium hirsutum*) competition in cotton. *Weed Science* **33**, 503–506.

Browning, J. A. (1974). Relevance of knowledge about natural ecosystems to development of pest management programs for agroecosystems. *Proceedings of the American Phytopathology Society* **1**, 191–199.

Buttel, F. H., Kennet, M. and Kloppenberg, J. (1985). *From Green Revolution to Biorevolution: Some Observations on the Changing Technological Bases of Economic Transformation in the Third World*. University of Chicago Press: Chicago.

Byrd, G. V. (1972). Common myna predation on wedge-tailed shearwater eggs. *Elepaio* **39**, 69–70.

Cameron, E. A. (1986). The gypsy moth *Lymantria dispar* L. (Lepidoptera, Lymantriidae) in the New World. *Melsheimer Entomology Serive* **36**, 9–19.

Casagrande, R. A. (1987). The Colorado potato beetle: 125 years of mismanagement. *Bulletin of the Entomology Society of America* **33**, 142–150.

Cisar, J. L. (1981). Control of *Veronica filiformis* in turf. M.S. thesis, Cornell University, Ithaca.

Cole, L. J. (1904). *The German carp in the United States*. In *Appendix 12 to the Report of the Commissioner of Fisheries*. 523–641. US Bureau of Fisheries, US Government Printing Office, Washington, DC.

Colwell, R. K., Norse, E. A., Pimentel, D., Sharples, F. E. and Simberloff, D. (1985). Letter to the Editor on Genetic Engineering in Agriculture. *Science* **229** 111–112.

Colwell, R. K., Barnthouse, L. W., Dobson, A., Taub, F. and Wetzler, R. (1986). *Letter to the Office of Science and Technology Policy, Executive Office of the President*. Washington, DC.

Courtenay, W. R. and Robins, C. R. (1975). Exotic organisms: an unsolved complex problem. *BioScience* **25**, 306–313.

Courtenay, W. R., Sahlman, H. F., Miley, W. W. and Herrema, D. J. (1974). Exotic fishes in fresh and brackish waters in Florida. *Biological Conservation* **6**, 291–302.

Courtenay, W. R., Hensley, D. A., Taylor, J. N. and McCann, J. A. (1984). Distribution of exotic fishes in the continental United States. In *Distribution, Biology, and Management of Exotic Fishes* (ed. W. R. Courtenay and J. R. Stauffer), pp. 41–77. Johns Hopkins University Press, Baltimore, MD.

Cross, E. G. (1969). Aquatic weed control using grass carp. *Journal of Fish Biology* **1**, 27–30.

DeVos, A., Manville, R. H. and Van Gelder, R. G. (1956). Introduced mammals and their influence on native biota. *Zoologica* **41**, 163–194.

DeWet, A. and Harlan, J. R. (1975). Weeds and domesticates: evolution in the man-made habitat. *Economics of Botany* **29**, 99–107.

Elton, C. S. (1958). *The Ecology of Invasions by Animals and Plants*. Methuen, London.

Feldhamer, G. A., Chapman, J. A. and Miller, R. L. (1978). Sika deer and white-tailed deer on Maryland's eastern shore. *Wildlife Society Bulletin* **6**, 155–157.

Flyger, V. (1960). Sika deer on islands in Maryland and Virginia. *Journal of Mammology* **41**, 140.

Gallun, R. L. (1977). Genetic basis of Hessian fly epidemics. *Annual New York Academy of Science* **287**, 223–229.

Georghiou, G. P. (1990). Overview of insecticide resistance. In *Managing Resistance to Agrochemicals: from Fundamental Research to Practical Strategies* (ed. M. B. Green, H. M. LeBaron and W. K. Moberg), pp. 18–41. American Chemical Society, Washington, DC.

Goeden, R. D. and Kok, L. T. (1986). Comments on a proposed 'new' approach for selecting agents for the biological control of weeds. *Canadian Entomology* **118**, 51–58.

Gopal, B. (1987). *Water Hyacinth*. Elsevier Science, Amsterdam, The Netherlands.

Greathead, D. J. (1973). Progress in the biological control of *Lantana camara* in East Africa and discussion of problems raised by the unexpected reaction of some of the more promising insects to *Sesamum indicum*. In *Proceedings of 2nd International Symposium on Biological Control of Weeds* (ed. by P. H. Dunn), pp. 89–92. London: Commonwealth Agricultural Bureaux, Slough.

Hagen, K. S., Allen, W. W. and Tassan, R. L. (1981). Mediterranean fruit fly: the worst may be yet to come. *Californian Agriculture* 5–7 March–April.

Halvorson, H. O., Pramer, D. and Rogul, M. (eds) (1985). *Engineered Organisms in the Environment: Scientific Issues*. American Society for Microbiology, Washington, DC.

Haselwood, E. L. and Motter, G. G. (1983). *Handbook of Hawaiian Weeds*. University of Hawaii Press, Honolulu.

Herbold, B. and Moyle. P. B. (1986). Introduced species and vacant niches. *American Nature* 128, 751–760.

Hokkanen, H. M. T. and Pimentel, D. (1989). New associations in biological control: theory and practice. *Canadian Entomology* 121, 828–840.

Holmes, M. (1987). *Control of the Medfly in California*. APHIS, USDA, Beltsville, MD.

Hopson, J. (1981). The green menace. *Science Digest* 89, 146–147.

Howarth, F. G. (1985). Impacts of alien land arthropods and mollusks on native plants and animals in Hawaii. In *Hawaii's Terrestrial Ecosystems: Preservation and Management* (ed. C. P. Stone and J. M. Scott), pp. 149–179 Cooperative National Park Resources Studies Unit, University of Hawaii, Honolulu.

Howarth, F. G. (1991). Environmental impacts of classical biological control. *Annual Review of Entomology* 36, 485–509.

Jackson, D. S. and Lee, B. G. (1985). Med-fly in California 1980-1982. *Bulletin of Entomology Society of America* 31, 29–37.

Javier, R. T., Sedarati, F. and Stevens, J. G. (1986). Two avirulent herpes simplex viruses generate lethal recombinats in vivo. *Science* 234, 746–798.

Johnson, P. (1985). Quiet invasion: Cordgrass establishes a beach head. *Oceans* 18, 62–64.

Kerr, A. (1987). The genetic basis for virulence, pathogenicity and host range in *Agrobacterium tumefaciens*. In *Plant Pathogenic Bacteria* (ed. E. L. Civerolo, A. Collmer, R. E. Davis and A. G. Gillaspie), pp. 377–387. Martinus Nijhoff, Dordrecht, The Netherlands.

Kresovich, S. (1987). *Introduced Crop Plants*. USDA-ARS Germplasm Resource Unit, Geneva, NY.

Krimsky, S. (1991). *Biotechnics and Society: The Rise of Industrial Genetics*. Praeger, New York.

Kuhn, L. W. and Peloquin, E. P. (1974). Oregon's nutria problem. *Proceedings of the Vertebrate Pest Conference* 6, 101–105.

Laycock, G. (1966). *The Alien Animals*. Natural History Press, New York.

Laycock, G. (1983). Mile-a-minute vine. *Audubon* 85, 30, 32.

Leonard, K. J. (1987). The host population as a selective factor. In *Populations of Plant Pathogens* (ed. M. S. Wolfe and C. E. Caten), pp. 163–179. Blackwell Scientific, Oxford.

Lindow, S. E. (1983). Methods of preventing frost injury caused by epiphytic ice-nucleation-active bacteria. *Plant Disease* 67, 327–333.

McCann, J. A. (1984). Involvement of the American Fisheries Society with exotic species, 1969–1982. In *Distribution, Biology, and Management of Exotic Fishes* (ed. W. R. Courtenay and J. R. Stauffer), pp. 1–7. The Johns Hopkins University Press, Baltimore, MD.

McKnight, T. (1964). Feral livestock in Anglo-America. *University of California Publication* 16, 1–87.

McNeil, J. N. (1985). The true armyworm, *Pseudaletia unipuncta* (Haw.) *(Lepidoptera: Noctuidae)*: a possible migrant species. In *The Movement and Dispersal of Agriculturally Important Biotic Agents* (ed. D. R. MacKenzie, C. S. Barfield, G. C. Kennedy, R. D. Berger and D. J. Taranto), pp. 433–441. Claitors, Baton Rouge, LA.

McWhorter, C. G. (1971). Introduction and spread of Johnsongrass in the United States. *Weed Science* 19, 496–500.

Mooney, H. A. and Drake, J. A. (1986). *Ecology of Biological Invasions of North America and Hawaii*. Springer Verlag, New York.

Muenscher, W. C. (1980). *Weeds*. 2nd edn: Cornell University Press, Ithaca.

NAS (1987). *Introduction of Recombinant DNA-engineered Organisms into the Environment: Key Issues*. National Academy of Sciences, Washington, DC.

NAS (1989). *Alternative Agriculture*. National Academy of Sciences, Washington, DC.

Neal, M. C. (1965). *In Gardens of Hawaii*. Bishop Museum Press. Lancaster, PA.

Nellis, D. W, and Everard, C. O. R. (1983). The biology of the mongoose in the Caribbean. *Studies in Fauna of Curacao Caribbean Islands* 64, 1–162.

OTA (1987). *New Developments in Biotechnology. 2. Background Paper: Public Perceptions of Biotechnology*. Office of Technology Assessment. US Congress, US Government Printing Office, Washington, DC.

OTA (1988). *New Developments in Biotechnology – Field-Testing Engineered Organisms: Genetic and*

Ecological Issues. Office of Technology Assessment. US Congress, OTA-BA-350. US Government Printing Office, Washington, DC.

Pammel, L. H. (1911). *Weeds of the Farm and Garden*. Orange Judd, New York.

Panem, S. (ed.) (1985). *Biotechnology. Implications for Public Policy*. The Brookings Institution, Washington, DC.

Patterson, D. T. (1982). Effects of shading and temperature on showy Crotalaria (*Crotalaria spectabilis*). *Weed Science* **30**, 692–697.

Patterson, D. T., Terrell, E. E. and Dickens, R. (1983). Cogongrass in Mississippi *Imperata cylindrica*, a major weed in agricultural crops, history and present distribution control. *MAFES Research Highlights, Missouri Agriculture Experimental Station* **46**, 1–2.

Penfound, W. T. and Earle, T. T. (1948). The biology of the water hyacinth. *Ecology Mongraphs* **18**, 447–472.

Pierce, B. A. (1983). Grass carp status in the United States: a review. *Environment Management* **7**, 151–160.

Pimentel, D. (1955). The control of the mongoose in Puerto Rico. *American Journal of Tropical Medicine and Hygiene* **41**, 147–151.

Pimentel, D. (1963). Introducing parasites and predators to control native pests. *Canadian Entomology* **95**, 785–792.

Pimentel, D. (1986). Biological invasions of plants and animals in agriculture and forestry. In *Ecology of Biological Invasions of North America and Hawaii* (ed. H. A. Mooney and J. A. Drake), pp. 149–162. Springer Verlag, New York.

Pimentel, D. (1987). Down on the farm: Genetic engineering meets technology. *Technical Review* **90**, 24–30.

Pimentel, D., Glenister, C., Fast, S. and Gallahan, D. (1984). Environmental risks of biological pest controls. *Oikos* **42**, 283–290.

Pimentel, D., Hunter, M. S., LaGro, J. A., Efroymson, R. A., Landers, J. C., Mervis, F. T., McCarthy, C. A. and Boyd, A. E. (1989). Benefits and risks of genetic engineering in agriculture. *Bioscience* **39**, 606–614.

Pimentel, D., Acquay, H., Biltonen, M., Rice, P., Silva, M., Nelson, J., Lipner, V., Giordano, S., Horowitz, A. and D'Amore, M., (1993). Assessment of environmental and economic costs of pesticide use. In *The Pesticide Question: Environment Economics and Ethics* (ed. D. Pimentel and H. Lehman), pp. 223–278. Chapman and Hall, New York.

Pimm, S. L. and Pimm, J. W. (1982). Resource use, competition, and resource availability in Hawaiian creepers. *Ecology* **63**, 1468–1480.

Pines, D. S. and Gerdes, G. L. (1973). Wild pigs in Monterey County, California. *Californian Fish and Game* **59**, 126–137.

Presnall, C. C. (1985). The present status of exotic mammals in the United States. *Journal of Wildlife Management* **22**, 45–50.

Ridings, W. H. (1985). Biological control of the stranglervine in citrus – a researcher's view. *Weed Science* **34**, 31–32.

Robbins, C. S. (1973). Introduction, spread and present abundance of the house sparrow in North America. *Ornithology Mongraphs* **14**, 3–9.

Robel, R. J. (1961). The effects of carp populations on the production of waterfowl food plants on a western waterfowl marsh. *Transactions of the North American Wildlife Natural Resources Conference* **26**, 147–159.

Robinson, W. L. and Bolen, E. G. (1984). *Wildlife Ecology and Management*. Macmillan, New York.

Robocker, W. C. (1974). *Life History Ecology, and Control of Dalmation Toad Flax*. Washington Agriculture Experimental Station Technical Bulletin 79, Pullman, WA.

Roots, C. (1976). *Animal Invaders*. Universe Books, New York.

Roush, R. T. and McKenzie, J. A. (1987). Ecological genetics of insecticide and acaracide resistance. *Annual Review of Entomology* **32**, 361–380.

Sailer, R. I. (1978). Our immigrant insect fauna. *Bulletin of the Entomology Society of America* **24**, 3–11.

Sailer, R. I. (1983). History of insect introductions. In *Exotic Plant Pests and North American Agriculture* (ed. C. L. Wilson and C. L. Graham), pp. 15–38. Academic Press, New York.

Schitoskey, F., Evans, J. and Lavae, G. R. (1972). Status and control of nutria in California. *Proceedings of the Vertebrate Pest Conference* **5**, 15–17.

Science (1991). 'Superbug' attacks California crops. *Science* **254**, 1445.

Sharp, W. M. (1957). Social and range dominance in gallinaceous birds – pheasants and prairie grouse. *Journal of Wildlife Management* **21**, 242–244.

Silver, J. (1924). The European hare (*Lepus europaeus* Pallas) in North America. *Journal of Agriculture Research* **28**, 1113–1117.

Sinclair, W. A. and Campana, R. J. (eds.) (1978). Dutch elm disease perspectives after 60 years. *Search* **8**, 5–52.

Smith, C. W. (1985). Impact of alien plant on Hawaii's native biota. In *Hawaii's Terrestrial Ecosystems Preservation and Management* (ed. C. P. Stone and J. M. Scott), pp. 180–250. Cooperative National Park Resources Studies Unit, University of Hawaii, Honolulu.

Southern, H. N. (1945). The economic importance of the house sparrow: a review. *Annals of Applied Biology* **32**, 57–67.

Stegman, L. C. (1938). The European wild boar in The Cherokee National Forest, Tennessee. *Journal of Mammology* **19**, 279–290.

Stotzky, G. and Babich, H. (1984). Fate of genetically-engineered microbes in natural environments. *Recombinant DNA Technical Bulletin* **7**, 163–188.

Strauss, H. S., Hattis, D., Page, G., Harrison, K., Vogel, S. and Caldart, C. (1986). Genetically engineered microorganisms. II. Survival, multiplication, and genetic transfer. *Recombinant DNA Technical. Bulletin* **9**, 69–88.

Threinem, C. W. and Helm, W. T. (1954). Experiments and observations to slow carp destruction of aquatic vegetation. *Journal of Wildlife Management* **18**, 247-251.

Tierkel, E. S., Arbona, G., Rivera, A. and deJaun, A. (1952). Mongoose rabies in Puerto Rico. *Public Health Report* **67**, 274–278.

Trevors, J. T., Barkay, T. and Bourquin, A. W. (1987). Gene transfer among bacteria in soil and aquatic environments: a review. *Canadian Journal of Microbiology* **33**, 191–198.

USDA (1971). *Common Weeds of the United States.* Dover, New York.

USDI (1987). *Spread, Impact, and Control of Purple Loosestrife (Lythrum salicaria) in North American Wetlands.* US Fish and Wildlife Service, US Department of Interior, Washington, DC.

Van Vuren, D. and Coblentz, B. E. (1987). Some ecological effects of feral sheep on Santa Cruz Island, California, USA. *Biological Conservation* **41**, 253-268.

Vance, D. R. and Westemeirer, R. L. (1979). Interactions of pheasants and prairie chickens in Illinois. *Wildlife Society Bulletin* **7**, 221–225.

Vietmeyer, N. (1986). Another invader: *Melaleuca*. *American Forestry* **92**, 25.

Weller, M. W. (1969). Potential dangers of exotic waterfowl introductions. *Wildfowl* **20**, 55–58.

Wiley, R. B., Schweizer, E. E. and Ruppel, E. G. (1985). Interaction of Kochia (*Kochia scoparia*) and *Rhizopus* sp. on sugarbeet (*Beta vulgaris*) germination. *Weed Science* **33**, 275–279.

Williams, C. S. and Hayes, R. M. (1984). Johnsongrass (*Sorghum halepense*) competition in soybeans (*Glycine max*). *Weed Science* **32**, 498–501.

Williams, M. C. (1980). Purposefully introduced plants that have become noxious or poisonous weeds. *Weed Science* **28**, 300–305.

Zeleny, L. (1976). *The Bluebird*. Indiana University Press, Bloomington.

3

Frequency and consequences of insect invasions

Joop C. van Lenteren

Introduction

A species is described as an invader when it colonizes and persists in an ecosystem in which it has never been before (Mooney and Drake, 1989). Invasions by insects seem to be rather well documented, but that is a false impression. It is only for two categories of insect species that documentation of successful invasions is complete: (*i*) for pest species (usually herbivores); and (*ii*) for natural enemies (predators and parasitoids of insect pests, or phytophagous insects used in weed control). This information relates mainly to agro-ecosystems and few or no data are available for insect invasions in other ecosystems. This might mean that such ecosystems are not frequently invaded, that the effect of invaders is not dramatic (but see Howarth, 1991; also, see Hopper, Chapter 6 and Andow *et al.*, Chapter 10), or that these other ecosystems are considered economically so unimportant that they receive no attention. The first view seems to be supported by the general literature on colonization, where the massive restructuring of natural ecosystems is usually regarded as the main cause for making invasions possible. Complex natural systems are replaced by a few plants and animals that are of direct use to humans (Mooney and Drake, 1986*a*).

This paper is structured as follows: first, some general facts about insect invasions are stated and then several case studies are provided to illustrate problems created by insect invasions. Following this the characteristics of the insect invaders and the systems which are invaded are discussed.

Finally, some specific questions related to the predictability of the effects of insect invasions are answered.

Insect invasions: some general facts

How many invaders establish?

Di Castri (1989), in an article on the history of unintentional biological invasions of plants and animals in the Old World, estimates that 5 out of 100 introduced species do become established (Fig. 3.1). The percentage establishment is much higher when species are purposefully introduced. From the literature on insect biological control it is known that 25–34% of the introduced arthropod species are able to become established (e.g. Hall and Ehler, 1979; van Lenteren, 1983). In weed control the success of establishment of natural enemies is still higher, 65% (Crawley, 1986). This information tells us that the process of identification and introduction of certain categories of species leads to an increased probability of establishment, so some characteristics of a successful invader and/or aspects for creating the proper conditions for establishment may be identified from this work.

Which part of the local insect fauna consists of invaders?

About 750 000 species of insects had been described worldwide around 1980 (Arnett, 1985). Estimates of the number of species in existence range from 2.5 to 30 million. The number of species known as insect colonizers does not exceed a few thousand, so the data on which we build our generalizations about invasions are very limited indeed.

Sailer (1983) made a compilation of historical insect introductions into the USA and cites 1554 successful introductions, 66% of western pale-arctic origin, 14% from South and Central America and the West Indies. If we make a further selection of these data we come to the following results for insect pests: 40% of the insect species that are pests in USA agriculture were accidentally introduced (57/148); many (25/57) came from Europe (Pimentel, 1986). Sailer (1983) believes that the pattern of human activities is largely responsible for this: (1) most commerce and travel was and still is with Europe; (2) many of the crops in Europe are the same as those in the USA; (3) similar climatic areas can be found on both continents; and (4) the seasons are synchronized,

which does not hold for e.g. South America and Africa. Even for recent years this reasoning seems to hold: 40 of the 68 new insect and mite pests introduced to the USA between 1970 and 1982 are of European origin (Sailer, 1983). About 13 new insect species establish in the USA yearly (Sailer, 1978). It is not surprising, therefore, that most of the really successful invaders are those that are able to cross major barriers because of their relationship with *Homo sapiens* (Ehrlich, 1986). In the USA, with 1554 successful insect introductions from a total number of 100 000 insect species (Danks, 1988), colonists make up 1.55% (Fig. 3.2). In the Netherlands insect colonizers make up 0.41% of the total number of insect species (82 colonizers out of 20 000 species; van Lenteren *et al.*, 1987) and for the UK the percentage is 0.82% (169 out of 20 553 species; Williamson and Brown, 1986).

When the number of successful insect invaders

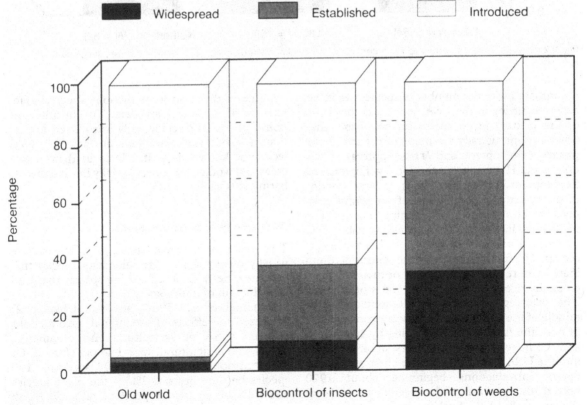

Fig. 3.1 Estimated percentage success for establishment of organisms invading the Old World (*N* unknown, Di Castri, 1989), organisms introduced for biological control of insects (*N* = 2295, Hall and Ehler, 1979) and organisms introduced for biological control of weeds (*N* = 627, Crawley, 1986).

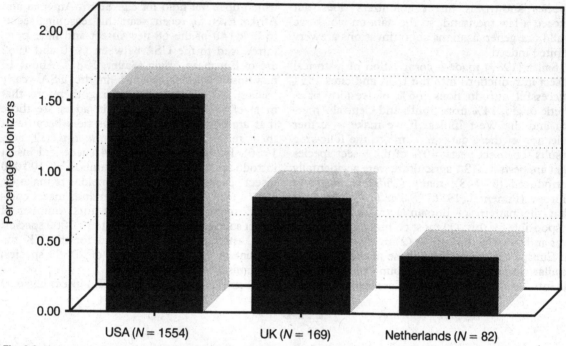

Fig. 3.2 Insect colonizers expressed as percentage of the total number of insect species (N = number of colonizing species).

is compared with the number of species per insect order occurring in the USA, it appears that some orders contain more successful colonizers than others: proportionally too many of certain phytophages (Homoptera and Thysanoptera), Coleoptera and Hymenoptera and too few Diptera and Lepidoptera (Simberloff, 1986). A large fraction of the systematic distribution of successful colonists can be attributed to the timing of commerce, plant introduction and biological control, and to differential intensity of study for different insect orders. The fact that beetle introductions predominated before 1900 is the result of their ability to cross the ocean in ship ballast and stored grain. The over-representation of Homoptera in the middle of the nineteenth century is a consequence of both the faster oceanic crossing by steamships and the huge increase in nursery stock imported into the USA from Europe. The burst of Hymenoptera introductions beginning about 1920 reflects the increasing biological control efforts (Sailer, 1978, 1983). Such information helps us to explain part of the invasion history; however, it does not provide the material needed to formulate

hypotheses that lead to predictions about characteristics of successful invaders. To illustrate this point: of the 212 accidentally introduced insects that have become major pests in the USA, 65% were not known as pest insects in their native areas and would, therefore, initially be classified as harmless (Calkins, 1983).

Negative effects of invasions

Two types of negative effects of invasions are usually distinguished: (a) colonizing species that become a pest; or (b) colonizing species that lead to extinction of native species.

Pimentel et al. (1989, also see Chapter 2) discussed the effects of intentional introductions into the USA of agricultural and ornamental plants, and agricultural, sport and pet animals. Of these introductions – some 7800 species – 160 species became pests; 2.2% of the plant species and 1.4% of the animal species: a rather high risk rate. They also remarked that insects introduced for biological control have generally had minimal or no environmental impacts and they relate that to greater ecological knowledge and established regulations.

Data on extinctions due to colonization have been provided by Simberloff (1981). He evaluated 10 papers that covered 850 plant and animal species introductions and concluded that less than 10% of the introductions caused species extinctions through habitat alteration, predation or competition (71 out of 850 introductions): introductions apparently tend to add species to a community, rather than cause extinctions, but there is a considerable risk that changes in the composition of biotae occur after an invasion. Predation was the principal cause of extinctions (Fig. 3.3), followed by habitat change, and competition caused only few extinctions. Interpretation of the effects of invasions leads to very different and conflicting conclusions, e.g. 'There is no invasion of natural communities without disturbance'(Fox and Fox, 1986) and 'Sufficient examples can now be assembled to indicate that invasion can proceed without continuing disturb-

ance' (Mack, 1985). However, I consider that extinction in 8.3% (71 out of 850) of the cases as a very high risk rate.

In a rather controversial article about the environmental impacts that result from purposeful introductions in classical biological control, Howarth (1991) lists factors that affect the degree of risk of an introduced organism on non-target organisms (also see, Andow *et al.*, Chapter 10). The list is comparable with the list of characteristics of successful invaders (see below). He states that two zygaenid moth species (a pest and a non-pest species) may have become extinct in Fiji as a result of the introduction of a tachinid parasitoid, *Bessa remota*. He further believes that the introduction of generalist predators and parasitoids of Lepidoptera has caused the extinction of at least 15 species (including 5 pest species) of larger native moths of Hawaii. Most of the other examples of insect introductions that he gives relate to decreases in numbers rather than extinctions.

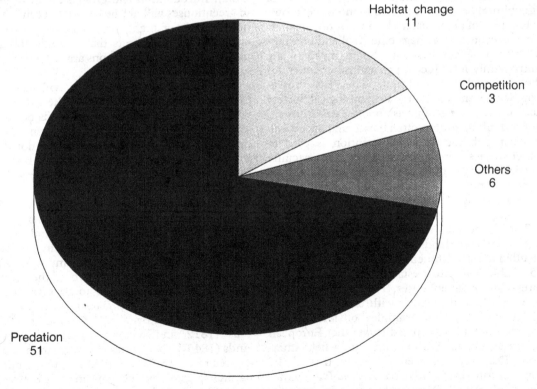

Fig. 3.3 Causes of extinction due to colonization based on analysis of 850 colonizations resulting in 71 cases of extinction (after Simberloff, 1981).

Precise predictions of the potential establishment and success of a given invading species cannot be made with the information shown above. What we need to know is: (1) how many insects arrive in new areas; (2) what proportion is able to establish; and (3) which characteristics of the invading species, of the system which is invaded, and of the interaction between the invader and the system, determine establishment (Mooney and Drake, 1986a)? Before addressing these questions I will first outline some case studies on insect invasions.

Case studies

Insect invasions into the Netherlands

Thousands of insects of some hundred different species invade the Netherlands every year. Natural invasions occur by insects that cross borders by active flight or are carried in by wind. Invasions of insects that are caused by man are also numerous. The important role Holland has as a producer of agricultural products is related to intensive import and export of plant material. Most of the invaders causing economic damage enter Holland unintentionally on plant material. Some invaders are purposefully introduced e.g. natural enemies for the biological control of pests. The Plant Protection Service and the Dutch Entomological Society keep track of cases of invasion and establishment, but this information is often biased and anecdotal: in their publications most attention is paid to either species that cause damage or a few peculiar species. Large numbers of invading species remain unnoticed because of their small size or lack of strong effects (negative or positive) on ecosystems.

The number of invasions of pest arthropods into the Netherlands since 1900, which have resulted in establishment, is estimated to be about 75 (Table 3.1). Most establishments occurred in forests and in greenhouses. The pest insects were often introduced together with their host plants. There are very few examples of foreign pest insects that naturally spread from other European countries and established in a Dutch field situation. The only documented case is that of the summer fruit tortrix *Adoxophyes orana* (van Lenteren *et al.*, 1987). Some other pest insects also immigrated from neighbouring countries, but

Table 3.1. Number of insect pest species that were introduced into the Netherlands since 1900

Crop	Number of new pest species	New pest problems
Arable farming	1	Few
Fruit orchards	2	Many
Vegetables (field)	0	None
Vegetables and ornamentals (greenhouses)	>35	Many
Forestry	17	Considerable
Ornamental trees	2	Considerable
Public parks etc.	6	Considerable
Food storage (granaries)	5	Considerable
Food storage (homes)	9	Considerable

these were first introduced unintentionally from elsewhere to Europe (e.g. the Colorado potato beetle, *Leptinotarsa decemlineata*, and the woolly apple aphid *Eriosoma lanigerum*, both from the USA). Introduction and establishment of insects in greenhouses will not be covered because greenhouses harbour highly artificial agro-ecosystems, and most of the insects that are able to survive under the artificial conditions are not able to do so in the field.

New introductions in forests will have taken place mainly since the end of the last century when reforestation in the Netherlands began with introduced tree species. Introductions of new deciduous tree species apparently did not lead to the introduction of harmful insects, but newly introduced species of coniferous trees were accompanied by 17 new pest species. Most of these pest species do not cause serious damage in the Netherlands, because of the small areas occupied by introduced tree species and the rich diversity in age and species composition of Dutch forests (P. Grijpma personal communication).

In Dutch apple orchards two of the 25 insect pest species are considered to be invaders. The woolly apple aphid (*Eriosoma lanigerum*) came from the USA via Britain (1787), the French west coast (1812) and Belgium (1843) to the Netherlands (1845). The species is now cosmopolitan. In its area of origin (West USA) it is under effective natural control by the parasitic wasp *Aphelinus mali*. This parasitoid was introduced into France in 1920 and afterwards to many other countries where it successfully controls the apple aphid. The parasitoid also established in Holland, but it is not able to reduce populations of aphids sufficiently, supposedly because of adverse climatological conditions.

Another current pest in apple orchards, the summer fruit tortrix *Adoxophyes orana* was found in the Netherlands in 1939 for the first time. Its origin is Central Europe, where it was described on birch trees in 1834. Until 1940 it was not regarded as a pest of orchards in West Europe. In the Netherlands it has been known since 1948 as a serious pest in apple and pear orchards. The causes of its spread and development to pest status are not known. Factors such as more intensive plantations and cultivation (crop protection, fertilization and pruning) are suggested, as well as a reduction of natural enemies as a result of intensive insecticide spray regimes. For integrated pest management in orchards I refer to Edland (see Chapter 4).

If an invader becomes established, control can be very expensive. An estimate for the annual costs of control on a selected number of crops of 4 of the 70 pests, which were introduced into the Netherlands, easily adds up to more than 15 million Dutch guilders (1 US Dollar = 2 Dutch guilders) (van Lenteren *et al.*, 1987).

International cases

For international case studies of insect invasions I refer to: Elton (1958), DeBach (1974), Hsiao (1985), Pimentel (1986), Jones (1987) and Herren and Neuenschwander (1991), all summarized by van Lenteren (1992).

Field introductions of genetically modified insects

In entomology there is some experience with the introduction of genetically modified arthropods into the field. Both, DeBach (1964) and Huffaker and Messenger (1976) warn against naivety in this area and state that: 'Where it is possible to attribute failure to a particular shortcoming of the natural enemy, such as the inability to tolerate high or low temperatures, it is preferable to exhaust sources of natural variation for remedying the defect before embarking on a program of artificial selection. . . . Only when it is established

that suitable strains are unavailable and that there are no alternative natural enemies, would collections from dissimilar regions or artificial selection for a desired trait seem warranted', (DeBach, 1964). Several attempts were made to overcome weaknesses and/or to improve parasitoids through selective breeding, but no one was able to demonstrate increased effectiveness in the field until the 1980s (see DeBach, 1964; Messenger *et al.*, 1976).

During the 1980s some successes with genetically modified natural enemies were obtained in the field, particularly with pesticide resistance. Resistance in several predatory mites against a number of organo-phosphorus compounds and one carbamate compound developed naturally (Hoyt, 1969; Meyer, 1975; Croft, 1976). Hoy (1983) obtained promising results with genetic improvement of predators: she successfully selected for resistance to two pesticides (permethrin and carbaryl) in the laboratory with the phytoseiid *Metaseiulus occidentalis*, and the effectiveness of the resistant strains has been shown in the field. At present several other selection programs with phytoseiids are aimed at attributes such as temperature tolerance, non-diapausing and increased fecundity (Hoy, 1987). Genetic improvement by laboratory selection for pesticide resistance has recently also been successful with parasitoids (Rosenheim and Hoy, 1988; Hoy *et al.*, 1990): a pesticide resistant laboratory strain of the walnut aphid parasitoid *Trioxys pallidus* was released in the field. The parasitoid was still found in the field after pesticide treatment and it had also dispersed to neighbouring fields.

The means to genetic engineering in natural enemies is still a long way off, as few genes that determine traits that lead to improved performance have been identified. It is expected that many such traits are under polygenic control, and are therefore difficult to transfer with current techniques (Beckendorf and Hoy, 1985). Based on very pragmatic reasons, the most obvious target for gene transfer is currently insecticide resistance as it is in many cases controlled by a single gene, and it has the added advantage that successfully transformed individuals can be identified by treatment with an insecticide. Attempts to introduce pesticide-resistance genes into natural enemies of insects in the UK and France have been unsuccessful until now.

Table 3.2. Factors influencing the invasion potential of animals (after Ehrlich, 1986, 1989)

Successful invaders	Unsuccessful invaders
1 Large native range	Small native range
2 Abundant in original range	Rare in original range
3 Vagile	Sedentary
4 Polyphagous	Mono- or oligophagous
5 Short generation times	Long generation times
6 Much genetic variability	Little genetic variability
7 Gregarious	Solitary
8 Fertilized female able to colonize alone	Fertilized female not able to colonize alone
9 Larger than most relatives	Smaller than most relatives
10 Associated with *H. sapiens*	Not associated with *H. sapiens*
11 Able to function in a wide range of physical conditions	Able to function only in a narrow range of physical conditions

The relatively small amount of success that has been obtained with genetically modified natural enemies has been in a field that I personally do not consider as one of the main topics in biological control, i.e. the development of pesticide resistance in natural enemies. The dilemma here is whether the optimal strategy for a biological-control research worker is either to search for other natural-enemy species or strains (which are abundantly available), or to start a breeding programme to improve the quality of the natural enemy already available. Although I believe genetic improvement to be an attractive and important area of research, I am astonished by the indiscriminate enthusiasm that exists for this technique. There is still a very long way to go before genetic manipulation can be used in order to improve natural enemies and we should be realistic in this respect: it is only one of many ways to help to control pests. For a more comprehensive discussion of this topic I refer to Pimentel (Chapter 2) and Whitten (Chapter 24).

Characteristics of invaders and the system being invaded

Characteristics of successful invaders

Several attempts have been made to list the characteristics of successful invaders. I will present here only one list (Table 3.2), which originates from Ehrlich (1986, 1989) and relates to characteristics for animal invaders. For plants similar lists have been made (see e.g. Bazzaz, 1986). Such lists are based on case studies of invasions.

Characteristics, as mentioned in Table 3.2, seem logical but there are so many exceptions that it is not justified to use these lists as reliable indicators of a species' potential to invade. Additions can be made to the list for specific groups of invaders. For herbivorous insects, for example, it has been found that polyphagous, external feeders (chewers and suckers) are much more successful colonizers of new plants than monophagous/oligophagous internal feeders (leaf-miners and gall-formers) (Strong *et al.*, 1984). Crawley (1986) and Pimm (1989) support the importance of several of the factors listed by Ehrlich (Table 3.2). They particularly stress a high reproductive rate as important for successful invaders. Levin (1989) emphasizes good dispersal capabilities as elementary for colonization success.

One category of insect invaders which deserves special attention is insect predators. Maynard Smith (1989), in discussing the causes of extinction, distinguishes a category of extinctions caused by man, with a subcategory of the introduction of foreign species. He concludes that of this subcategory the functional group of predators – especially if they are of an ecologically unfamiliar type – have a stronger effect on the biota (i.e. have a higher chance to eliminate species) than competitors. His conclusion is in line with that of Simberloff (1981), who provided data to show that it is mainly predators that cause extinction (see section on negative effects of invasions). Competitive exclusion does occur, but it requires

Table 3.3. Criteria for pre-introductory evaluation of natural enemies (after van Lenteren, 1986*a*)

1   Seasonal synchronization with host
2   Internal synchronization with host
3   Climatic adaptation
4   No negative effects
5   Host specificity
6   Great reproductive potential
7   Good density responsiveness

great similarity in resource utilization (Maynard Smith, 1989), and the chance that a species will succeed decreases significantly with the morphological similarity between the invader and the nearest congeneric species already present (Pimm, 1989). Pimm (1989) predicts that, as a result of competition, the surviving species will be morphologically more dissimilar than one would expect by chance. He also states that the effect of an invader on the ecosystem is greater if the invader arrives at a place where natural enemies are absent, than at a place where natural enemies are present. This statement is abundantly supported in the biological control literature.

Recent examples given in the literature on invasions (e.g. Crawley, 1986) emphasize how far we still are from being able to predict properly the success of anticipated invasions. But, I have already indicated that for some groups of invaders, e.g. organisms introduced for biological control, establishment rates are much higher than for others.

This higher establishment rate is the result of two developments: (1) the formulation of criteria to evaluate the effectiveness of natural enemies before they are introduced; and (2) an estimation of potential negative effects of natural enemies on the local fauna; i.e. to predict the probability of establishment and the reduction effect on its target organisms, and to prevent unwanted side effects (see also Greathead, Chapter 5; Aeschlimann, Chapter 7; and Blossey, Chapter 8). The use of such criteria during the past century is the main reason that virtually no problems were created with importations. The list with evaluation criteria that I developed (Table 3.3) was based on ideas expressed earlier in the biological control literature and on ecological properties that were supposed to characterize predators and parasitoids with a strong reduction effect (van Lente-

ren, 1980, 1983, 1986*a*). A procedure for selection of natural enemies with these pre-introductory evaluation criteria is given in van Lenteren (1986*b*). Such lists have also been made to evaluate herbivorous insects for the biological control of weeds (e.g. Harris, 1973; and Blossey, Chapter 8). For the biological control of insects the criteria list has already proven its usefulness, both for the evaluation of natural enemies for use in the field (e.g. Pak, 1988) as well as for use in protected crops (e.g. van Lenteren and Woets, 1988; Minkenberg, 1990).

Ehrlich's list and my own have little similarity. This is mainly the result of attempting to prevent negative effects with the introduction of natural enemies. In particular, polyphagy is not appreciated in biological control, which is in line with Pimm's (1989) notion that the potential negative effect of an invader is greater when it is polyphagous. In addition, matching of the natural enemy with the climate and with its specific host increase the establishment rate. Several of the criteria used for the evaluation of natural enemies can also be applied for risk assessment of genetically modified organisms, although they have to be fine tuned to the specific situation.

Characteristics of the system being invaded

Some locations seem to be invaded more easily than others. During the past three decades much attention has been paid to the difference between islands and mainland with regard to resistance to invasions. It was hypothesized that islands are more prone to invasions because the island biotae are shaped by less stringent selective pressures and, also, that island species are less successful as invaders. Simberloff (1986) evaluated the available data and had to conclude that there is as yet no material to support the hypothesis that the 'environmental resistance' in islands is less than on the continents. To test the view that mainland species are somehow stronger and/or island communities possess less biotic resistance, one would need information on the success and failure rates for a large number of related species, some from islands and some from mainlands, when introduced to both island and mainland target areas. Reliable data to perform such a test are not available, but data on introductions of insect genera widely used in biological control do not support the hypothesis that the mainland is more resistant to invaders than islands (DeBach, 1974; Simberloff, 1986).

Simberloff (1986) has designed the following alternative hypothesis. For any species of insect not found at some site (either island or mainland), there is an intrinsic probability that a propagule will initiate establishment, and this probability does not depend on the other species (competitors, predators, etc.) found on that site. Rather, it depends on the biology of the species and the availability of suitable habitat on the site. Since native insular entomofaunas tend to be small, the introduced species that establish themselves constitute a large fraction of the total entomofauna.

Another often expressed opinion is that disturbed areas are particularly vulnerable to invasions because their simplified biotae are less resistant (Elton, 1958; and later many others, e.g. Sharples, 1983; Pimm, 1989). According to Simberloff (1986) only anecdotal evidence is available to support this idea, although Orians (1986) states that Elton's (1958) insight into the importance of disturbance favouring invasions into ecological communities has generally been upheld by subsequent research. Simberloff's alternative hypothesis reads as follows. Patterns of successful insect introduction on disturbed and undisturbed sites do not depend primarily on different amounts of biotic resistance. Rather, each potential invader has a probability of successfully colonizing each site, and this probability rests largely on the nature of its habitat requirements and habitat availability at the site, and only secondarily on the other species that are present. Baker's (1986) data on plant invasions seem to support Simberloff's view that environmental resistance does not play an important role in the explanation of establishment, but Crawley (1986) provides material that supports the idea that disturbed areas are more vulnerable to plant invasions. A critical reanalysis of these data seems to be of high priority! It is refreshing to experience that Simberloff is not simply adhering to earlier expressed, seemingly attractive ideas concerning introductions: it prevents us from formulating dogmas that hamper a good understanding and application of ideas related to invasions.

Other factors influencing establishment

Several other factors influence the success of establishment (see e.g. Beirne, 1975; van Lenteren, 1986b; Tiedje et al., 1989):

season and environmental conditions at time of introduction;
number of organisms introduced;
frequency of introductions;
method of introduction (released in sensitive or relatively insensitive stage; e.g. natural enemy as adult (sensitive) or pupa (insensitive).

Small invading populations are supposed to establish less easily than larger ones (Ehler and Hall, 1982; Pimentel et al., 1989; Pimm 1989; Simberloff, 1989; Williamson, 1989). This might be the result of the Allee effect: effects that occur at 'undercrowding' conditions where e.g. mates are difficult to find, inability to find resources, greater vulnerability to predators, lack of genetic variability, etc. (Allee et al., 1949; Williamson, 1989). Evidence for this idea is presented in Fig. 3.4.

Predicting the probability of establishment

Attempts to make predictions of the potential invasive success of a given organism have generally taken the form of lists of attributes – genetic, physiological and ecological – that are most often associated with successful invaders (see Table 3.2). Many of the ideas come from qualitative natural history-type observations and retrospection. There has been very little experimentation utilized in the study of biological invasions (Mooney and Drake, 1986b, 1989). Most ecologists who published studies during the last decade that concerned invasions remain very sceptical about the possibility of making correct predictions on establishment and negative effects on the local fauna and flora.

Ecologists can predict with near certainty that a pest insect introduced from elsewhere will become a pest in the country of introduction, if the crop it feeds on in the country of origin is also widely grown in the country of introduction (Pimentel et al., 1989). But making predictions is much more difficult if related to non-pest species: for example, of the 212 introduced insects that have become major pests in the USA, 65% were not pests in their native ecosystems. The very serious problems created by, for example, the cottony cushion scale, the cassava mealybug and the Colorado potato beetle, were not expected at all on the basis of the knowledge of their biology in their respective endemic areas. Predictions based on the role these organisms played in their country of origin would have been completely inaccurate.

Cases of good intent in purposeful introductions that ultimately have bad consequences stem from a narrow view of the potential good that can come from the introduction. Frequently, the desired feature of the organism is considered in isolation from the total impact that the organism will have on the target system as well as on those who depend on that system for a variety of purposes. The potential effects of release must be considered in a total system context (Mooney and Drake, 1989). Often short-term economical considerations predominate over long-term ecological reflections.

With regard to genetically modified organisms Regal (1986) states: 'An organism with new biological properties will be an exotic element in nature whether its origins are the laboratory or a distant continent. Ecologists may not be able to make predictions about genetically engineered species in abstract terms; however, on a case-by-case basis, when specific details are available, ecologists with broad field experience working together with molecular biologists, should be able to give advice that can improve the predictions considerably. Modern, carefully planned introductions for biological control have been relatively safe . . .'. Also Levin (1989) states that if introductions have been based on proper ecological studies, they usually have not created problems. But where ecological information has been lacking or faulty, deliberate introductions of insects other than natural enemies have led to major and sometimes catastrophic ecological consequences.

In general, attributes likely to increase the probability of establishment are known, but we are not in a position to make accurate predictions about individual cases. One of the elements that make accurate prediction so difficult is the interaction between chance and timing of an invasion. Rare chance events that occur at just the right time may

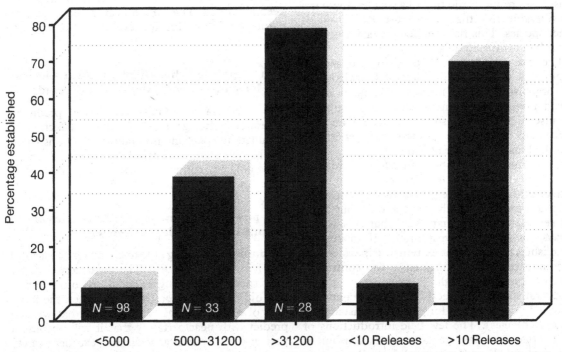

Fig. 3.4 Establishment success rate of natural enemies introduced for biological control of insects in Canada (after Beirne, 1975).

well be the cause of major, long-term structural and dynamic changes in ecological communities, (examples of which can be found in Crawley, 1989).

I am not in a position to comment on the probability of lateral transfer of modified genes in insects, as for this taxonomic group there is no published information available. Lateral transfer and exchange of genetic material between species is known to occur in micro-organisms (Tiedje *et al.*, 1989) and plants, but has not yet been demonstrated for insects (Howarth, 1991).

Conclusions

What do case studies tell us?

First of all, one cannot conclude that an insect which is harmless in its country of origin, will not cause problems in the country where it is introduced. Review of a large number of animal and plant introductions indicates that of the unintentional introductions 7% become a pest species and even of the intentionally introduced organisms 1–2% become pests. Although colonizations in most cases result in addition of a species, there is a considerable risk that they cause extinction of other species. This risk is much lower for arthropods than for vertebrates.

The risk of negative effects of an introduction diminishes with greater ecological knowledge of the organism (e.g. no negative effects of introduced insect natural enemies). This is hopeful, because it means that some criteria for an ecological risk analysis of invaders and genetically modified insects are already known.

Sometimes, the density of an introduced insect can be greatly reduced by the introduction of its native natural enemies (e.g. cottony cushion scale, cassava mealybug, woolly apple aphid), but there are several examples that show that this is not always possible (e.g. gypsy moth). If an invader establishes, eradication is extremely expensive and seldom successful. Annual costs for control of colonists can be very high.

Several species of insects have been genetically modified (e.g. selection for pesticide resistance in natural enemies). The few field introductions of these organisms have resulted in establishment and dispersal. The effect of such introductions on their conspecifics has not yet been evaluated.

Can we predict whether an insect species will be able to establish in a new area?

The wide range of available information on introduction of new organisms to a habitat does not lead to easy generalization: the outcome is extremely unpredictable (e.g. Levin and Harwell, 1986). Whether the introduced organism can survive and become established, and whether an established population may cause harmful effects on other populations depends greatly on subtle details of the particular situation. Even predictions about the probability of establishment are difficult to make. With increasing knowledge about the invader and about the ecosystem in which it will be introduced, the predictability increases. For the best studied insect invaders – herbivores of weeds – percentage establishment is on average 65%, without causing negative side effects.

Which characteristics distinguish a successful colonist from one which will fail?

A number of characteristics of both successful invaders and the system that is being invaded have been identified. They are, however, too general in nature to be useful for evaluation of specific cases.

Can we predict what effect an insect species will have on the ecosystems that it invades?

The effect of a colonist cannot be predicted accurately. In general, negative effects are obtained in 2% of intentional introductions when all organisms are considered. Insect introductions for biological control of arthropods and weeds have not resulted in negative effects.

Is it possible to provide characteristics of harmful and harmless colonizers?

We are far from being able to characterize harmful and harmless colonizers. The conditions leading to a species becoming harmful are often subtle and difficult to foresee. It is only when the invasion of a pest species is concerned that we can predict with near certainty that it will become a pest in the new area as well. The excellent safety record of purposeful introductions of insect natural enemies, however, tells us that thorough ecological studies of candidates for importation are likely to prevent problems later.

Acknowledgments

Thanks are due to J. C. van Loon for improving the text, and to Marianne Bergeman for assistance with editing.

References

Allee, W. C., Emerson, A. E., Park, O., Park, T. and Schmidt, K. P. (1949). *Principles of Animal Ecology*. Saunders, Philadelphia.

Arnett, R. H. (1985). *A Handbook of the Insects of America North of Mexico*. Van Nostrand Reinhold Company, New York.

Baker, H. G. (1986). Patterns of plant invasion in North America. In *Ecology of Biological Invasions of North America and Hawaii*, (ed. H. A. Mooney and J. A. Drake, pp. 44–57. Springer-Verlag, New York.

Bazzaz, F. A. (1986). Life history of colonizing plants: some demographic, genetic, and physiological features. In *Ecology of Biological Invasions of North America and Hawaii* (ed. H. A. Mooney and J. A. Drake), pp. 96–110. Springer-Verlag, New York.

Beckendorf, S. K. and Hoy, M. A. (1985). Genetic improvement of arthropod natural enemies through selection, hybridization or genetic engineering techniques. In *Biological Control in Agricultural IPM Systems* (ed. M. A. Hoy and D. C. Herzog), pp. 167–187. Academic Press, New York.

Beirne, B. P. (1975). Biological control attempts by introductions against pest insects in the field in Canada. *Canadian Entomologist* 107, 225–236.

Calkins, C. O. (1983). Research on exotic insects. In *Exotic Plant Pests and North American Agriculture* (ed. C. L. Wilson and L. Graham), pp. 321–359. Academic Press, New York.

Crawley, M. J. (1986). The population biology of invaders. *Philosophical Transactions Royal Society London, B* 314, 711–731.

Crawley, M. J. (1989). Chance and timing in biological invasions. In *Biological Invasions: A Global Perspective*, (ed. J. A. Drake), pp. 407–423. Wiley, Chichester.

Croft, B. A. (1976). Establishing insecticide-resistant phytoseid mite predators in deciduous tree fruit orchards. *Entomophaga* 21, 383–399.

Danks, H. V. (1988). *Insects of Canada: a synopsis prepared for delegates to the XVIIIth International Congress of Entomology (Vancouver, 1988). Biological Survey of Canada Document Series No. 1*, Ontario.

DeBach, P. (ed.) (1964). *Biological Control of Insect Pests and Weeds*. Chapman and Hall, London.

DeBach, P. (1974). *Biological Control by Natural Enemies*. Cambridge University Press, Cambridge.

Di Castri, F. (1989). History of biological invasions with special emphasis on the old world. In *Biological Invasions: A Global Perspective*, (ed J. A. Drake), pp. 1–30. Wiley, Chichester.

Ehler, L. E. and Hall, R. W. (1982). Evidence for competitive exclusion of introduced natural enemies in biological control. *Environmental Entomology* 11, 1–4.

Ehrlich, P. R. (1986). Which animal will invade? In *Ecology of Biological Invasions of North America and Hawaii* (ed. H. A. Mooney and J. A. Drake), pp. 79–95. Springer-Verlag, New York.

Ehrlich, P. R. (1989). Attributes of invaders and the invading process: vertebrates. In *Biological Invasions: A Global Perspective* (ed. J. A. Drake), pp. 315–328. Wiley, Chichester.

Elton, C. S. (1958). *The Ecology of Invasions by Animals and Plants*. Methuen, London.

Fox, M. D. and Fox, B. J. (1986). The susceptibility of natural communities to invasion. In *Ecology of Biological Invasions: an Australian perspective* (ed. R. H. Groves and J. J. Burdon), pp. 57–66. Australian Academy of Science, Canberra.

Hall, R. W. and Ehler, L. E. (1979). Rate of establishment of natural enemies in classical biological control. *Bulletin Entomological Society America* 25, 280–282.

Harris, P. (1973). The selection of effective agents for the biological control of weeds. *Canadian Entomologist* 105, 1495–1503.

Herren, H. R. and Neuenschwander, P. (1991). Biological control of cassava pests in Africa. *Annual Review of Entomology* 36, 257–283.

Howarth, F. G. (1991). Environmental impacts of classical biological control. *Annual Review of Entomology* 36, 485–509.

Hoy, M. A. (1983). Opportunities for genetic improvement of mites as biological control agents. In *Biological Control of Pests by Mites* (ed. G. L. Cunningham, M. A. Hoy and L. Knudson), pp. 141–147. Division of Agricultural and Natural Resources, University of California, USA.

Hoy, M. A., (1987). Developing insecticide resistance in insect and mite predators and opportunities for gene transfer. In *Biotechnology in Agricultural Chemistry*, (ed. H. M. LeBaron, R. O. Mumma, R. C. Honeycutt and J. H. Duesing), pp. 125–138. American Chemical Society Symposium Series No. 334.

Hoy, M. A., Cave, F. A., Beede, R. H., Grant, J., Krueger, W. H., Olson, W. H., Spollen, K. M., Barnett, W. W. and Hendricks, L. C. (1990).

Release, dispersal, and recovery of a laboratory-selected strain of the walnut aphid parasite *Trioxys pallidus* (Hymenoptera: Aphidiidae) resistant to Azinphosmethyl. *Journal of Economic Entomology* **83**, 89–96.

Hoyt, S. C. (1969). Integrated chemical control of insects and biological control of mites on apple in Washington. *Journal of Economic Entomology* **62**, 74–86.

Hsiao, T. H. (1985). Ecophysiology and genetic aspects of geographic variations of the Colorado Potato Beetle. In *Proceedings of the Symposium on the Colorado Potato Beetle, XVIIth International Congress of Entomology* (ed. D. N. Ferro, and R. H. Voss), pp. 63–77. University of Massachusetts, Amherst.

Huffaker, C. B. and Messenger, P. S. (eds) (1976). *Theory and Practice of Biological Control*. Academic Press, New York.

Jones, R. E. (1987). Behavioural evolution in the cabbage butterfly (*Pieris rapae*). *Oecologia* **72**, 69–76.

van Lenteren, J.C. (1980). Evaluation of control capabilities of natural enemies: does art have to become science? *Netherlands Journal of Zoology* **30**, 369–381.

van Lenteren, J. C. (1983). The potential of entomophagous parasites for pest control. *Agriculture, Ecosystems and Environment* **10**, 143–158.

van Lenteren, J. C. (1986a). Parasitoids in the greenhouse: successes with seasonal inoculative release systems. In *Insect Parasitoids* (ed. J. K. Waage and D. J. Greathead), pp. 341–374. Academic Press, London.

van Lenteren, J. C. (1986b). Evaluation, mass production, quality control and release of entomophagous insects. In *Biological Plant and Health Protection* (ed. J. M. Franz.), pp. 31–56. Fischer, Stuttgart.

van Lenteren, J. C. (1992). Insect invasions: origins and effects. In *Ecological Effects of Genetically Modified Organisms*, pp. 59–90. Netherlands Ecological Society, Amsterdam.

van Lenteren, J. C. and Woets, J. (1988). Biological and integrated pest control in greenhouses. *Annual Review of Entomology* **33**, 239–269.

van Lenteren, J. C., Woets, J., Grijpma, P., Ulenberg, S. A. and Minkenberg, O. P. J. M. (1987). Invasions of pest and beneficial insects in the Netherlands. *Proceedings Koninklijke Nederlandse Akademie van Wetenschappen*, **C 90**, 51–58.

Levin, S. A. (1989). Analysis of risk for invasions and control programs.In *Biological Invasions: a Global Perspective* (ed. J. A. Drake), pp. 425–435. Wiley, Chichester.

Levin, S. A. and Harwell, M. A. (1986). Potential ecological consequences of genetically engineered organisms. *Environmental Management* **10**, 495–513.

Mack, R. N. (1985) Invading plants: their potential contribution to population biology. In *Studies in Plant Demography: A Festschrift for John L. Harper* (ed. J. White), pp. 127–141. Academic Press, London.

Maynard Smith, J. (1989). The causes of extinction. *Philosophical Transactions of the Royal Society London B* **325**, 241–252.

Messenger, P. S., Wilson, F. and Whitten, M. J. (1976). Variation, fitness and adaptability of natural enemies. In *Theory and Practice of Biological Control* (ed. C. B. Huffaker and P. S. Messenger), pp. 233–254. Academic Press, New York.

Meyer, R. H. (1975). Release of carbaryl-resistant predatory mites in apple orchards. *Environmental Entomology* **4**, 49–51.

Minkenberg, O. P. J. M. (1990). On Seasonal Inoculative Biological Control. Ph.D. thesis, Wageningen Agricultural University.

Mooney, H. A. and Drake, J. A. (1986a). Preface in *Ecology of Biological Invasions of North America and Hawaii* (ed. H. A. Mooney and J. A. Drake), pp. v–vii. Springer-Verlag, New York.

Mooney, H. A. and Drake, J. A. (1986b). *Ecology of Biological Invasions of North America and Hawaii*. Springer-Verlag, New York.

Mooney, H. A. and Drake, J. A. (1989). Biological invasions: a SCOPE program overview. In *Biological Invasions: a Global Perspective* (ed. J. A. Drake), pp. 491–508. Wiley, Chichester.

Orians, G. H. (1986). Site characteristics favoring invasions. In *Ecology of Biological Invasions of North America and Hawaii* (ed. H. A. Mooney and J. A. Drake), pp. 133–148. Springer-Verlag, New York.

Pak, G. A. (1988). Selection of *Trichogramma* for inundative biological control. Ph.D. thesis, Wageningen Agricultural University.

Pimentel, D. (1986). Biological invasions of plants and animals in agriculture and forestry. In *Ecology of Biological Invasions of North America and Hawaii* (ed. H. A. Mooney, and J. A. Drake), pp. 149–162. Springer-Verlag, New York.

Pimentel, D., Hunter, M. S., LaGro, J. A., Efroymson, R. A., Landers, J. C., Mervis, F. T., McCarthy, C. A. and Boyd, A. E. (1989). Benefits and risks of genetic engineering in agriculture. *Bioscience* **39**, 606–614.

Pimm, S. L. (1989). Theories of predicting success and impact of introduced species. In *Biological Invasions: a Global perspective* (ed. J. A. Drake), pp. 351–367. Wiley, Chichester.

Regal, P. J. (1986). Models of genetically engineered organisms and their ecological impact. In *Ecology of Biological Invasions of North America and Hawaii*, (ed. H. A. Mooney and J. A. Drake), pp. 111–129. Springer-Verlag, New York.

Rosenheim, J. A. and Hoy, M. A. (1988). Genetic improvement of a parasitoid biological control agent: artificial selection for insecticide resistance in *Aphytis melinus* (Hymenoptera: Aphelinidae). *Journal of Economic Entomology* **81**, 1539–1550.

Sailer, R. I. (1978). Our immigrant insect fauna. *Bulletin of the Entomological Society of America*, **24**, 3–11.

Sailer, R. I. (1983). History of insect introductions. In *Exotic Plant Pests and North American Agriculture* (ed. C. Graham and C. Wilson), pp. 15–38. Academic Press, New York.

Sharples, F. E. (1983). Spread of organisms with novel genotypes: thoughts from an ecological perspective. *Recombinant DNA Technical Bulletin* **6**, 43–56.

Simberloff, D. (1981). Community effects of introduced species. In *Biotic Crises in Ecological and Evolutionary Time* (ed. M. H. Nitecki), pp. 53–81. Academic Press, New York.

Simberloff, D. (1986). Introduced insects: a biogeographic and systematic perspective. In *Ecology of Biological Invasions of North America and Hawaii* (ed. H. A. Mooney and J. A. Drake), pp. 3–27. Springer-Verlag, New York.

Simberloff, D. (1989). Which insect introductions succeed and which fail? In *Biological Invasions: a Global Perspective* (ed. J. A.Drake), pp. 61–75. Wiley, Chichester.

Strong, D. R., Lawton, J. H. and Southwood, R. (1984). *Insects on Plants: Community Patterns and Mechanisms*. Blackwell, Oxford.

Tiedje, J. M., Colwell, R. L., Grossman, Y. L., Hodson, R. E., Lenski, R.E., Mack, R. N. and Regal, P. J. (1989). The planned introduction of genetically engineered organisms: ecological considerations and recommendations. *Ecology* **70**, 298–315.

Williamson, M. H. (1989). Mathematical models of invasion. In *Biological Invasions: a Global Perspective*, (ed. J. A. Drake), pp. 329–350. Wiley, Chichester.

Williamson, M. H. & Brown, K. C. (1986). The analysis and modelling of British invasions. *Philosophical Transactions of the Royal Society London B* **314**, 505–522.

4

Integrated pest management (IPM) in fruit orchards

Torgeir Edland

Introduction

The agro-ecosystems in Fennoscandia are characterized by relatively few and moderate plant protection problems. The number of pests and plant diseases is smaller than in warmer regions. Due to a shorter growing season and lower temperatures, many multivoltine insects develop only one generation a year. The beneficial fauna, primarily the many predaceous bugs (anthocorids and mirids) and mites (phytoseiids), have also proved to be highly efficient as control measures.

Pest control systems have been developed for several agricultural crops, which require minimum inputs of pesticides. The programmes for many greenhouse cultures, based on biocontrol, and those for fruit crops, using integrated approaches, have been especially successful.

Development of IPM in Norwegian fruit orchards

In the 1950s and early 1960s a spray programme that consisted of 6–8 annual applications was commonly used against insects and mites. This heavy routine spray programme was expensive and caused undesirable environmental impact.

An extensive research programme on IPM, which started in the 1960s, provided important information on the biology and ecology of major pests and their natural enemies. Based on this new knowledge, the use of pesticides could be reduced. By 1974, farmers who had adopted IPM in their orchards, used only 0–3 pesticide applications a

year without significant damage to the crop (Edland, 1989a). During recent years, common use of scab-warning devices has resulted in further reduction of spray programmes. Thus, in 1988 the average use of insecticides and acaricides was 1.1 sprays and of fungicides 3.5 sprays, for the whole season in many orchards (Edland, 1989b). This is less than 20% of what is being used in integrated fruit production in some other European countries (Prinoth, 1990).

In spite of the success achieved by sound IPM programmes, we still face important pest problems every year. Some cause serious damage to the current programmes and undesirable environmental impacts.

'Old pest problems'

Natural outbreaks of various phytophagous insects are common in Fennoscandia. Most of them are usually initiated in natural habitats, and subsequently extend to agricultural crops. As they are not 'man-made', their existence will probably continue in the future with the same pattern as in the past, unless some extraordinary environmental changes occur. Such changes could result from long-lasting improvements in the effectiveness of native natural enemies through habitat manipulation (Ehler, 1990), and the introduction of new association biocontrol agents as proposed by Hokkanen and Pimentel (1989).

The winter moth (*Operophtera brumata* (L.))
For more than a hundred years, regular outbreaks at 12–15 year intervals in the fruitgrowing areas of

western Norway have been reported. These outbreaks, which each last for 3 to 4 years, cause severe economic damage (Edland, 1971). Both deciduous forest and unsprayed orchards may be completely defoliated. The massive use of insecticide in such periods causes secondary outbreaks of other pests.

In Nova Scotia the winter moth was introduced more than 50 years ago, and severe damage to deciduous trees occurred 10–20 years later. In the 1950s, the tachinid fly *Cyzenis albicans* (Fallen) and the ichneumonid wasp *Agrypon flaveolatum* (Gravenhorst) were introduced and effectively established. After 6–7 years the winter moth population was again reduced to a very low level (Embree, 1971). In Europe, however, the association of the same host and parasites, has not shown the same success.

In England the winter moth was studied by Varley *et al.*, (1973). They found that the larval parasite *C. albicans* was dominant, while the pupal parasite *Cratichneumon culex* (Müller), an ichneumonid wasp, also appeared to be an important natural enemy. Nevertheless, in this ecosystem these parasites do not seem to be satisfactory biocontrol agents, since their host is still an important pest of fruit and deciduous forest trees, thus occasional outbreaks occur.

During severe outbreaks in Norway in the 1960s, a number of different parasites was reared from various winter moth populations. Surprisingly, *C. albicans* in association with *A. flaveolatum* were the most common and abundant natural enemies in the important fruit area of Hardanger, in western Norway. In addition, at least two different pupal parasites were commonly found, one of them probably being *C. culex*.

In spite of the presence of these parasites, a new severe outbreak of the winter moth occurred in the Hardanger area in late 1970s, and recent observations indicate that another is to be expected in the middle of this decade. Evidently, a parasite that is a very efficient biocontrol agent in one area, may be quite worthless in another, when associated with competitively superior species and other interspecific constraints (Ehler, 1990).

The apple fruit moth (*Argyresthia conjugella* Zeller)

In many areas of Fennoscandia this is regarded as a key apple pest. Its natural and preferred host plant, however, is the commonly distributed and wild-growing mountain ash *Sorbus aucuparia* L. But, in those years when berry production is too small to support all the egglaying females, a number of them emigrate to apple for oviposition. The injury to the apple crop may then be severe, even resulting in total destruction of the crop.

In order to protect the apple crop from damage by this moth, two annual applications of insecticides were common. This changed drastically in the 1970s, however, when a reliable forecasting system for the apple fruit moth was developed (Edland, 1974). This system has made it possible to avoid spraying in years when no attack is expected, which gives a reduction in pesticide application of 60–90% (Edland, 1989a).

'New pest problems'

Every year new and potentially damaging pests are introduced into Fennoscandia, mostly on infested plant material.

Exotic pests

During recent years several consignments of fruit imported to Norway from the USA and southern Europe, have been infested with important pests, e.g. San José scale (SJS) *Quadraspidiotus perniciosus* (Comstock), woolly aphid *Eriosoma lanigerum* (Hausmann) and an American leaf miner, probably *Phyllonorycter elmaella* Doganlar & Mutuura. In the absence of their natural enemies we consider all of them as potentially very destructive for our fruit production, if they become established. In Denmark, the SJS survived successfully for several years before it was eradicated. The woolly aphid has been introduced and has established in Norway four times during the last century, but through intensive use of insecticides it was eradicated.

In 1991 two exotic fruit pests were reported to be established in two of the most important fruit areas in Norway, the European cherry fruit fly *Rhagoletis cerasi* L., and the apple leaf midge *Dasineura mali* (Keiffer) (Edland, 1991). Intensive efforts, which represented different strategies and control measures, were performed during 1992 in an attempt to eradicate them.

Pesticide-resistant strains

Many insect and mite species, formerly considered as non-pests, have become serious pests after they developed strong resistance to many widely used insecticides (Kremer, 1971; Riedl et al., 1992).

Any new pests or diseases established are likely to cause an increased need for pesticide use. Such excessive use is always harmful to most natural enemies and will consequently result in further trouble in the control of native pests. A typical example of this is the increasing mite problem in various crops, against which special biocontrol agents may be an appropriate measure.

Benefits and risks of introducing biocontrol agents

Organophosphate-resistant strains of phytoseiids in the control of phytophagous mites

Utilization of natural enemies, especially some predaceous mites, has become a fundamental factor for IPM in integrated fruit production in Europe. In controlling harmful mite species, various species/strains of phytoseiids are widely used.

Since the organophosphate (OP)-resistant strain of *Typhlodromus pyri* Scheuten was introduced into England in 1977 for use in IPM programmes in apple orchards, it has provided excellent control of both fruit tree red spider mite (*Panonychus ulmi* Koch) and rust-mites (*Aculus* spp.), and has virtually abolished the need for chemical treatments (Solomon, 1988). In 1987–8 this strain, mass-produced at East Malling Research Station, Kent, UK, was introduced and successfully established in a commercial fruit orchard in Norway. In 1990, it survived very well after a full-dosage treatment of azinphosmethyl, applied against the apple fruit moth, and in the subsequent year it gave significantly better control of spider mites than did any of the best new acaricides being tested in heavily infested experimental plots. Biocontrol agents that tolerate important pesticides, which sometimes are needed to control certain pests, are of great benefit in the effort to produce large yields of high quality.

However, such introductions of biocontrol agents, on living plants or plant products, involve great risks of spreading dangerous pests and dis-

eases. Special precautions should be taken to prevent the introduction of e.g. fire blight, woolly aphid and apple leaf midge.

As already stated, phytoseiid mites have become a very important biocontrol agent in the European integrated fruit production. Leaves and twigs that harbour these natural enemies are transported over long distances. Unfortunately, in some countries little care is taken to prevent new and harmful pests and diseases being spread by importation of plant material. Thus, in 1990 in one of the Scandinavian countries, some large fruit growers imported several truckloads of apple branches or twigs from the Netherlands. The plant material, which contained OP-resistant phytoseiids, was placed directly on the fruit trees in several orchards. In these areas the apple leaf midge has now become widespread, thus showing that guidelines and requirements for practical procedures, which could prevent introduction and establishment of new and harmful organisms are most important.

Hymenopterous parasites for the control of *Dasineura mali*

If the efforts in eradicating the newly established leaf midge by intensive use of insecticides should fail, we must search for alternative solutions. Different strategies of biocontrol should be considered. As pointed out by Ehler (1990) naturally occurring parasites and predators may sometimes exploit exotic pests and give acceptable biocontrol. But in cases when control is unsatisfactory, exotic natural enemies could provide a sound solution.

In 1979–85 Trapman (1988) investigated the incidence, biology, natural enemies and control of *D. mali* in the Netherlands. In some experimental orchards the pest had a rapid build-up, the percentage of infected shoots increased from about 1 in 1979 to about 90 in 1982–3. Two larval parasites were associated with the pest, the minute *Platygaster demades* Walker, which was sometimes able to parasitize up to 83% of the spring generation midge larvae, and the much larger *Torymus* sp., which reduced the summer generations by up to 94%.

On the basis of this knowledge, we may expect that the introduction of these parasites and other natural enemies could be a sound strategy for resolving the problem. However, such introduction cannot be conducted without risks. It may, for instance, be very difficult to isolate the parasites from the host, and also any hyperparasites from the parasites. Therefore, this approach may involve a great risk of introducing more adverse types of the pest, e.g. pesticide resistant strains, and undesirable hyperparasites, which may greatly reduce the potential of various natural enemies.

In any case, rearing of such parasites must be performed in isolation, preferably through several seasons before the natural enemies are released, to prevent spread of exotic noxious organisms into new areas.

Habitat manipulation, a potential strategy for the control of *Argyresthia conjugella*

During the last 30 years, extensive studies have been conducted in Norway on our key apple pest *Argyresthia conjugella* Zeller, its preferred hostplant *Sorbus aucuparia* L. and its natural enemies, of which the braconid *Microgaster politus* Marshall is the dominant parasite. Based on this knowledge, and a better understanding of the relationship between the three trophic levels, a new approach for a permanent solution of the pest problem has been suggested, this involves just a simple modification of the natural habitats (Edland, 1989c).

Several parasites of *Argyresthia* may severely reduce the moth population. For instance, the common and widely distributed *Microgaster* sometimes parasitizes up to 100% of its host larvae over large areas. Nevertheless, *Argyresthia* still occurs as an important pest in our fruit-growing areas. In the following paragraphs, I will try to explain why this is so.

In the current ecological situation, the number of *Argyresthia* is largely dependent on the amount of available food, i.e. the number of *Sorbus* berries that develop each year. Since only one larva develops in a single berry, the number of *Argyresthia* can never be higher than the total number of berries. Thus, in years when *Sorbus* has no berries in a certain area, no *Argyresthia* larvae are produced, and consequently there will be no parasites in this area. Through evolutionary mechanisms, however, both the host and its parasites have adapted to meet such 'catastrophic' situations. Both groups survive because a small number

(2–20%) of their pupae always hibernate for two or more seasons. These survivors have thus even prevented the extinction of the species locally.

During the season that follows a year with little or no berry crop, *Sorbus* normally produces very large numbers of berries. On each tree there may be hundreds of clusters, each of them containing 100–200 single berries. For the *Argyresthia* survivors there is now an excess of food, and the female moths fly from tree to tree, or from cluster to cluster, distributing their eggs at random. The females spend most of their energy on oviposition, and each of them probably produces more than 100 eggs. In contrast, the survivors of *Microgaster* and other parasites meet a completely different situation. At emergence, they will find an extreme shortage of food, and they have to spend all their time and energy in searching out individual berries for their host. Consequently, very few *Argyresthia* larvae are parasitized. Under such circumstances the rate of increase will be very different for the host and parasite. *Argyresthia* multiplies rapidly and within a few years it becomes so numerous that some of the moths will emigrate to apple for oviposition. The parasites, on the other hand, increase slowly in number, as long as the available food is the limiting factor. Thus, as indicated in Fig. 4.1(a) they will not exert sufficient influence on the *Argyresthia* population in time to prevent emigration and damage to apple. Similar relationships apply to many other associations of different trophic levels (Andrewartha and Birch, 1961).

In more southerly countries in Europe the apple fruit moth (*A. conjugella*) is a common and widely distributed insect in all areas where *Sorbus aucuparia* is common. However, in those areas it very seldom occurs as a pest on apple.

How can a species behave so differently, being a serious pest in one region and entirely harmless in another? The explanation of this may be found in the availability of food. In the southern areas, where *Sorbus* normally produces an abundant and regular crop every year, *Argyresthia* always finds enough food to maintain its population at a reasonably high level. Under such conditions *Microgaster* and other natural enemies will also operate and reproduce sufficiently every year to regulate their host below the level where *Argyresthia* is forced to emigrate. Hence, apple avoids infestation, as indicated in Fig. 4.1(b).

During the last few decades, various exotic cultivars of *Sorbus aucuparia* have been introduced into Fennoscandia, and planted as ornamentals in parks and private gardens. Many of these cultivars produce large numbers of berries every year, even when the wild-growing native type is completely fruitless. If such trees were planted out in sufficient number in the surroundings of the orchards, we could expect an important change in the current relationships between the three trophic level association. By this simple manipulation of the natural habitat, a sufficient equilibrium between *Argyresthia* and its parasites will most probably take place at a level that prevents emigration, and damage to apple. A permanent solution to this important pest problem could then be achieved.

In 1991 the Agricultural Research Council of Norway allocated a grant of 0.75 million NOK for research and application of the habitat modification. In cooperation with other research institutions, ten different exotic clones, in addition to one native, were selected for the project. Annual inspections, since 1970, have shown all of them to be yearly cropping types. By means of meristem cultures, 25 000 plants were produced and further cultivated in nurseries. The largest plants were planted out in three different areas in southern Norway in spring 1992, with the rest to be planted within 2 years.

During the project period we wish to clarify whether all the 11 clones will produce satisfactory berry crops every year, and in all areas. Furthermore, attractiveness of all of the clones for

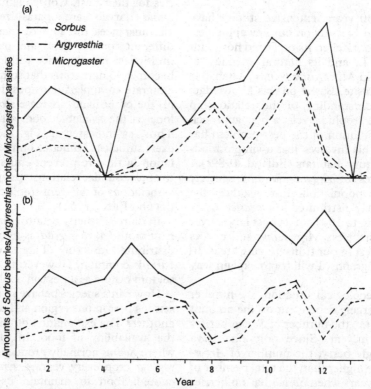

Fig. 4.1 Relationships between the three trophic level association of the mountain ash *Sorbus aucuparia*, the apple fruit moth *Argyresthia conjugella* and its dominant parasite *Microgaster politus*. Generalized diagram based on actual field data (Edland, unpublished).

(a) The annual berry production of *Sorbus* varies greatly. In year 4, 7 and 13 the amount of available food is insufficient for *Argyresthia* and the moths emigrate to apple for oviposition.

(b) *Sorbus* produces an abundant and regular crop every year, resulting in a sufficient equilibrium between *Argyresthia* and *Microgaster*, which prevents emigration and damage to apple.

oviposition by *Argyresthia* will be investigated, as well as any repellent effect on its parasites.

If this approach gives the expected result, several benefits will be achieved. There will be one important pest less to fight, all sprays against this target insect can cease, which will mean considerable savings in application costs, and detrimental side effects on natural enemies and other undesirable enviromental impacts can be avoided. Eventually, further pesticide reduction will be possible.

No serious risks seem to be involved. As the exotic *Sorbus aucuparia* types are already established in Norway, they do not need to be imported from other countries with the risk of the introduction of new and harmful organisms. Also, the argument sometimes being expressed, that introduction of exotic plants and animals into new and natural habitats means unfortunate pollution of nature, is a moot point, as it would apply to all forms of introduction, including biocontrol agents and crop plants.

Conclusion

Many programmes, employing various biocontrol agents, have proved to be successful and safe for pest control. Therefore, well assessed and appropriate biocontrol programmes seem to be a sound approach for plant protection of many agricultural crops.

However, any type of control measure should be considered only when needed. In many cases, therefore, the best solution would be to prevent the needs, rather than to search for the most effective control measures.

In this connection an important factor is preventing or retarding introduction and spread of harmful organisms from one region to another. International phytosanitary regulations have been established in order to achieve this, and to harmonize national requirements as much as possible. Unfortunately, as the regulations are often in strong conflict with current political convictions and statements for more liberal commerce and trade, it may sometimes be very difficult to maintain sufficiently safe restrictions.

Since the needs very often are created by man, it might be advisable to analyse the connections between the cause and effect. Then we can choose appropriate solutions, which offer permanent advantages.

When this is considered the following points should be carefully evaluated:

Prohibition of importation of specific plants and plant products into a new area, from a region where a serious disease or pest species/strain is known to be commonly distributed.

Significant improvements of the plant inspection services in both exporting- as well as importing-countries, to ensure a minimum of undesirable organisms being spread.

More attention should be paid to pesticide resistance, in order to prevent or retard the spread of destructive organisms, and to encourage better utilization of resistant biocontrol agents, where this is desirable.

Careful selection of biocontrol agents with regard to their control potential, specificity, environmental requirements etc., with the object of improving their benefits and reducing the risks of failure or adverse effects.

Preparation of appropriate and viable guidelines for safe procedures and other requirements concerned with the introduction of biocontrol agents, to prevent spread and establishment of destructive organisms (See Greathead, Chapter 5).

Consideration of the possibilities of applying favourable habitat modification as a solution for permanent control /regulation of pests and diseases.

If all countries would support the implementation of these statements and requirements, significant, world-wide advantages could be realized.

References

Andrewartha, H. G. and Birch, L. C. (1961). *The Distribution and Abundance of Animals*. University of Chicago Press, Chicago.

Edland, T. (1971). Wind dispersal of the winter moth larvae *Operophtera brumata* L. (Lep., Geometridae) and its relevance to control measures. *Norsk entomologisk Tidsskrift* **18**, 103–105.

Edland, T. (1974). Prognosis for attack of apple fruit moth (*Argyresthia conjugella* Zell.). A preliminary report on methods and results. (English summary.) *Gartneryrket* **64**, 524–532.

Edland, T. (1989*a*). Integrated pest management in Norwegian orchards. *Noragric Occasional Papers, Series C, Development and Environment No. 3*: pp. 65–74.

Edland, T. (1989*b*). Kvantifisering av kjemikalbruken nytta mot soppsjukdomar og skadedyr i frukthagar. *Aktuelt fra SFFL* **3**, 297–305.

Edland, T. (1989c). Kan vi i framtida utelate all sprøyting mot rognebærmøll? *Gartneryrket* **79**, 17.

Edland, T. (1991). Eit nytt skadedyr for Noreg: Eplebladgallmygg registrert i Telemark. *Gartneryrket* **81**, 16–17.

Ehler, L. G. (1990). Introduction strategies in biological control of insects. In *Critical Issues in Biological Control* (ed. M. Mackauer, L. E. Ehler and J. Roland), pp. 111–134. Intercept, Andover, Hants.

Embree, D. G. (1971). The biological control of the winter moth in eastern Canada by introduced parasites. In *Biological Control* (ed. C. B. Huffaker), pp. 217–226. Plenum Press, New York.

Hokkanen, H. M. T. and Pimentel, D. (1989). New associations in biological control: theory and practice. *Canadian Entomologist* **121**, 829–840.

Kremer, F. W. (1971). Change of species dominance among pests of intensively protected pome fruit crops in Italy. *Pflanzen-schutz-Nachrichten 'Bayer'* **24**, 232–238.

Prinoth, H. (1990). Auswertung von Betriebsheften auf Pflanzenschutzmassnahmen. *Obstbau Weinbau* **27**, 184–185.

Riedl, H., Barnett, W. W., Beers, E., Brunner, J. F., Burts, E., Croft, B. A., Shearer, P. W. and Westigard, P. H. (1992). Current status, monitoring and management of insecticide and miticide resistance on deciduous tree fruits in the Western United States. *Acta Phytopathologica et Entomologica Hungarica* **27**, 535–544.

Solomon, M. G. (1988). Managing predators in apple orchards in the U. K. *Proceedings XVIII International Congress of Entomology*, Vancouver, July 1988, p. 323.

Trapman, M. (1988). Appelbladgalmug populatie-opbouw en natuurlijke regulatie. *De Fruitteelt* **78**, 34–35.

Varley, G. C., Gradwell, G. R. and Hassel, M. P. (1973). *Insect Population Ecology, an Analytical Approach*. Blackwell Scientific Publication, Oxford.

Part II CLASSICAL BIOCONTROL

5

Benefits and risks of classical biological control

David J. Greathead

Introduction

Biological control by introduction and permanent establishment of exotic natural enemies of pests has been practised for over 100 years. Although a few introductions of beneficial insects had been made earlier, the introduction of an Australian ladybird (*Rodolia cardinalis* Mulsant) into California to control the cottony cushion scale (*Icerya purchasi* Maskell) on citrus in 1888–9 (Caltagirone and Doutt, 1989), is generally considered to mark the beginning of the practice of biological control as an effective pest control strategy. This operation was not only highly successful in controlling the pest but also was widely publicized with the result that economic entomologists in a number of countries were soon importing ladybirds for the control of a wide range of pests in what Lounsbury (1940) referred to as the 'ladybird fantasy'. Few of these introductions were successful, consequently practitioners began to study the ecology and population dynamics of pest natural enemy systems and a more scientific approach to biological control developed (Waage and Greathead, 1988), with a consequent increase in the success rate. However, the introduction of natural enemies remains a largely empirical activity that depends to a large extent on the knowledge and insight of the practitioner.

There are other ways in which natural enemies can be applied as pest control agents; by augmentative or inundative releases of native or exotic agents, which includes formulation of pathogens as biological pesticides (see Huber, Chapter 18; and Waage, Chapter 9) and conservation or enhancement of the action of native species of natural enemies (see Edland, Chapter 4). Because the introduction method was the first to achieve an outstanding success, it is now referred to as 'classical biological control'.

These other uses of biological control agents do not necessarily involve the introduction of exotic species or anticipate their permanent establishment. However, if a species is introduced for use in augmentative biological control or as a biological pesticide it may become established even when this is not the intended objective. Thus, all introductions of exotic biological control agents should be subject to the same scrutiny as introductions for classical biological control.

Organisms that have been proposed, or used, as biological control agents include vertebrate and invertebrate predators, parasitic insects and nematodes, as well as pathogenic micro-organisms – protozoa, fungi, bacteria and viruses. Introductions of mongooses (Hinton and Dunn, 1967) and other generalist vertebrate predators (see Pimentel, Chapter 2), chiefly onto oceanic islands during the nineteenth century for rodent control, were an ecological disaster because these animals are opportunists and take the most readily available prey. Not only did they frequently fail to provide adequate rodent control but they attacked chickens and native fauna, and thereby contributed to extinction of some endemic species. Although other generalist predators continued to be introduced, including the giant toad (*Bufo marinus* (L.)) last introduced successfully into Queensland in 1935–6 (Wilson, 1960) and predatory snails (*Gonaxis* spp. and *Euglandina rosea*

Ferrusac) introduced onto islands in the Pacific and Indian Oceans during 1950–80 (Mead, 1979), often with unfortunate side effects, the introduction and liberation of such organisms is no longer recommended by responsible biological control practitioners and, indeed, is forbidden in some countries.

There remain special circumstances where generalists can be acceptable as biological control agents. Thus, predatory fish are used for mosquito control (Gerberich and Laird, 1985) and can be applied safely in wells and ornamental ponds from where they are unable to escape into natural waterbodies. The restricted fauna of such habitats also ensures that mosquito larvae are the major component of the prey. Another exceptional situation is the use of herbivorous fish to control submersed waterweeds in ponds, reservoirs and irrigation canals. The Chinese grass carp (*Ctenopharyngodon idella* Cuv. & Val.) is attractive for this purpose because it requires very specific conditions for reproduction that are not met in the places where it is used, and so the numbers of fish can be controlled and the population will die out eventually. Also, in many situations sterile grass carp hybrids are used to avoid the possibility of reproduction (van Zon, 1981).

On the whole, the types of organisms used as biological control agents are now limited to species belonging to groups of organisms that experience has shown tend to have very restricted host ranges and do not readily exhibit switches in host preferences. These are principally: specialist arthropod predators such as Coccinellidae, predatory mites, etc.; insect parasitoids; certain species of nematodes specialized as parasites of insects and weeds; strains of pathogenic micro-organisms with a narrow host range; herbivorous arthropods that exhibit a high degree of host plant specificity; and harmless competitors and antagonists (invertebrates or micro-organisms), these preclude the establishment of harmful pest species if applied prior to pest attack. So far specialized insects have been most widely used in classical biological control, chiefly against insect pests and weeds.

This paper examines the benefits and risks of biological control introductions, with special reference to those made for classical biological control, and discusses action that is being taken to develop guidelines for the safe introduction of biological control agents.

Targets for classical biological control

The suitability of different pests as targets for biological control depends on many factors; viz., the origin of the pest, the group of organisms to which it belongs, and the existence of suitable natural enemies.

Traditionally, introduced pest species are considered prime targets for biological control using introduced natural enemies, on the premise that the pests have arrived accidentally in small numbers with a low probability that they were accompanied by their natural enemies, especially if the pest is scarce in its area of origin. However, native or other long established pests may also be good targets, if it can be shown that their importance as pests varies in different parts of their range and that this is due to the action of natural enemies present in some areas but not in others. Good prospects for biological control may also exist when a pest species has a close relative in another region that has effective natural enemies (Carl, 1982).

Pests must also belong to groups of organisms that have effective natural enemies that are also safe to other organisms. Few vertebrate pests satisfy these requirements; their predators are mostly opportunists, their parasites are nonspecific or have stages affecting unrelated nontarget organisms, and many of their diseases are also not specific. An exception, which has been exploited, is the myxoma virus from South American relatives of the European rabbit (*Oryctolagus cuniculus* (L.)), which provided effective control of rabbits in Europe and Australia (Fenner and Ratcliffe, 1965) until the pathogenicity of the virus became attenuated.

Similarly, pest molluscs have not proved to be good targets. Their parasites tend to have vertebrates as their final hosts and the pathogens discovered so far are not sufficiently host specific. Predators have been used but do not meet current standards of specificity and were not very successful (Mead, 1979).

Many insect pests present ideal targets and are the subject of the majority of successful biological controls reported to date. Some 66% of the successes relate to Homoptera – aphids, plant hoppers, scale insects, mealybugs, etc. The success rate with other insect orders is lower (Table 5.1) but some excellent results have been obtained for

Table 5.1. Relative success rate of classical biological control of insect pests by order

Order	Success rate (%)
Homoptera	66
Lepidoptera	18
Coleoptera	7
Diptera	5
Hymenoptera	1
Heteroptera	1
Others	2

From BIOCAT Database.

Table 5.2. Summary of classical biological control results using insect agents to control insect pests and weeds

	Insect pests N	Weeds N
Introductions	4769	692
Establishments	1445	443
Target pests	543	115
Good controls	421	73
Countries or islands	196	55

From BIOCAT database (updated Aug. 1992) (insects) and Julien (1992) (weeds).

most of them. Success rates are higher for insects that feed in exposed positions; those with larvae that bore into plant tissue may escape natural enemies and consequently success rates against these are lower. Soil pests present the most difficult targets.

Many introduced weeds have proved to be good targets for biological control, especially dominant species which have a major impact in pastures, rangeland or in perennial crops. To date, most effort has been applied to perennial weeds (Julien, 1992). Annual weeds of arable land provide a less attractive target because in many instances a large number of species are present and all of them must be controlled at the same time. Also, because of their short life span and the unstable environment in which they grow they do not provide ideal targets for long-term biological control, so host-specific strains of pathogens are being developed as biological herbicides for short-term suppression of the seedlings of these weeds (Evans, 1991).

There are as yet few examples of biological control of plant pathogens (Deacon, 1988; Risbeth, 1988) and none are strictly classical biological control (see also Défago *et al.*, Chapter 12; Weller *et al.*, Chapter 13; Fokkema, Chapter 15; and Jensen and Wolffhechel, Chapter 16). However, cultures and commercial preparations for inoculative application are now being shipped from country to country.

Benefits of classical biological control

To date (August 1992), the IIBC has gathered 4769 records of the introduction of insect para-

sitoids and predators for the control of insect pests (beginning with the 1880s introduction of *Rodolia cardinalis*), which are held in a database (BIOCAT). There are also records of the introduction of other categories of natural enemies, including pathogens for the biological control of animal pests, but these have not been catalogued. Julien (1992) has catalogued over 690 records of introductions of agents for the control of weeds. Thus, overall there have been more than 5500 introductions of beneficial species for biological control. The records for insect natural enemies and weeds are summarized in Table 5.2.

Although only about one in four introduced biological control agents have become established (see van Lenteren, Chapter 3) and only a proportion of these make an important contribution to control of pests, overall some 165 species of insect pests (BIOCAT) and 35 species of weeds have been controlled in at least one country (Julien, 1992). The pests that have been controlled include many of major importance. Thus, prickly pear cactus (*Opuntia* spp.) was controlled over 24 million hectares in Queensland and New South Wales by the introduction of two insect species (Mann, 1970). Further, a number of successful biological controls have been repeated in other countries. For example, the cottony cushion scale has been controlled in at least 30 countries (BIO-CAT) and the control of prickly pear cactus has been repeated in some 16 countries (Julien, 1992). In these instances the cost of control was minimal because the development work had already been done.

Unfortunately, economic assessments of classical biological control are scarce. Too often,

Table 5.3. Some economic assessments of classical biological control programmes (US $ millions)

Pest	Region	Savings[a]	Costs of control programme
Cassava mealybug *Phenacoccus manihoti*	Africa (1984–2003)[a]	96.0	14.8
Rhodes grass scale *Antonina graminis*	Texas (1974–8)	194.0	0.2
Skeleton weed *Chondrilla juncaea*	Australia (1975–2000)	13.9	3.1
Wood wasp *Sirex noctilio*	Australia (1975–2000)	0.8	8.2
White wax scale *Ceroplastes destructor*	Australia (1975–2000)	0.09	1.4
Two-spotted mite *Tetranychus urticae*	Australia (1975–2000)	0.9	0.9
Potato tuber moth *Phthorimaea operculella*	Zambia (1974–1980)	0.09	0.04
Alfalfa weevil *Therioaphis trifolii*	USA (1954–1986)	77.0	1.00
Water fern *Salvinia molesta*	Sri Lanka (1987–2112)	0.5	0.22

[a]The years in parentheses are those of the period used by economists in calculating the discounted benefits shown in column 3 as annual 'savings'.

because the result is so obvious, cost/benefit analysis has been considered unnecessary. However, some rigorous assessments exist and show that the return on investment can be very high (Table 5.3), because control continues indefinitely without the need for further expenditure to maintain it. This characteristic of classical biological control makes it unattractive as a commercial investment. Consequently, most programmes have been carried out by governments or international organizations using public funds. On the other hand the farmer does not bear any direct financial or labour costs, which makes the method especially attractive for resource-poor farmers in the developing countries.

In many instances, classical biological control has been used where alternative controls were not feasible. The recent control of the South American cassava mealybug (*Phenacoccus manihoti* Matile-Ferrero) in 34 affected African countries (Herren and Neuenschwander, 1991) is noteworthy. Chemical control of mealybugs is difficult, too costly on a subsistence foodcrop and

inadvisable in countries where the young shoots are gathered as food. Weeds of pasture and rangeland are also prime targets because of the low unit value of the land, which makes herbicides uneconomic, and because of the extent of the infestation to be controlled.

Classical biological control, and other biological control techniques, also have the advantage that their impact is more or less restricted to the target pest and so have minimal effects on populations of non-target organisms or on humans, since there are no toxic residues to pollute the environment or to contaminate food.

Thus, classical biological control is an economical and sometimes the only practicable control method for major introduced pests. Because of its achievements and potential for the future, given the likelihood of an increasing number of invasions by alien pests as the volume of international trade and travel increases and journey times are reduced, the regulation of biological control introductions should be designed so as to encourage its safe use while avoiding unnecessary restriction.

Risks associated with introduced biological control agents

Background

All introduced organisms present a potential hazard, and proposals for introductions should be subject to scrutiny, but the greatest risk relates to those introductions made in the expectation that the biological control agent will become permanently established. Experience with invading species has shown that eradication is seldom feasible, or indeed possible, once an organism has spread beyond its point of entry. Thus, it is important that biological control introductions are made only after an evaluation of the advantages, disadvantages and risks. The advantages have been reviewed and it is shown that very considerable economic benefit can result from successful biological control. The possible negative effects are discussed in the following paragraphs (see also Hopper, Chapter 6).

Direct damage to non-target species of economic importance and native fauna and flora

Damage to non-target organisms is the most serious potential risk from the introduction of biological control agents. The record of classical biological control has been remarkably good during the century since the introduction of *Rodolia cardinalis* into California. There are a few records of insects introduced for weed control feeding on useful plants or crops but the majority of these are not serious and chiefly refer to species introduced without screening (Greathead, 1971; Harris, 1990). The most often cited is the attack on sesame (*Sesamum indicum* L.) at Serere Research Station in Uganda by *Teleonemia scrupulosa* Stål. The lace bug had been widely introduced into other tropical countries for control of lantana (*Lantana camara* L.) without untoward side effects since it was first introduced into Hawaii in 1902 before screening procedures were developed. Investigation showed that although the insect oviposited on sesame plants, few of the resulting nymphs survived to become adults and these laid no eggs (Davies and Greathead, 1967). Further, it was found that only a few cultivars were at all attractive and unfortunately, it was one of these that was widely planted on the Research Station at that time (Greathead, 1973; and unpublished observations). Thus, no long-term harm resulted from this introduction.

Howarth (1991) attempted to gather evidence of damage to native fauna and flora, however, most of the examples he cites do not stand up to scrutiny of the original reports and are largely based on speculation, or are totally unsubstantiated. The most convincing evidence relates to predatory snails introduced onto Pacific islands for control of the giant African snail, *Achatina fulica*. These generalist predators on snails are implicated in the decline in numbers of native endemic snails.

There has also been occasional interference between biological control of insects and weeds that could have been avoided had more thought been given to the possibility in the past; e.g., the coccinellid, *Cryptolaemus montrouzieri* Muls., widely introduced to control mealybugs has been implicated in reducing the impact of *Dactylopius* spp. introduced for control of *Opuntia* spp. in South Africa and Mauritius (Greathead, 1971).

In spite of past lack of foresight, the number of adverse side effects has been remarkably small and generally of minor importance. Now biological control practitioners avoid generalist predators, and such side effects are unlikely to recur in the future.

Screening programmes are designed to assess the risk that non-target organisms will be attacked and provide the data for a risk assessment. However, the level of acceptable risk must be decided by the authorities in the importing country, but it is incumbent on those who propose introductions to provide a dossier detailing possible adverse effects should the introduction be permitted, in order to enable an informed decision to be made.

Indirect environmental damage

Reduction in the abundance of pests may leave empty niches that can be occupied by other noxious species. For example, destruction of stands of major weeds may leave bare ground, which could be colonized by another weed (examples cited by Harris, 1990), or result in soil erosion. These possibilities must be assessed by the importing country, which should consider measures to ensure that the pest is replaced by useful or other non-controversial species. Thus, pasture grasses may be seeded to replace a weed. However, prior analysis should have been carried out to assess the economic value of a proposed biological control programme, which should not have been sanctioned unless the removal of the pest problem was expected to result in economic or environmental benefits of value to the community.

Conflicts of interest

Conflicts of interest may arise over the status of targets for control. For example, graziers in Australia desired biological control of the weed they call Paterson's Curse (*Echium plantagineum* L.) but another group led by beekeepers, who know the plant as Salvation Jane, objected and delayed action until a new law was in place (Cullen and Delfosse, 1985). This enabled a decision to be made in favour of biological control after an assessment, which showed that the benefits to the nation out-weighed the disadvantages. Dilemmas that arise from conflicts of interest may constrain the application of biological control but do not strictly relate to the safety of biological control, although they must be considered by authorities charged with the task of deciding whether to grant permission for the release of exotic organisms into the environment. Such problems should be settled in advance of approval of a biological control programme on the basis of where the greater benefit lies for the community as a whole.

Environmental pollution

Biological control agents do not normally pollute the environment. However, following introduction, successful agents may become very abundant and become a nuisance temporarily. Thus, there are anecdotal reports of biological control agents swarming under street lights (Harris, 1990) and there have been instances when coccinellids have become so abundant that they were a temporary nuisance (Greathead, unpublished observations). For example, successful weed control agents may leave large amounts of decaying biomass, which may be offensive. These problems are transient, and if the public is informed of their temporary nature and the long-term benefits which will ensue, complaints can be minimized.

Damage from contaminants

Biological control agents for shipment should be screened to ensure that they are free from their own natural enemies (which might limit their effectiveness), and from contamination with plant and animal pathogens. Further, they should be packed using sterile or otherwise inert materials free of contaminating organisms. As with host-specificity screening, cost/benefit considerations should be taken into account so that precautions are appropriate and not excessive. Thus, elimination of all pathogens from insects can be expensive and ineffective if the pathogen is widespread and non-specific, so that the agent will soon become reinfected in the receiving country. The shipping authority should be responsible for taking sensible precautions, but the receiving authority should specify any special needs and conditions. For example, intermediate quarantine in a third country may be specified to ensure that major diseases of the crop host of the pest are eliminated. Where possible, agents should be shipped as resting stages (insects) or *in vitro* cultures (pathogens) and the inclusion of living host plant or animal material avoided.

As an added precaution, on receipt, all shipments should be opened for inspection in quarantine or other secure premises where the escape of the contents can be prevented and all packaging and other extraneous material destroyed or sterilized promptly. Any contaminated shipments, or suspect material of the control agent, should also be destroyed at once.

Concern of neighbouring countries

Natural enemies are not constrained by political boundaries and may spread out of the country in which they are released into neighbouring countries in the same ecological region. Importing countries should be aware of this and, where appropriate, inform their neighbours or a regional organization acting on behalf of their governments of the intention to introduce biological control agents and thus provide an opportunity for them to voice their objections.

Conclusion

Thus, the dangers of biological control introductions chiefly relate to the safety of the biological control agent to non-target organisms and the avoidance of contaminants. The first can be minimized by host specificity screening and the latter by judicious screening prior to shipment. Issues relating to the level of acceptable risk and conflicts of interest are the province of the importing country alone.

Protocols for host specificity screening tests

The potential dangers posed to crops by introduced herbivorous organisms have long been recognized. Consequently, it has been a requirement for many years that agents for weed control should be rigorously screened to ensure, as far as possible, that they will not damage crops and other useful plants. Thus, safety procedures for weed control agents began to be developed early this century, and the procedures developed over the years provide a guidance for devising screening procedures that are being introduced for other categories of biological control agents. Further, concern for native flora and fauna has now resulted in many countries requiring the screening of endangered native species as well as economically important non-target organisms. Therefore, safety screening of all biological control agents proposed for introduction is now being demanded by many countries. Because of the pioneering work done on weed control agents and the greater experience gained with them, procedures for screening organisms that attack plants will be outlined first.

All screening tests on candidate biological control agents should employ the strain that would be released so as to ensure that intraspecific variation does not invalidate the tests. Likewise, it is advisable to use in tests the same strain of the pest that is to be controlled to avoid the possibility of incompatibility, which might lead to failure.

Weeds

Initially, screening of insect agents for weed control was done by carrying out 'no-choice' feeding (starvation) tests and oviposition tests on a range of economically important plant species, in which

the test agents were held isolated in small containers with the test plant until they died or fed (oviposited) on the plant offered. It was soon realized that under the highly artificial conditions of these tests where insects are confined in small containers, abnormal results are frequent and lead to the rejection of species that are host-plant specific under field conditions. Accordingly, more realistic procedures were developed and protocols for their use were debated by participants in successive International Symposia on Biological Control of Weeds. The outcome has been the adoption of agreed protocols by the International Institute of Biological Control (IIBC) and CSIRO (and proposed for adoption by the USDA) for the selection of test plants and the conduct of screening tests. These incorporate a centrifugal testing sequence developed during the 1960s and 1970s (Wapshere, 1989), which seeks to determine the host range of the candidate agent by tests on varieties of the target weed, species in the same genus as the weed, genera in the same tribe, and so on, until only negative results are obtained. As a check on these results other plants selected on the following criteria are also tested (see Blossey, Chapter 8).

The degree of specificity that needs to be demonstrated and the level of risk that is acceptable depend on the importance of the weed and the presence of non-target species closely related to it in the environment where the weed is to be controlled. Thus, less specificity can be tolerated for cactus insects in countries such as Australia where there are no native cacti than for natural enemies of wild blackberries, because other species and hybrids of the genus *Rubus* are cultivated for soft-fruit production (Field and Bruzzese, 1985). However, agents are usually rejected if there is evidence that other plant species are at risk.

The nature of the screening tests depends to an extent on the life cycle of the agent and the stage, or stages, which select the host plant. For insects, feeding tests are carried out except for those species where the ovipositing parent is the sole determinant of where the offspring will feed, e.g. gall makers and leaf miners. Similarly, oviposition tests are carried out for all species that oviposit in or on the host plant. Both tests are usually carried out as multiple choice tests in the presence of the weed host as control, using excised plant parts or potted plants, as appropriate, for species that fail no-choice screening. Where positive results are obtained, further tests are carried out to determine whether the candidate can complete its life cycle on the plants in question. These may be further supplemented with field experiments in the area of origin, in which test plants are exposed in a stand of the weed infested by the candidate and observations are made on the candidate's host range under natural conditions (see Blossey, Chapter 8). Sometimes it is also possible to set up a garden of test plants that include examples of the target weed imported from its invasion area, in order to carry out outdoor screening – provided that there are no restrictions on importing noxious plants back into the area of origin (Harley and Forno, 1992). These additional tests may demonstrate host specificity in the early stages of host finding, which are circumvented under cage conditions, e.g. the mite for the control of gorse (*Ulex europaeus* (L.)) now released in New Zealand is never found on non-hosts outdoors but does attack some varieties of French beans (*Phaseolus vulgaris* (L.)) in captivity (Hill, 1987).

Recently, fungi have been successfully used in weed control and the protocols developed for insects have been adapted for safety testing of pathogens (Evans, 1992). Infectivity tests are carried out in which infective stages are applied to test plants in the laboratory and, when possible, plants are exposed to natural infection in the field.

These procedures appear to provide a reliable guide to the performance of weed-control agents because no unanticipated attacks on non-target plants have been reported for agents released after being screened using the above protocols. However, the realization that confinement can lead to results that are not found in the field and thus to the rejection of agents that are host specific under natural conditions, has prompted a reassessment of screening protocols (Harris and McEvoy, 1992). Many practitioners now favour giving precedence to field observations and experiments as well as evaluation of ecological and phenological data in host specificity assessment in line with thinking on the screening of control agents for arthropod pests where abnormal results from laboratory tests are more frequent.

Arthropod pests

Until recently, control agents for arthropod pests were not rigorously tested, it being deemed sufficient that they were not known to attack economically important insects (bees, silkworms, lac insects etc.) or other beneficial species. However, more rigorous requirements are being introduced in some countries to protect native fauna. There are, as yet, no agreed procedures and methodology is being developed on a case-by-case basis.

For a few insect parasitoids and predators, the protocols developed for weed agents can be adapted and used, but under cage conditions many more of these insects, than herbivorous insects, behave abnormally. Since entomophagous species respond to a complex of chemical and physical cues from the environment, host plant and pest hosts, key determinants of host specificity may occur at any of these levels and be eliminated in highly simplified test environments. Consequently, greater reliance on other measures of host range will be required in making safety assessments of entomophagous insects, viz. field studies in the country of origin to determine the natural host range with special regard to the determination of the environmental and biological factors that delimit the niche occupied by the candidate insect. Unfortunately, test insects, being mobile, cannot be used in the type of field experiments that can be used to screen test plants in the field. Therefore, these observations may be supplemented by laboratory studies on host identification behaviour, olfactory cues used to locate the hosts, developmental parameters including nutritional requirements and modelling of the life cycle and phenology in relation to that of species at risk (Goldson and Phillips, 1991).

Micro-organisms and parasites (e.g. nematodes, see Ehlers and Peters, Chapter 11) for insect control can be screened with the same procedures as used for micro-organisms for weed control; i.e. by carrying out infectivity tests and following them when necessary with in-depth studies on the course of infection in the laboratory. However, as with insect control agents, screening in the field is not feasible and recourse has to be made to study of the epidemiology of the pathogen in relation to the ecology of non-target species of insects which may be at risk.

Micro-organisms

Antagonists, principally micro-organisms used for the control of plant diseases, can be screened using the same principles as for pathogens to be used for weed control to ensure that they are not pathogenic to plants. Similarly, competitors can be screened in this way when they are micro-organisms (see Fokkema, Chapter 15; and Jensen and Wolffhechel, Chapter 16).

Vertebrate pests

Pathogens of vertebrates present a potential risk to human and animal health. Therefore, it is strongly advised that World Health Organization (WHO) should be consulted before they are considered as candidate biological control agents for vertebrate pests.

Conclusion

Because screening cannot usually include all species at risk, for cost or logistic reasons, and the precise behaviour of an introduced species cannot be predicted from pre-introduction studies, no absolute guarantee of safety can be given; therefore, the degree of specificity required in a control agent should depend on circumstances and consequently each proposed introduction should be judged on its merits. A comprehensive assessment should be made by the regulatory authority, which should include a review of the economic importance of the pest, the benefits and disadvantages of alternative control measures and their costs, as well as the potential benefits and risks of the introduction of a biological control agent. Only when this is done is it possible to make an informed decision on where the greater benefit lies and to determine if any possible negative effects are sufficiently offset by the advantages.

Current international action to ensure the safety of imported biological control agents

The growing interest in biological control in response to concern over the side effects of chemical pesticides has attracted the attention of agencies concerned with the safety of biological control and a desire to regulate its use. Thus, conserva-tionists require assurance that introductions will not affect rare or endangered species, some pesticide regulatory bodies wish to extend registration requirements to include biological control agents (appropriate for formulated biological pesticides but not classical biological control agents), and some of those charged with regulating the release of genetically modified organisms wish to include all 'alien' germplasm, natural or engineered. Plant quarantine departments often have the responsibility for permitting the introductions of agents for crop pest control and many do not have the knowledge to make informed decisions, consequently there is a tendency to play safe and refuse permission. In a number of countries legislation is proposed or under discussion for the regulation of biological control by one or more of these groups (see Klingauf, Chapter 28).

For these reasons the International Organization for Biological Control (IOBC) and the IIBC approached the Food and Agriculture Organization of the United Nations (FAO) to urge it to take the initiative in the development of guidelines for governments that were considering the regulation of biological control so as to try to ensure that any such legislation promoted the safe use of biological control and did not unnecessarily impede or prevent its adoption. The FAO responded by commissioning the IIBC to engage Professor Michael Way to assist in the preparation of a discussion document to be considered by an expert consultation with a view to the development of guidelines or a code of conduct. The experts met in September 1991 and agreed a draft Code of Conduct for the Import and Release of Biological Control Agents. They also recommended that FAO should commission the preparation of a series of guidelines on procedures for the implementation of the Code. The draft Code (FAO, 1992) has been circulated for comment and, if endorsed, an agreed text will be submitted for approval by FAO member countries (see Waage, Chapter 9).

The draft Code provides a basis for the adoption of national regulations or legislation that would ensure the safe importation and release of biological control agents, avoid irresponsible action, and promote the responsible use of biological pest control. Interest in the document has already been shown by a number of countries, which are considering the development of national policy on the conduct of biological control. It is to be hoped that the action by FAO will preempt inappropriate legislation and promote the orderly development of rules for the movement of biological control agents between countries.

Conclusion

Classical biological control has now been practised for over a century and many important insect pests and weeds have been controlled. It has proved to be an economical and environmentally benign solution to severe pest problems, particularly those caused by invading noxious organisms. The considerable body of experience and the knowledge gained has been used to improve the performance and safety of classical biological control. Thus, properly carried out by experienced practitioners, classical biological control has few risks that cannot be assessed so that rational decisions can be made by countries as to the benefits and risks of importing and releasing agents proposed for introduction. However, increasing concern among conservationists and regulatory bodies over the safety of introducing species into new areas has led the Food and Agriculture Organization of the United Nations to draft a code of conduct for the introduction of agents for classical biological control.

Acknowledgments

Sections of this paper have been adapted from the discussion document prepared for FAO. I am grateful to Professor M. J. Way and to Dr G. C. M. Schulten of FAO for their agreement to my use of this material.

References

Caltagirone, L. E. and Doutt, R. L. (1989). The history of the vedalia beetle importation to California and its impact on the development of biological control. *Annual Review of Entomology* **34**, 1–16.

Carl, K. P. (1982). Biological control of native pests by introduced natural enemies. *Biocontrol News and Information* **3**, 191–200.

Cullen, J. M. and Delfosse, E. S. (1985). *Echium plantagineum*: catalyst for conflict and change in Australia. *Proceedings of the VI International Symposium of Biological Control of Weeds, Vancouver, Canada, 1984*, pp. 249–292.

Davies, J. C. and Greathead, D. J. (1967). Occurrence of *Teleonemia scrupulosa* on *Sesamum indicum* Linn. in Uganda. *Nature* **213**, 102–103.

Deacon, J. W. (1988). Biocontrol of soil-borne plant pathogens with introduced inocula. *Philosophical Transactions of the Royal Society of London B* **318**, 249–264.

Evans, H. C. (1991). Biological control of tropical grassy weeds. In *Tropical Grassy Weeds* (ed. F. W. G. Baker and P. J. Terry), pp. 52–72. CAB International, Wallingford.

Evans, H. C. (1992). Pathogen–weed relationships: the practice and problems of host range screening. *Abstracts, VIII International Symposium on Biological Control of Weeds, Canterbury, New Zealand, February, 1992*, p. 62.

FAO (Food and Agriculture Organization) (1992). *Report of the Expert Consultation on Guidelines for the Introduction of Biological Control Agents.* FAO, Rome.

Fenner, F. and Ratcliffe, F. N. (1965). *Myxomatosis.* Cambridge University Press, Cambridge.

Field, R. P. and Bruzzese, E. (1985). Biological control of blackberries: resolving a conflict in Australia. *Proceedings of the VI International Symposium of Biological Control of Weeds, Vancouver, Canada, August 1984*, pp. 341–349.

Gerberich, J. B. and Laird, M. (1985). Larvivorous fish in the biocontrol of mosquitoes, with a selected bibliography of recent literature. In *Integrated Mosquito Control Methodologies, vol. 2* (ed. M. Laird and J. W. Miles), pp. 47–76. Academic Press, London.

Goldson, S. L. and Phillips, C. R. (1991). Biological control in pasture and lucerne and the requirements for further responsible introduction of entomophagous insects. *Bulletin of the Entomological Society of New Zealand* **10**, 63–74.

Greathead, D. J. (1971). A review of biological control in the Ethiopian Region. *Commonwealth Institute of Biological Control Technical Communication, no. 5.*

Greathead, D. J. (1973). Progress in the biological control of *Lantana camara* in East Africa and a discussion of problems raised by the unexpected reaction of the more promising insects to *Sesamum indicum. Commonwealth Institute of Biological Control, Miscellaneous Publications*, No. 6, pp. 89–92.

Harley, K. L. S. and Forno, I. W. (1992). *Biological Control of Weeds. A Handbook for Practitioners and Students.* Inkata Press, Melbourne.

Harris, P. (1990). Environmental impact of introduced biological control agents. In *Critical Issues in Biological Control* (ed. M. Mackauer, L. E. Ehler and

J. Roland), pp. 289–300. Intercept, Andover, Hants.

Harris, P. and McEvoy, P. (1992). Analysis and management of risk in introducing insects for biological weed control. *Abstracts, VIII International Symposium on Biological Control of Weeds, Canterbury, New Zealand, February*, p. 21.

Herren, H. R. and Neuenschwander, P. (1991). Biological control of cassava pests in Africa. *Annual Review of Entomology* **36**, 257–283.

Hill, R. (1987). *The Biological Control of Gorse* (Ulex europaeus *L.*) In *New Zealand an Environmental Impact Assessment*. DSIR Entomology Division, Christchurch.

Hinton, H. E. and Dunn, A. M. S. (1967). *Mongooses Their Natural History and Behaviour*. Oliver & Boyd, Edinburgh.

Howarth, F. G. (1991). Environmental impacts of biological control. *Annual Review of Entomology* **36**, 485–509.

Julien, M. H. (1992). *Biological Control of Weeds. A World Catalogue of Agents and Their Target Weeds*, 3rd edn. CAB International, Wallingford.

Lounsbury, C. P. (1940). The pioneer period of economic entomology in South Africa. *Journal of the Entomological Society of Southern Africa* **3**, 9–29.

Mann, J. (1970). *Cacti Naturalised in Australia and Their Control*. Department of Lands, Brisbane.

Mead, A. R. (1979). *Economic Malacology with Particular Reference to* Achatina fulica. Academic Press, London.

Rishbeth, J. (1988). Biological control of air-borne pathogens. *Philosophical Transactions of the Royal Society of London B* **318**, 265–281.

van Zon, J. C. J. (1981). Status of the use of grass carp (*Ctenopharyngodon idella* Val.). *Proceedings of the Vth International Symposium of Biological Control of Weeds, Brisbane, 1980*, pp. 249–264.

Waage, J. K. and Greathead, D. J. (1988). Biological control: challenges and opportunities. *Philosophical Transactions of the Royal Society of London B* **318**, 111–128.

Wapshere, A. J. (1989). A testing sequence for reducing rejection of potential biological control agents for weeds. *Annals of Applied Biology* **114**, 515–526.

Wilson, F. (1960). A review of the biological control of insects and weeds in Australia and Australian New Guinea. *Commonwealth Institute of Biological Control Technical Communication, no. 1*.

6

Potential impacts on threatened and endangered insect species in the United States from introductions of parasitic Hymenoptera for the control of insect pests

Keith R. Hopper

Introduction

Introductions of parasitic wasps have often led to economic control of introduced insect pests (Clausen, 1978). Biological control practitioners argue that such introductions have fewer and less severe detrimental impacts than alternative controls (DeBach and Rosen, 1992). However, others have challenged this viewpoint (Howarth, 1983, 1991) with the argument that introduced biological control agents can drive non-target populations extinct, especially if these are already rare. Others have responded that the impact on conservation of insects species is unlikely if introductions are done carefully (Samways, 1988). Unfortunately, this controversy has given rise to more heat than light, being plagued by excessive generalization, lack of hard data on adverse impacts, and lack of concrete recommendations about how to reduce the risks of introductions. The lack of data on adverse impacts arises, in part, because many biological control projects lack the funds for evaluating the impact of the introduced agent on the target species, let alone on non-target species. The low level of funding, relative to the complexity of the problems involved, also means that methods to reduce risks will have to rely

heavily on extant knowledge, if biological control introductions are to continue. In this chapter, I will address the following questions: How can one predict the potential impacts on threatened and endangered insect species from introductions of parasitic Hymenoptera for the control of insect pests in the continental United States?; and, What are the potential impacts from selected introductions? I have chosen a narrow scope of impacts and types of introductions to avoid over-generalization.

My viewpoint is that of a researcher actively engaged in exploration for and introduction of natural enemies of insect pests with the desire and necessity of doing this with minimum impact on non-pest species. The draft guidelines developed by the Agricultural Research Service, United States Department of Agriculture, state specifically:

II. B. 1. d. Protection against entry of arthropods and other organisms inimical to native or other introduced beneficial arthropods.

Provisions are required for the elimination of all obligatory hyperparasites; facultative hyperparasites should also be eliminated except under circumstances where potential benefits can be shown to outweigh potential detriments. Likewise, the potential effect of the biological control agent on non-target organisms, e.g. biological control agents for weeds, predators, pollinators, endangered species, etc., should be considered; if a parasite attacks hosts in taxonomic groups which include primarily beneficial species (e.g. Coccinellidae), host specificity studies are required to determine the safety of beneficial species. In addition, precautions should be taken that any entomopathogens capable of severely affecting beneficial arthropods are eliminated prior to release of imported arthropods in the United States.

The degree of risk involved as to potential detrimental effects of a natural enemy of a pest that may also attack non-target organisms should be weighed against the potential beneficial effects of the natural enemy proposed for release. (Coulson *et al.*, 1991.)

My approach will be to review the biology of the species listed as threatened or endangered by the United States Fish and Wildlife Service (1991), to discuss methods of host range measurement and prediction, review the biology of parasitic wasps proposed for introduction against insect pests in the United States, estimate the potential for impacts given the available data and theory, and suggest areas where data and theory could be improved.

Threatened and endangered insects in the United States

In 1989, 436 insect species were proposed for classification as threatened or endangered, but insufficient data were available for classification of 427, and only 9 were considered eligible for immediate listing (see Table 6.1, United States Fish and Wildlife Service, 1989). The distribution of species across orders does not reflect taxonomic diversity or necessarily reflect distribution of danger of extinction. Given the lack of data, I have concentrated on the species actually listed as threatened or endangered. As of 1991, there were 18 insect species so listed (see Table 6.2, United States Fish and Wildlife Service, 1991). I have drawn the information on their biologies discussed below from Lowe *et al.* (1990) and Moseley (1992) and references cited therein.

Of the seven species of beetles listed as threatened or endangered, three are cave dwellers threatened by pollution and lowering of the water table, one is a specialist herbivore whose habitat is being destroyed, one is a burying beetle whose

Table 6.1. Distribution across orders of taxa proposed for inclusion on threatened and endangered list (from USFWS 1989)

Order	Total	Number of species	
		Immediate listing	Insufficient data
Collembola	2	0	2
Ephemeroptera	12	0	12
Odonata	21	1	20
Plecoptera	9	0	9
Orthoptera	26	0	26
Hemiptera	5	0	5
Homoptera	2	0	2
Neuroptera	2	0	2
Coleoptera	210	4	206
Mecoptera	2	0	2
Diptera	10	0	10
Lepidoptera	66	1	65
Trichoptera	61	2	59
Hymenoptera	8	1	7
Total	436	9	427

habitat is disappearing, and two are predatory tiger beetles whose habitats are being disturbed.

Thirteen species or subspecies of butterflies (7 lycaenids, 4 nymphalids, an hesperid and a papilionid) and a sphingid moth are listed as threatened or endangered. Almost all have limited-distribution host plants that are threatened by some form of land development, and almost all are univoltine on annual plants or shrubs far from agricultural areas. Two taxa may be affected by pest control: the Kern primrose sphinx moth (*Euproserpinus euterpe*) whose only habitat occurs next to a wheat field, the Schaus swallowtail (*Heraclides aristodemus ponceanus*) which occurs in areas where aerial applications of insecticide are made for mosquito control. It is worth noting that insecticide has been applied to eradicate Mediterranean fruit fly (*Ceratitis capitata*) in the San Francisco Bay area, where there are several taxa of endangered butterflies.

Finally, one species of naucorid (creeping water bugs) is threatened by destruction of the only hot spring in which it occurs.

What makes these species rare? What is likely to preserve them from extinction? The answer to the first question is that most of them are rare because suitable habitat is rare. The answer to the second

Table 6.2. Biology of currently listed threatened and endangered species in the United States

Taxon	Status	Distribution	Food	Phenology	Threat
Coleoptera					
Cincindela dorsalis dorsalis (Cincindelidae), Northeastern Beach Tiger Beetle	T	Atlantic coast beaches	Carrion, small arthropods	Biennial, larvae overwinter, adults Jun–Aug	Habitat disruption (beach recreation), collecting
Cincindela puritana (Cincindelidae), Puritan Tiger Beetle	T	New England beaches and nearby cliffs	Carrion, small arthropods	Biennial, larvae overwinter, adults Jun–Jul	Habitat destruction (dams, cliff stabilization, urbanization) and disruption (beach recreation), collecting
Desmocerus californicus dimorphus (Cerambycidae), Valley Elderhorn Longhorn Beetle	T	Sacramento-San Joaquin River, California	*Sambucus* spp.	2 yr cycle, adults Mar–Jun	Riparian habitat destruction
Elaphrus viridis (Carabidae), Delta Green Ground Beetle	T	Jepson Prairie, Fairfield, California	Collembola spp.	Univoltine, adults Feb–May	Vernal pool destruction
Nicrophorus americanus (Silphidae), American Burying Beetle	E	Oklahoma, New England	Carrion	—	Reduction of leaf litter depth
Rhadine persephone (Carabidae), Tooth Cave Ground beetle	E	Limestone caves, Austin, Texas	Cave cricket eggs	—	Pollution, lowering of water table
Texamaurops reddelli (Pselaphidae), Kretchmarr Cave Mold Beetle	E	Limestone caves, Austin, Texas	Omnivorous, but prefers fungi	—	Pollution, lowering of water table
Heteroptera					
Ambrysus amargosus (Naucoridae), Ash Meadows Naucorid	T	Hot spring, Ash Meadows, Nevada	Aquatic plants	—	Habitat destruction (urbanization)
Lepidoptera					
Hesperia leonardus montana (Hesperidae), Pawnee Montane Skipper	T	Upper South Platte River canyon	*Bouteloua gracilis*	Univoltine, larvae Oct–Jun, pupae Jun–Aug, adults Aug–Sep	Habitat destruction (reservoir construction)
Apodemia mormo langei (Lycaenidae), Lange's Metalmark	E	6 ha of Antioch dunes on San Joaquin river, California	*Eriogonum nudum* var. *auriculatum*	Univoltine, larvae fall–winter, pupae spring–summer, adults summer	Sand mining, industrial development
Callophrys mossii bayensis (Lycaenidae), San Bruno Elfin	E	San Bruno Mts., San Francisco, California	*Sedum* spp.	Univoltine, larvae Apr–Jun, pupae Jun–Feb, adults Feb–Apr	Habitat destruction (urbanization)

Table 6.2. (cont.)

Taxon	Status	Distribution	Food	Phenology	Threat
Euphilotes battoides allyni (Lycaenidae), El Segundo Blue	T	80 ha El Segundo sand dunes, California	*Eriogonum parvifolium*	Univoltine, adults Jun–Sep	Habitat destruction (urbanization)
Euphilotes enoptes smithi (Lycaenidae), Smith's Blue	E	Coastal dunes, Monterey Co., California	*Eriogonum* spp.	Univoltine, larvae Jun–Aug, pupae Aug–Jun, adults Jun–Sep	Habitat destruction (coastal development)
Glaucopsyche lygdamus palosverdesensis (Lycaenidae). Palo Verdes Blue	E	Palos Verdes Peninsula, California	*Astragalus trichopodus* var. *lonchus*	Univoltine, pupae overwinter	Habitat destruction (urbanization)
Incarcia icarioides missionensis (Lycaenidae), Mission Blue	E	San Francisco, San Mateo Counties, California	*Lupinus* spp.	Univoltine, larvae winter, pupae spring, adults Apr–May	Habitat destruction (urbanization)
Lycaeides argyrongnomon lotis (Lycaenidae), Lotis Blue	E	Mendecino Co. peat bog	*Lotus formosissimus*, clover	Univoltine, larvae winter, pupae spring, adults Jun–Jul	Habitat change (succession), power line corridor maintenance
Boloria acrocnema (Nymphalidae), Uncompahgre Fritillary	E	Mountains of southwestern Colorado	*Salix reticulata* ssp. *nivalis*	Biennial, eggs 1st year, larvae 2nd year, adults summer 2nd year	Collecting
Euphydryas editha bayensis (Nymphalidae), Bay Checkerspot	T	Grasslands on serpentine soil, San Francisco Bay Counties, California	*Plantago erecta*, *Orthocarpus densiflora*	Univoltine	Habitat destruction (urbanization)
Speyeria zerene hippolyta (Nymphalidae), Oregon Silverspot	T	Coastal dunes, Oregon	*Viola adunca*	Univoltine, larvae winter, pupae Jun–Jul, adults Jul–Sep	Habitat destruction (urbanization)
Neonympha mitchelli mitchelli (Nymphalidae), Mitchell's Marsh Satyr	E	Tamarack bogs, Michigan, Indiana	*Carex*, *Scirpus* spp.	Univoltine, larvae overwinter, adults Jun–Jul	Habitat change (succession)
Heraclides aristodemus ponceanus (Papilionidae), Schaus Swallowtail	E	Florida Keys	*Amyris elimifera*, *Xanthoxium sheara*	Larvae Jun–Jul, pupae Jul–Apr, adults May–Jun	Habitat destruction (urbanization), insecticides for adult mosquitoes
Euproserpinus euterpe (Sphingidae), Kern Primrose Sphinx Moth	T	Upper Kern River, California	*Oenothera contorta* var. *epilobioides*	Univoltine, larvae spring–sum., pupae fall–winter, adults Feb–Mar	Habitat destruction (agriculture), oviposition on introduced, unsuitable, host plant

question is usually preservation of habitat in the face of land development (Collins and Thomas, 1991). However, mortality from introduced parasitic Hymenoptera would be a dangerous additional stress. The question is how likely is such mortality. To answer this question we must determine whether these threatened or endangered insect species are likely to be included in the host range of introduced parasitic Hymenoptera.

Determination and prediction of host range

Five processes can affect host range of parasitic Hymenoptera (Vinson, 1976). First, parasitoids must find the habitat and substrate (often, but not always, a host plant) on which the insect of interest occurs. This process may be affected by the light level, temperature, and humidity of the habitat and the colour, odour, and taste of the host plant (Barbosa and Letourneau, 1988). Secondly, parasitoids must find the insect of interest on its substrate. Finding an insect requires synchrony of phenology of adult parasitoids and the insect of interest, as well as recognition of shape, size, colour, odour, and/or taste. Thirdly, parasitoids must accept the insect as a suitable host. This involves evaluation of the stage, location, shape, size, colour, odour, and taste of the insect of interest (Gross, 1993). Fourthly, parasitoids must overcome the defences of the insect of interest. These may be physical barriers (e.g. bark or soil in which the insect hides itself, webbing secreted by the insect, a tough cuticle, and urticating hairs), behavioural responses (e.g. kicking, biting, spitting, dropping from the host plant), chemical composition (e.g. sequestered secondary plant compounds), and physiological responses (e.g. encapsulation). Finally, parasitoids must regulate the insect of interest for proper development. Knowledge concerning this last process may provide important information about host range. Parasitoids that kill their hosts in the stage attacked (idiobionts, e.g. egg and pupal endoparasitoids, and many larval ectoparasitoids) are likely to have broader more malleable host ranges than parasitoids that allow further development of the host (koinobionts, e.g. many egg-larval, larval, and larval pupal endoparasitoids) because of the

intimate interaction with host development necessary for the latter.

Although the processes described above are no doubt under genetic control, change in host range after introduction does not require genetic change. Indeed, the evidence for genetic change in host range is weak (Hopper et al., 1993) (although evidence of the genetic basis for variation in resistance to encapsulation in parasitoids of Drosophila spp. is mounting (Bouletreau, 1986; Carton et al., 1989).

One can classify host ranges according to the potential impact on threatened or endangered species: (1) not possible that endangered taxa could be included, e.g. morphological, physiological limitations; (2) possible, but unlikely, that threatened or endangered taxa would be included, e.g. the family not known to be attacked; (3) likely that endangered taxa would be included, e.g. a closely related species is known to be in the host range; and (4) endangered taxa known to be in host range. It is possible that even for levels 3 an 4, the impact would be low because of poor overlap in space and time between the biological control agent and the threatened or endangered species. However, given the potential risk and current law, decision makers will probably avoid such introductions.

Several methods are available for determining whether a taxon is likely to be in the host range of a parasitoid: (1) comparison with rearing records from field collected hosts; (2) laboratory exposure to the taxon; (3) inference from the biology of attack; and (4) some combination of 1–3 for a related parasitoid species. Unfortunately, each of these methods has drawbacks. In laboratory screening, one cannot normally test threatened or endangered taxa because their collection is prohibited. However, some are subspecies of species that are relatively abundant, and furthermore, one could test other species in the same genus. A more serious drawback to laboratory screening is shown by the complexity revealed in the above listing of processes and factors that affect host range. Thus, whether a parasitoid will or will not attack a given species in the laboratory does not mean that it will or will not attack in the field. Therefore, host-range determinations from field-host records are likely to be more reliable, if a sufficient variety of habitats and potential hosts are covered, because such determinations integrate all of the processes and factors listed above. The problem is that field-host records for an introduction candidate may be difficult and expensive to obtain. First, although it is relatively easy to determine the species that parasitize a given insect, it is much harder to determine all the hosts of a parasitoid because of the large number of candidates. Secondly, most rearing records are for economically important, abundant, and/or striking hosts. Lastly, without some biological interpretation, host records are not predictive, i.e. they do not show how host range could change.

Despite these drawbacks, some conclusions may be drawn from our present knowledge of mechanisms that affect host range and the methods available for its determination. First, parasitoids that attack hosts in specialized locations or at particular seasons are unlikely to attack hosts in different places or times. Secondly, koinobionts are less likely to change host range than idiobionts. Thirdly, change in host range is not necessarily or even frequently genetic so that one can probably infer future host range from past behaviour. Lastly, to save effort, one should, where possible, estimate poorly known host ranges from those of closely related parasitoid species.

Current and proposed introductions into the United States

At least 22 insect pests are currently targets for introduction of biological control agents into the United States (Table 6.3). Five orders are represented, but one could argue that parasitoids of Coleoptera and Lepidoptera are the most troublesome as far as threatened/endangered species are concerned. This list of targets, which I complied from a search of the Agricultural Research Service (ARS) project data file and from queries of quarantines and university researchers, is probably incomplete. (During the research for this chapter, I was surprised to find that there is no central list of natural enemies that have been proposed for introduction, although efforts were under way to develop such a list (Coulson, 1992).) Nonetheless, there are too many species to discuss in detail here. Thus, I will discuss parasitoids of five pests which are current targets at my laboratory: *Bemisia tabaci*, *Cydia pomonella*, *Diuraphis noxia*, *Lymantria dispar*, and *Yponomeuta malinellus*. The

Table 6.3. Target pests for biological control introductions in the United States

Coleoptera
Anthonomus grandis (Curculionidae), Bollweevil
Crioceris asparagi (Chrysomelidae), Asparagus Beetle
Diabrotica spp. (Chrysomelidae), Rootworms
Epilachna varivestis (Coccinellidae), Mexican Bean Beetle
Leptinotarsa decemlineata (Chrysomelidae), Colorado Potato Beetle
Pyrrhalta luteola (Chrysomelidae), Elm Leaf Beetle
Oulema melanopis (Chrysomelidae), Cereal Leaf Beetle

Diptera
Anastrepha suspensa (Tephritidae)
Anastrepha ludens (Tephritidae), Mexican Fruit Fly

Hemiptera
Lygus lineolaris (Miridae), Tarnished Plant Bug
Adelphocoris lineolatus (Miridae), Alfalfa Plant Bug

Homoptera
Aphis spiraecola (Aphididae)
Aphis gossypii (Aphididae), Cotton Aphid
Bemisia tabaci (Aleyrodidae), Sweetpotato Whitefly
Diuraphis noxia (Aphididae), Russian Wheat Aphid
Schizaphis graminum (Aphididae), Greenbug
Unaspis euonymi (Diaspididae), Euonymus Scale
Psylla pyricola (Psyllidae), Pear Psylla

Lepidoptera
Lymantria dispar (Lymantriidae), Gypsy Moth
Ostrinia nubilalis (Pyralidae), European Corn Borer
Cydia pomonella (Tortricidae), Codling Moth
Yponomeuta malinellus (Yponomeutidae), Apple Ermine Moth

information on pest and parasitoid biologies is from unpublished data and from Doane and McManus (1981), Lopez-Avila (1986), Dijkerman *et al.* (1986), Hughes (1988) and Araya *et al.* (1990) and references cited therein.

Parasitoid introductions could follow four possible scenarios with three outcomes: (1) the parasitoid cannot include any threatened or endangered species in its host range, thus there is no impact; (2) the parasitoid could include a threatened or endangered species in its host range, but the parasitoid and threatened or endangered species do not co-occur in space or time, thus there is no impact; (3) the parasitoid includes one or more threatened or endangered species in its host range, but preference is low and/or parasitoid abundance in habitats of these hosts is low, thus there is small impact; and (4) the parasitoid includes one or more threatened or endangered

Table 6.4. Biology of parasitic Hymenoptera proposed for introduction or being introduced for biological control of insect pests in the United States by the European Biological Control Laboratory

Species	Origin	Distribution	Host plant	Phenology	Natural enemies	Development	Phenology	Host range
Bemisia tabaci (Homoptera: Aleyrodidae), Sweetpotato Whitefly	Indian subcontinent	California, Arizona, New Mexico, Texas, Alabama, Florida	Leguminosae, Compositae, Malvaceae, Solanaceae, Euphorbiaceae, Convolvulaceae, Cucurbitacea, Labiatae, Verbenaceae, Cruciferae, Amaranthaceae, Rosaceae, Moraceae, Oleaceae, Gramineae, Capparidaceae, Chenopodiaceae, Tiliaceae, Umbelliferae and others	11–15 generations/yr	Aphelsoma spp., Encarsia aspididiotiola, E. deserti, E. shafeei, E. formosa, E. lutes, E. mohyuddini, E. partenopea, E. smithi, E. sublutee, Eretmocerus aligarhensis, E. corni, E. mundus, Pteroptrix bemisiae, Bulgarialeurodes cotesii	Koinobiont, nymphal-pupal endoparasitoids	Multivoltine	Homoptera: Aleyrodidae; occasionally Diaspididae, Aphididae; rarely noctuid eggs
Diuraphis noxia (Homoptera: Aphididae), Russian Wheat Aphid	Eurasia	18 western states	>100 spp. Gramineae	Multivoltine, anholocyclic	Aphelinus asychis, A. varipes (Aphelinidae), Aphidius spp. (Aphidiidae), Diaeretiella rapae, Praon volucre, Ephedrus plagiator (Braconidae)	Koinobiont, nymphal endoparasitoids	Multivoltine	Homoptera: Aphididae
Lymantria dispar (Lepidoptera: Lymantriidae), Gypsy Moth	Eurasia	Northeast and Atlantic seaboard	>300 tree and shrub species	Univoltine (adults, summer; egg 8–9 months)	Casinaria tenuiventris (Ichneumonidae)	Koinobiont, larval endoparasitoid	—	L. dispar, L. mathura
					C. nigripes (Ichneumonidae)	Koinobiont, larval endoparasitoid	—	Lasiocampidae, Lymantriidae, Notodontidae
					Glyptapanteles flavicoxis (Braconidae)	Koinobiont, larval endoparasitoid	Multivoltine, 2 generations/yr	Lymantria obfuscata, L. dispar
					Glyptapanteles liparidis (Braconidae)	Koinobiont, larval endoparasitoid	Multivoltine; requires overwintering host not found in US	Lasiocampidae, Lymantriidae
					Glyptapanteles porthertriae (Braconidae)	Koinobiont, larval endoparasitoid	Multivoltine, 1 generation per year on gypsy moth, must find other hosts for other generations	Arctiidae, Lasiocampidae, Lymantriidae

Pest species	Origin	Distribution	Host plants	Generations	Parasitoid	Parasitoid biology	Voltinism	Parasitoid host range
Cydia pomonella (Lepidoptera: Tortricidae), Codling Moth	Eurasia	Apple growing areas	Apple, crabapple, pear, quince, walnut and other fruits	Up to 3 generations/yr	*Trichomma enecator* (Ichneumonidae)	Koinobiont, bivoltine, larval endoparasitoid	Bi- or trivoltine	Lepidoptera: Eucosmidae, Gelechiidae, Geometridae, Lymantriidae, Noctuidae, Psychidae, Pyralidae, Tortricidae
					Pristomerus vulnerator (Ichneumonidae)	Koinobiont, larval endoparasitoid	Bivoltine	Lepidoptera: Aegeriidae, Gelechiidae, Lasiocampidae, Lymantriidae, Oecophoridae, Phycitidae, Pyralidae, Tineidae, Tortricidae
					Ephialtes spp. (Ichneumonidae)	Idiobiont, larval (in cocoon) ectoparasitoid	Uni- or bivoltine	Lepidoptera: Tineidae, Tortricidae
Yponomeuta malinellus (Lepidoptera: Yponomeutidae), Apple Ermine Moth	Eurasia	Northwest Washington state, British Columbia	Apple trees	Univoltine	*Ageniaspis fuscicollis* (Encyrtidae)	Koinobiont, egg–larval endoparasitoid	Univoltine	Yponomeutidae
					Diadegma armillata (Ichneumonidae)	Koinobiont, larval or larval–pupal endoparasitoid	Multivoltine	Microlepidoptera, one generation/yr on *Yponomeuta*, rest on other species
					Herpestomus brunnicornis (Ichneumonidae)	Idiobiont, pupal endoparasitoid	Univoltine	Maybe only *Yponomeuta*, maybe also *Cacoecia muriana* (Tortricidae) and *Plutella maculipennis* (Plutellidae)

species in its host range, its preference for them is high or parasitoid abundance in their habitats is high, thus there is a large impact.

To determine the potential impact of introductions of parasitoids against the targets discussed here, I used the following reasoning. If the parasitoid's host range did not include related taxa, e.g. from the same family, and the parasitoid was unlikely to expand its host range, i.e. was a koinobiont with a systematically restricted host range, I put the parasitoid in scenario 1 above, i.e. no impact potential. If the parasitoid's host range include related taxa, or the parasitoid was likely to expand its host range, i.e. was an idiobiont with a wide host range, but was unlikely to occur in same time and place as the threatened or endangered taxa, I put the parasitoid in scenario 2, i.e. little or no impact potential. If the parasitoid's host range included related taxa or the parasitoid was likely to expand its host range and the parasitoid and threatened or endangered taxa were likely to co-occur, I put the parasitoid in scenarios 3–4, i.e. potential impact so do not introduce.

For *Bemisia tabaci*, all the species of Hymenoptera proposed for introduction or being introduced are koinobiont parasitoids of whiteflies, occasionally of diaspidid scales and aphids (Table 6.4). There have also been some reports of attack on noctuid eggs. Thus, all the candidates should follow scenario 1 with no potential for impact on threatened or endangered species. (In some cases, males of these parasitoids are hyperparasitic on females. Although this may affect their usefulness in biological control, it does not affect their impact on threatened or endangered species.) For *Cydia pomonella*, two of the species of Hymenoptera proposed for introduction are koinobiont parasitoids of microlepidoptera, noctuoids, and bombycoids, and one is an idiobiont parasitoid of microlepidoptera (Table 6.4). Given the broad host ranges of these species, tests with species in the same families or genera as the threatened or endangered Lepidoptera would be useful, especially for the idiobiont. However, the target host does not occur on the same or related host plants or in close spatial proximity to the threatened and endangered species. Thus, the candidates should follow scenarios 1 or 2 with no potential impact on threatened or endangered species. For *Diuraphis noxia*, all the species of Hymenoptera proposed for introduction or being introduced are koino-

biont parasitoids of aphids (Table 6.4). Thus, all the candidates should follow scenario 1 with no potential for impact on threatened or endangered species. For *Lymantria dispar*, all the species of Hymenoptera proposed for introduction or being introduced are koinobiont parasitoids of noctuoid and bombycoid moths (Table 6.4). None are known to attack members of the Papilionoidea, Hesperioidea, or Sphingoidea. So that, again, all the candidates should follow a scenario 1 with no potential for impact on threatened or endangered species. For *Yponomeuta malinellus*, one of the species of Hymenoptera being introduced is an koinobiont of the target genus, one is a koinobiont of microlepidoptera, and one is an idiobiont either of the target genus, or of tortricids and plutellids as well. For the latter two species, tests with species in the same families or genera as the threatened or endangered Lepidoptera would be useful, especially for the idiobiont. However, the target host does not occur on the same or related host plants or in spatial proximity to the threatened and endangered species. Thus, the candidates should follow scenarios 1 or 2 with no potential impact on threatened or endangered species. It is worth noting that *Agria mamillata* Pandelle (Diptera: Sarcophagidae), which also attacks *Y. malinellus*, has not been introduced because its potential host range is considered too broad.

Discussion

The analysis of impacts given here was intentionally limited to parasitic Hymenoptera, threatened and endangered species, the continental United States, and impacts of biological control introductions only. This was in part because the goal was mainly to illustrate the approach and to suggest that often potential impact can and is being assessed from currently available data. However, it was also because a wider taxonomic or geographical scope would bring up more controversial issues. For example, if biological control practitioners were required to assess and avoid impact on all insect species, rather than those that are threatened and endangered, it is likely that biological control introductions would have to stop. Such impacts are always possible and yet their magnitudes are extremely difficult or impossible to predict or measure without much greater resources than are currently available for the conduct of biological control introductions. Furthermore, a full assessment would require comparison of the impact of biological control with that of other control measures or no control at all.

Some have argued that the irreversibility of introductions puts them in a class separate from other control measures. The development of control agents with conditionally lethal genes and the very fact that species can be driven to extinction suggests that introductions may not be so irreversible. However, the important point is not whether introductions are irreversible, but whether their impact is irreversible. Allele effects at small population sizes (Hopper and Roush, 1993) and multiple domains of attraction for complex systems (Barbosa and Schultz, 1987) among other phenomena show that other controls may be just as irreversible in their impact as biological control introductions.

It appears that limited habitat and habitat destruction are the most important hazards for threatened or endangered taxa. Current biological knowledge, in particular of host ranges of parasitoids can, and is, being used to assess their potential impacts on threatened and endangered taxa. The introductions examined here have low potential for impact on threatened or endangered taxa. Nonetheless, we need to know more about possible shifts in parasitoid host ranges for certain introductions and we need to know more in general about the mechanisms that affect host range.

References

Araya, J. E., Quiroz, C. and Wellso, S. G. (1990). Pest status and control of the Russian wheat aphid, *Diuraphis noxia* (Mordvilko) (Homoptera: Aphididae), a review. *Purdue University Agricultural Experiment Station Bulletin* No. 588.

Barbosa, P. and Schultz J. C. (eds) (1987). *Insect Outbreaks*. Academic, San Diego.

Barbosa, P. and Letourneau, D. K. (eds) (1988). *Novel Aspects of Insect-plant Interactions*. Wiley, New York.

Bouletreau, M. (1986). The genetic and coevolutionary interactions between parasitoids and their hosts. In *Insect Parasitoids* (ed. J. Waage and D. Greathead), pp. 169–200. Academic, Orlando, FL.

Carton, Y., Capy, P. and Nappi, A. J. (1989). Genetic variability of host-parasite relationship traits: utilization of isofemale lines in a *Drosophila simulans* parasitic wasp. *Genetics Selection Evolution* 21, 437–46.

Clausen, C. P. (ed.) (1978). *Introduced Parasites and Predators of Arthropod Pests and Weeds: A World Review*. USDA Agr. Handbook 480. USDA; Washington, DC.

Collins, N. M. and Thomas, J. A. (1991). *The Conservation of Insects and Their Habitats. 15th Symposium of the Royal Entomological Society of London*. Academic, New York.

Coulson, J. R. (1992). Documentation of classical biological control introductions. *Crop Protection* 11, 195–205.

Coulson, J. R., Soper, R. S. and Williams, D. W. (eds) (1991). *Proceedings of the USDA-ARS Workshop on Biological Control Quarantine: Needs and Procedures* (14–17 January, Baltimore, Maryland). USDA, ARS, Washington, DC.

DeBach, P. and Rosen, D. (1992). *Biological Control by Natural Enemies*, 2nd edn. Cambridge University Press, Cambridge.

Dijkerman, H. J., de Groot, J. M. B. and Herrebout, W. M. (1986). Parasitoids of the genus *Yponomeuta* (Lepidoptera, Yponomeutidae) in the Netherlands. *Proceedings of the Koninklijke Nederlandse Akademie van Wetenschappen C* 89, 379–98.

Doane, C. C. and McManus, M. L. (1981). The gypsy moth: research toward integrated pest management. *USDA Forest Service and APHIS Technical Bulletin*, No. 1584.

Gross, P. (1993). Insect, behavioral and morphological defenses against parasitoids. *Annual Review of Entomology* 38, 251–273.

Hopper, K. R., Roush, R. T. (1993). Mate finding, dispersal, number released, and the success of biological control introductions. *Ecological Entomology* 18, 321–331.

Hopper, K. R., Roush, R. T. and Powell, W. (1993). Managing the genetics of biological control introduction. *Annual Review of Entomology* 38, 27–51.

Howarth, F. G. (1983). Classical biological control: panacea or Pandora's box. *Proceedings of the Hawaiian Entomology Society* 24, 239–244.

Howarth, F. G. (1991). Environmental impacts of classical biological control. *Annual Review of Entomology* 36, 485–509.

Hughes, R. D. (1988). A synopsis of information on the Russian wheat aphid, *Diuraphis noxia* (Mordwilko). *CSIRO Division Entomology Technical Paper No. 28*.

Lopez-Avila, A. (1986). Natural enemies. In Bemisia tabaci – *a Literature Survey* (ed. M. J. W. Cock), pp. 27–35. Commonwealth Agricultural Bureau, International Institute of Biological Control, Ascot, UK.

Lowe, D. W., Matthews, J. R. and Moseley, C. J. (eds) (1990). *The Official World Wildlife Fund Guide to Endangered Species of North America*, vol. 2, pp. 1061–1102. Beacham, Washington, DC.

Moseley, C. J. (ed.). (1992). *The Official World Wildlife Fund Guide to Endangered Species of North America*, vol. 3, pp. 1469–1480. Beacham, Washington, DC.

Samways, M. J. (1988). Classical biological control and insect conservation: are they compatible? *Environmental Conservation* 15, 348–354.

United States Fish and Wildlife Service (1989). *Endangered and threatened wildlife and plants.* 50 CFR 17.11 and 17.12, 30 November 1988.

United States Fish and Wildlife Service (1991). *Endangered and threatened wildlife and plants.* 50 CFR 17.11 and 17.12, 15 July 1991.

Vinson, S. B. (1976). Host selection by insect parasitoids. Annual Review of Entomology 21, 109–133.

7

Lessons from post-release investigations in classical biological control: the case of *Microctonus aethiopoides* Loan (Hym., Braconidae) introduced into Australia and New Zealand for the biological control of *Sitona discoideus* Gyllenhal (Col., Curculionidae)

Jean-Paul Aeschlimann

Introduction

Species of the genus *Sitona* Germar (Col., Curculionidae) have been recorded on all continents and all develop at the expense of a number of Leguminosae. In their larval stages, they attack the root system and may be especially injurious to the nodules (Aeschlimann, 1986), whereas the long-lived adults consume the stems and foliage of their host plants. The four palaearctic representatives which comprise the *S. humeralis* group of species all depend upon *Medicago* spp. as host plants (Aeschlimann, 1984). One of these, *S. discoideus* Gyllenhal, was accidentally introduced into the southern hemisphere, where it has become a pest of economic importance to both perennial cultivated lucerne and volunteer annual *Medicago* species used for pasture throughout southeastern

Australia and subsequently in the south island of New Zealand.

As a consequence of the importance of *S. discoideus*, surveys were carried out from 1973 to 1985 over most of the Mediterranean basin (Aeschlimann, 1980) with the aim of identifying efficient natural enemies of the various stages of *S. discoideus* that would be suitable for deliberate introduction into Australia. These investigations demonstrated that the parasitoid *Microctonus aethiopoides* Loan (Hym., Euphorinae) was the most promising biological control agent for use against adult *Sitona* weevils. Based on these findings, several biotypes of this natural enemy were imported between 1975 and 1979 from various parts of the western Mediterranean region to Australia for mass-rearing in quarantine and subsequent field release against *S. discoideus*

Table 7.1. Biotypes of *Microctonus aethiopoides* introduced against *Sitona discoideus* in southeastern Australia

Country of origin	Host species	Release period	Success rate[a]
Morocco	*Sitona discoideus*	1977–79	+++
France	*Sitona humeralis*	—	−
Greece	*Sitona bicolor*	1979–81	+

[a]Using a scale: −, no success; +, established but of limited impact and dispersal; +++, high incidence and dispersal.

(Aeschlimann, 1983*a*). Initial establishment followed by spreading was recorded in Australia 2 years after the start of the release programme (Aeschlimann, 1983*a*).

This paper summarizes the findings from 1981 to 1984 while monitoring the subsequent incidence and dispersal of *M. aethiopoides* throughout southeastern Australia (States of Victoria, New South Wales, and the Australian Capital Territory). These results are analysed and compared with data obtained during release and follow-up studies made in the same period in South Australia, and towards the end of the 1980s in New Zealand, which used the same species of natural enemy for biological control purposes. The comparison between the different approaches adopted, illustrates some of the shortcomings inherent in such investigations and leads to formulation of several suggestions for the improvement of the assessment process of classical biological control programmes against introduced pest species.

Experimental approach

The various techniques used to sample, the special containers designed to rear the various stages of *S. discoideus*, all the devices developed to multiply, release, and recover their parasitoids have been described in detail elsewhere (Aeschlimann, 1979, 1983*a*; Cullen and Hopkins, 1982; Goldson *et al.*, 1990). Both incidence and dispersal of *M. aethiopoides* in the southeastern Australian region were assessed on a number of occasions each year for 4 successive years by collecting weevils from *Medicago* spp. sites in one or several

series of 10×25 strokes of a sweep-net (30 cm diameter). Each such sample was then divided into lots of 100–200 adults, each was placed in a special rearing box with double bottom for the emergence of larval parasitoids in the laboratory, and maintained at constant conditions ($21 \pm 2°C$; $70 \pm 10\%$ rel. hum.; 10 h photophase per day). When emergence was complete, each curculionid in each sample was dissected to detect the presence of any dead braconids.

Dispersal was followed by collecting regular weevil samples (as described above) from fields of *Medicago* spp. located at increasing distances from the release centres, and in various directions, any adult parasitoids caught during sweeping were also recorded.

The data obtained during the 4 years (1981 to 1984) of monitoring *M. aethiopoides* in southeastern Australia are presented below.[1] They will serve to illustrate some important aspects of particular concern for future biological control operations.

Biotypical differences

Three biotypes of *M. aethiopoides* were forwarded successively from the Mediterranean region to CSIRO Canberra for mass propagation in quarantine, as listed in Table 7.1. In 1976–7, a first rearing was successfully initiated based on a biotype that originated from south of the Haut-Atlas in southern Morocco. In 1979, however, an attempt at starting another culture with a biotype obtained from southern France failed during the first two generations, whereas a third biotype

[1] The 1982 to 1984 results were collected mainly by J. A. Cavanaugh (present address: Division of Clinical Sciences, John Curtin School of Medical Research, Australian National University, Canberra ACT 2601, Australia), under the supervision of D. T. Briese, CSIRO Division of Entomology, GPO Box 1700, Canberra ACT 2601, Australia.

collected in Central Greece succeeded in the same year.

Consistent differences in morphology, melanization, bionomics, and even, from the biochemical point of view, in the pattern of non-specific esterases have been described by Aeschlimann (1983*b*), not only between *M. aethiopoides* populations from various biogeographic origins, but also between cohorts of individuals that occur sympatrically but on different host weevil species. More recently, morphological and morphometric analyses carried out by Adler and Kim (1985) and Sundaralingam (1986), in particular, have confirmed the above findings. The relationships between biotypes of *M. aethiopoides* and their various host species in the curculionid genera *Hypera* Schönherr and *Sitona* Germar could only be clarified after the taxonomy of the *S. humeralis* Stephens group of species had been elucidated by Roudier (1980). Unfortunately, this only occurred at the end of the importation and rearing programme in Australia (Aeschlimann, 1984). It is now recognized that the taxon '*M. aethiopoides*' in fact represents a whole range of biotypes, each of which is associated with a particular host. Failure of the Greek biotype to perform successfully in Australia may therefore be attributed to the fact that it originated from another host, *S. bicolor* Fahraeus, which occupies a similar niche and in effect replaces *S. discoideus* in the Balkans (Aeschlimann, 1984). The situation was similar with regard to the second biotype, obtained from *S. humeralis* in southern France, and for which the climatic conditions that prevail in Canberra may have posed an additional problem for parasitoids collected at the end of the European winter. Failure to take note of this fine-tuned host preference and an inadequate evaluation of the potential host range undoubtedly accounted, to a large extent, for various failures reported so far for a number of previous attempts at introducing *M. aethiopoides* into North America against different species of *Hypera* and *Sitona* (encapsulation, reduced oviposition, shift in sex ratio, etc. resulting in non-establishment; cf. Aeschlimann, 1983*b*).

As a consequence, most of the incidence on *Sitona* populations recorded so far from Australian fields of *Medicago* spp. must be attributed to the Moroccan biotype, which adapted rapidly to the new climatic conditions, developed high rates of parasitization (over 60% on several occasions), and was able to disseminate very efficiently (cf. Table 7.2; Aeschlimann and Carl, 1987). With a maximum of up to 30% parasitization in only four instances out of 4 years, and a very limited spread after establishment (cf. Table 7.3), the Greek biotype was obviously less appropriate for the Australian situation. In addition, it seems that the 'belated' performance of this Greek biotype may have been due to a contamination of the culture units by Moroccan females, which most likely took place at some stage in 1980 (J. A. Cavanaugh, personal communication).

The restriction of a particular parasitoid – or of some of its biotypes or siblings – to a certain host has recently been well documented in several instances (cf. Chow and Mackauer, 1991; Debach and Rosen, 1991; Hughes *et al.*, 1994). Such strong adaptive liaison with one host species out of a range of field-recorded hosts was clearly demonstrated under laboratory conditions, in terms of drastically depressed reproductive performance and distinctly lower sex ratio in Hymenopteran parasitoids of aphids by Hughes *et al.*, (1994). These were the main reasons evoked for the failure of *Aphelinus varipes* Foerster (Hym., Aphelinidae) to establish on alternate cereal aphid hosts in Australia, during the first part of a pre-emptive integrated control programme directed against the Russian wheat aphid, *Diuraphis noxia* Mordvilko (Hom., Aphididae; Aeschlimann and Hughes, 1992). When offered its original *D. noxia* hosts, even after 20 generations under laboratory conditions, the allegedly 'oligophagous' Ukrainian *A. varipes* immediately recovered its earlier reproductive potential (Hughes *et al.*, 1994). It could also be argued that the rate of encapsulation generally represents a good measure of the adaptation of an endoparasitoid to its host. As a consequence, it is suggested that candidate biotypes that elicit some degree of rejection in their target host should be considered unsuited for classical biological control purposes.

Post-release monitoring

The parasitization rates recorded during 1979–84 for the various releases of *M. aethiopoides* made in southeastern Australia (Tables 7.2 and 7.3) show

Table 7.2. Recovery and dispersal of *Microctonus aethiopoides* (Moroccan biotype) on *Sitona discoideus* in Australia

Locality and state	Release period	Recovery at release centre		Dispersal as measured from the release centre			
		Year	Parasit. (%)	Year	Distance	Direction	Parasit. (%)
Yanco, NSW	1977	1979	61.8	1981	1 km	W	2.6–59.1
		1981	9.1		3 km	S	1.2–8.3
					5 km	N	71.4
					7 km	N	0.7
					14 km	N	5.9
					19 km	SE	0.0–16.7
		1982	42.0	1984	9 km	NW	37.5–55.7
		1983	58.0		11 km	SE	10.2
					12 km	SE	43.4
					18 km	N	10.9
Coolac, NSW	1977–79	1979	49.1	1981	12 km	SW	20.8
					17 km	NW	3.1
		1981	8.2–51.9	1983	54 km	E	4.0
					80 km	E	2.0–5.0
					85 km	E	1.0
					120 km	E	1.4
		1984	14.0	1984	20 km	NW	14.0
					90 km	W	57.9
					120 km	E	1.4–2.4
Cowra, NSW	1977–79	1981	0.0	1982	19 km	NW	0.5
					20 km	NW	1.0
				1984	10 km	W	17.4
					15 km	NW	11.0
					20 km	NW	18.0
					26 km	N	1.4
					30 km	NW	0.6
					34 km	N	6.7
					40 km	NW	4.8
					48 km	NW	2.0

that no recovery at one point of liberation need not necessarily imply that the release was unsuccessful. Some unidentified factors obviously prevented recovery at Cowra, for instance (Table 7.2), but the parasitoid was nevertheless found in substantial numbers 3–7 years later at a series of locations at 10 to 48 km distance from this initial release centre.

It is thus essential that monitoring is continued for several seasons after release to avoid any premature conclusions. This risk is best illustrated by the following two examples:

(a) at Dubbo where low numbers of natural enemies of *Sitona* were only detected 3 years after the beginning of the liberation programme (Table 7.3);
(b) at Fyshwick, where, after having achieved high parasitization rates in the second year, the braconid disappeared in the following year (Table 7.3).

Based on the experience with Moroccan *M. aethiopoides* in Australia, 5–7 years post-release monitoring can be assumed to be sufficient, whereas 3–5 years was not enough to assess adequately the establishment of the Greek biotype.

Monitoring the dispersal

The second part of Tables 7.2 and 7.3 (last four columns of each) lists all the cases in which evidence of dispersal was recorded (field samples that yielded no parasitoid in laboratory rearings are not mentioned here). Initial dispersal of some 10 km per year was reported from southeastern Australia during the first year following establishment (Hopkins, 1982; Aeschlimann, 1983a). In the second and third years (i.e. 1981 and 1982),

Table 7.3. Recovery and dispersal of *Microctonus aethiopoides* (Greek biotype) on *Sitona discoideus* in Australia

Locality and state	Release period	Recovery at release centre		Dispersal as measured from the release centre			
		Year	Parasit. (%)	Year	Distance	Direction	Parasit. (%)
Rochester, VIC	1980	1981	0.0	1982	4 km	NW	4.0
		1982	8.0–21.5	1984	5 km	NW	29.7
		1984	11.7		7 km	NW	8.7
Benalla, VIC	1980–81	1981	0.0				
Cooma, NSW	1979–80	1981	0.0				
Jugiong, NSW	1979	1981	0.0	1984	5 km	NW	19.7
		1984	14.0		10 km	N	34.0
					20 km	N	14.0–15.0
Cootamundra, NSW	1979	1981	0.0–31.8	1981	1 km	NE	3.2–10.8
Fyshwick, ACT	1981	1983	30.0				
		1984	0.0				
Dubbo, NSW	1981	1981	0.0				
		1984	0.9				

dispersal proceeded at the same rate, the entomophage reached fields situated up to 20 km from the release points (Table 7.2). During the fourth year, however, the spread increased considerably, and the braconid was found at sites located up to 50–120 km, and in different directions from the liberation centres (Table 7.2), which represents some 30 to 100 km per breeding season. This overall dispersal activity seems to be the result of an optimal combination of both active and passive (within the host insect) spread (Aeschlimann, 1983a). Dispersal rates of the same order of magnitude have already been reported from a number of beneficial Hymenoptera introduced for classical biological control purposes in widely different families in addition to the Braconidae (present study), i.e. Aphelinidae, Aphidiidae, Encyrtidae, Ichneumonidae, and even Mymaridae (Herren *et al.*, 1987; Quednau, 1990). This feature also needs to be taken into consideration when designing follow-up programmes to measure the success of a release project, and in particular for the selection of control plots located at appropriate distances from the points of liberation.

Incidence on host populations

It is difficult to evaluate adequately the impact of a parasitoid species such as *M. aethiopoides*, which is multivoltine (4–6 generations per year) with some overlap between generations. Large discrepancies in the parasitization rates may, therefore, be recorded during the same season at any one site due to ill-timed occasional samplings (cf. also Hopkins 1982, 1989; Goldson *et al.*, 1990). In the course of the 1981 winter for instance, these proportions fluctuated from 0.0 to 31.8% for the material collected at Cootamundra (Table 7.3), and from 8.2 to 51.9% at Coolac during the 1981 spring (Table 7.2). The incidence of the natural enemy may therefore be classified either as extremely low towards the end of winter, when almost all the individual parasitoids within a cohort have synchronously abandoned their host to pupate in the soil litter, or as unexpectedly high just before the new generation starts to emerge at a time when old adult weevils, most of which are parasitized, are still present in the field.

To try to assess the efficacy of the introduced parasitoid in southeastern Australia, the sampling success over all sites considered throughout the States of Victoria, New South Wales and the Australian Capital Territory (cf. Fig. 7.1) was pooled for the years 1979 to 1984. This is admittedly an indirect and very crude indication of the overall abundance of *S. discoideus* in *Medicago* spp. fields. The mean numbers of adult weevils collected per sampling effort, i.e. per series of 250 sweep-net strokes over the vegetation are presented in Table 7.4. As calculated over all sites and sampling dates during each calendar year (i.e. during the total life span of one adult *S. discoideus*

Table 7.4. Result of average sampling effort per site in southeast Australia 1979–84 (mean no. *Sitona discoideus* caught per sampling)

Year	Total no. adult *S. discoideus* caught	No. sites and collecting dates considered	Average no. adult *S. discoideus*/effort
1979	13321	15	888.1
1980	5916	21	281.7
1981	9133	67	136.3
1982	3459	33	104.8
1984	6405	60	106.8

generation), this average sampling success shows a clear decline from 1979 to 1984 (cf. last column of Table 7.4). As discussed earlier (Aeschlimann, 1980), 50 strokes of a sweep-net may be assumed to be sufficient to collect the whole curculionid fauna (imagos in particular) occurring on 1 m² of normal-sized *Medicago* spp. Taking this relationship into account, 1979 was the only year in which *S. discoideus* reached a population density that was likely to cause economic damage to cultivated lucerne (i.e. well over 100 specimens/m²). The mean numbers calculated for 1980 were still of the order of 60 individuals/m² at which some damage is likely to be observed. For 1981 and the following years, however, the average population density remained below 20 adults/m², i.e. at a level that can be considered as having no noticeable impact on plant yield. Investigations on the *Sitona* weevil were discontinued in southeastern Australia at the end of 1984. The data obtained so far showed that the entomophage had spread, and that it most probably occupied the whole of its potential range of distribution in the new area. Post-release studies were unable to provide any evidence for the

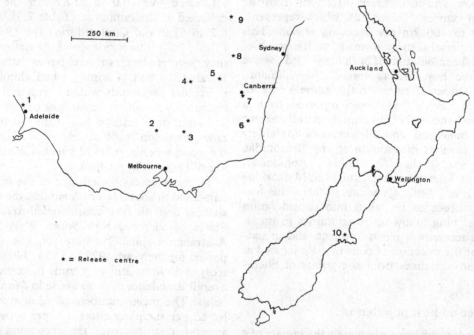

Fig. 7.1 Map showing the release centres of *Microctonus aethiopoides* in Australia and New Zealand. 1, Adelaide area (South Australia); 2, Rochester (Victoria); 3, Benalla (Victoria); 4, Yanco (New South Wales); 5, Coolac-Jugiong-Cootamundra (New South Wales); 6, Cooma (New South Wales); 7, Fyshwick (Australian Capital Territory); 8, Cowra (New South Wales); 9, Dubbo (New South Wales); 10, Christchurch area (Canterbury, New Zealand).

role played by *M. aethiopoides* in decreasing the population densities of its host.

The situation in South Australia, where the curculionid is essentially a pest of annual *Medicago* spp., appears to be different (Hopkins, 1989). In spite of being well established and achieving substantial levels of parasitization, at least comparable to the highest numbers shown in Table 7.2, *M. aethiopoides* is not considered an effective natural enemy of the *Sitona* weevil. Here also, results accumulated so far did not allow for a causal relationship to be established between the levels of visible damage to annual medics in sheep pastures, the persistence of *Sitona* weevil populations, and the abundance of its entomophage.

By contrast, the comprehensive approach followed by the group of research workers involved in the same project in New Zealand had to face a totally different situation. The Moroccan *M. aethiopoides* biotype was imported directly from Australian quarantine. Soon after release it seemed to exert little influence and was also considered inadequate for the control of host populations on cultivated lucerne (Goldson and Proffitt, 1986; Stufkens *et al.* 1987; Goldson and Muscraft-Taylor, 1988). A few years later, however, detailed investigations carried out by Goldson *et al.* (1990) revealed that a switch in the behaviour of the entomophage had occurred, in the Canterbury area at least, which enabled some of the summer generations to develop on aestivating adult weevils and, as a consequence, to suppress effectively the economic damage to lucerne fields on the South Island. An analysis by Barlow and Goldson (1990), Barlow (1993) and Barlow and Goldson (1994), based on the development and use of a simulation model, confirmed that the observed levels of parasitization were indeed responsible for the drastic decline in *S. discoideus* importance in New Zealand, in spite of it being very similar to the levels recorded for the same biotype in southeastern Australia.

Because no specific foreign exploration, subsequent quarantine or further specificity testing (cf. Goldson *et al.*, 1992) were required in New Zealand, all funds available for the project could be devoted to the development of a detailed release and recovery programme, complemented by an intense sampling effort at a few localities, and the design of a sophisticated simulation model. This model not only provided clues to answer some of the questions raised elsewhere, but it also suggested that a timely application of insecticides could be compatible with the most efficient activity of *M. aethiopoides*, should additional inputs be required to achieve satisfactory control of the *Sitona* weevil (Barlow and Goldson, 1994). This example clearly shows the benefits that may arise from thorough monitoring of the release and establishment phase of biological control programme. The post-release monitoring phase of a such programme should undoubtedly be considered equally important and receive as much attention, time and funding as the initial phase of prospection, importation and liberation.

This argument will be further emphasized using one of the best documented, and most impressive modern successes in classical biological control, i.e. the Australian programme directed against skeleton weed, *Chondrilla juncea* Linnaeus (Asteraceae). Introductions of the most effective natural enemies from the Mediterranean region were initiated in the early 1970s; some 10 years later, Groves and Cullen (1981) concluded that successful control had been obtained in southeastern Australia at a high level of efficiency. Groves and Cullen (1981) also demonstrated that the 95% reduction of the *C. juncea* population densities was mainly due to the near suppression of the predominant form of the weed, which had been highly susceptible to the introduced strain of the rust fungus *Puccinia chondrillina* Bubak and Sydenham. The two other forms of *C. juncea* present in Australia have, however, remained totally immune to this first strain of the rust fungus. Investigations aimed at identifying new strains of *P. chondrillina* (in Anatolia) virulent against the other two forms of skeleton weed that persist in Australia are continuing, as well as the follow-up studies in this continent, and a number of additional releases of natural enemies, which have occurred in the meantime both in Australia and elsewhere (Julien, 1992). This extended period of research has allowed the reproductive system of the weed, and the extremely high specificity mechanisms of its main natural enemies to be analysed and better understood (Chaboudez, 1989). This example, therefore, considerably reinforces the claim made above for an in-depth understanding of the relationships between target organisms and the biological control agents to be introduced, as despite the incomplete level of control achieved in the *C. juncea* project, the economic benefits for Australia from this programme are still enormous (Tisdell, 1991).

Conclusions

The following points arising from the discussion of the *S. discoideus* example in the southern hemisphere need to be emphasized for consideration in classical biological control in the future.

1. The relationships between biotypes of natural enemies considered for importation and the various forms of their target species must be investigated in great detail before any releases are made into the new environment.
2. It is crucial that post-release follow-up studies be thoroughly continued for a minimum period of 5–7 years.
3. When sampling programmes to monitor dispersal are designed, the spread potential of up to 100 km/ year for parasitic Hymenoptera must be taken into consideration.
4. Observations on presence/absence, and on the incidence of natural enemies must be supported by an adequate analytical model, developed in parallel, to corroborate field results.

Acknowledgments

The author would like to express his gratitude to D. Briese (Canberra), and R. Groves (Montpellier) for useful critical comments on early stages of this manuscript.

References

Adler, P. H. and Kim, K. C. (1985). Morphological and morphometric analyses of European and Moroccan biotypes of *Microctonus aethiopoides* (Hymenoptera: Braconidae). *Annals of the Entomological Society of America* 78, 279–283.

Aeschlimann, J. P. (1979). Sampling methods and construction of life tables for *Sitona humeralis* populations (Col., Curculionidae) in Mediterranean climatic areas. *Journal of Applied Ecology* 16, 405–415.

Aeschlimann, J. P. (1980). The *Sitona* (Col.: Curculionidae) species occurring on *Medicago* and their natural enemies in the Mediterranean region. *Entomophaga* 25, 139–153.

Aeschlimann, J. P. (1983a). Sources of importation, establishment and spread in Australia, of *Microctonus aethiopoides* Loan (Hymenoptera: Braconidae), a parasitoid of *Sitona discoideus* Gyllenhal (Coleoptera: Curculionidae). *Journal of the Australian Entomological Society* 22, 325–331.

Aeschlimann, J. P. (1983b). Notes on the variability of *Microctonus aethiopoides* Loan (Hymenoptera: Braconidae: Euphorinae). *Contributions of the American Entomological Institute* 20, 329–335.

Aeschlimann, J. P. (1984). Distribution, host plants, and reproductive biology of the *Sitona humeralis* Stephens group of species (Coleoptera, Curculionidae). *Zeitschrift für Angewandte Entomologie* 98, 298–309.

Aeschlimann, J. P. (1986). Rearing and larval development of *Sitona* spp. (Coleoptera, Curculionidae) on the root system of *Medicago* spp. plants (Leguminosae). *Zeitschrift für Angewandte Entomologie* 101, 461–469.

Aeschlimann, J. P. and Carl, K. P. (1987). Les tactiques curatives à l'aide d'ennemis naturels. In *Protection Intégrée: Quo Vadis?* (ed. V. Delucchi). *Parasitis* 86, 167–192.

Aeschlimann, J. P. and Hughes, R. D. (1992). Collecting *Aphelinus* spp. (Hymenoptera: Aphelinidae) in Southwestern CIS for pre-emptive biological control of *Diuraphis noxia* (Homoptera: Aphididae) in Australia. *Journal of Hymenoptera Research* 1, 103–105.

Barlow, N. D. (1993). Applications and challenges in pest population modelling. In *Individuals, Populations and Patterns in Ecology* (ed. S. Leather, A. Watt, A. Dervor and N. Kidd). Intercept, Andover.

Barlow, N. D. and Goldson, S. L. (1990). Modelling the impact of biological control agents. *Proceedings of the 43rd New Zealand Weed and Pest Control Conference*, pp. 282–283.

Barlow, N. D. and Goldson, S. L. (1994). A modelling analysis of the successful biological control of *Sitona discoideus* Gyllenhal (Coleoptera: Curculionidae) by *Microctonus aethiopoides* Loan (Hymenoptera: Braconidae) in New Zealand. *Journal of Applied Ecology* (in the press).

Chaboudez, P. (1989). Modes de reproduction et variabilité génétique des populations de *Chondrilla juncea* L.: implications dans la lutte microbiologique contre cette mauvaise herbe. Ph.D. thesis, Université de Montpellier II, France.

Chow, A. and Mackauer, M. (1991). Patterns of host selection by four species of aphidiid (Hymenoptera) parasitoids: influence of host switching. *Ecological Entomology* 16, 403–410.

Cullen, J. M. and Hopkins, D. C. (1982). Rearing, release and recovery of *Microctonus aethiopoides* Loan (Hymenoptera: Braconidae) imported for the control of *Sitona discoideus* Gyllenhal (Coleoptera: Curculionidae) in southeastern Australia. *Journal of the Australian Entomological Society* **21**, 279–284.

Debach, P. and Rosen, D. (1991). *Biological Control by Natural Enemies*, 2nd edn. Cambridge University Press, Cambridge.

Goldson, S. L. and Proffitt, J. R. (1986). The seasonal behaviour of the parasite *Microctonus aethiopoides* and its effects on Sitona weevil. *Proceedings of the 39th New Zealand Weed and Pest Control Conference*, pp. 122–125.

Goldson, S. L. and Muscroft-Taylor, K. E. (1988). Inter-seasonal variation in *Sitona discoideus* Gyllenhal (Coleoptera: Curculionidae) larval damage to lucerne in Canterbury and the economics of insecticidal control. *New Zealand Journal of Agricultural Research* **31**, 339–346.

Goldson, S. L., Proffitt, J. R. and McNeill, M.R. (1990). Seasonal biology and ecology in New Zealand of *Microctonus aethiopoides* (Hymenoptera: Braconidae), a parasitoid of *Sitona* spp. (Coleoptera: Curculionidae), with special emphasis on atypical behaviour. *Journal of Applied Ecology* **27**, 703–722.

Goldson, S. L., McNeill, M. R., Phillips, C. B. and Proffitt, J. R. (1992). Host specificity testing and suitability of the parasitoid *Microctonus hyperodae* (Hym.: Braconidae, Euphorinae) as a biological control agent of *Listronotus bonariensis* (Col.: Curculionidae) in New Zealand. *Entomophaga* **37**, 483–498.

Groves, R. H. and Cullen, J. M. (1981). *Chondrilla juncea*: the ecological control of a weed. In *The Ecology of Pests. Some Australian Case Histories* (ed. R. L. Kitching and R. E. Jones), pp. 6–17. CSIRO Australia, Melbourne.

Herren, H. R., Neuenschwander, P., Hennessey R. D. and Hammond, W. N. C. (1987). Introduction and dispersal of *Epidinocarsis lopezi* (Hym., Encyrtidae), an exotic parasitoid of the cassava mealybug, *Phenacoccus manihoti* (Hom., Pseudococcidae), in Africa. *Agriculture, Ecosystems and Environment* **19**, 131–144.

Hopkins, D. C. (1982). Establishment and spread of the Sitona weevil parasite *Microctonus aethiopoides* in South Australia. *Proceedings of the 3rd Australasian Conference on Grassland Invertebrate Ecology*, 177–182.

Hopkins, D. C. (1989). Widespread establishment of the Sitona weevil parasite, *Microctonus aethiopoides* and its effectiveness as a control agent in South Australia. *Proceedings of the 5th Australasian Conference on Grassland Invertebrate Ecology*, pp. 49–54.

Hughes, R. D., Hughes, M. A., Aeschlimann, J. P., Woolcock, L. T. and Carver, M. (1994). An attempt to anticipate biological control of *Diuraphis noxia* (Hom., Aphididae) in Australia. *Entomophaga* **39** (in the press).

Julien, M. H. (1992). *Biological Control of Weeds. A World Catalogue of Agents and their Target Weeds*, 3rd edn, pp. 1–186. CAB International Wallingford.

Quednau, F. W. (1990). Introduction, permanent establishment, and dispersal in eastern Canada of *Olesicampe geniculatae* Quednau and Lim (Hymenoptera: Ichneumonidae), an important biological control agent of the Mountain ash sawfly, *Pristiphora geniculata* (Hartig) (Hymenoptera: Tenthredinidae). *Canadian Entomologist* **122**, 921–934.

Roudier, A. (1980). Les Sitona Germar 1817 du groupe du *Sitona humeralis* Stephens 1831. *Bulletin de la Société Entomologique de France* **85**, 207–217.

Stufkens, M. W., Farrell, J. A. and Goldson, S. L. (1987). Establishment of *Microctonus aethiopoides*, a parasitoid of the sitona weevil in New Zealand. *Proceedings of the 40th New Zealand Weed and Pest Control Conference*, pp. 31–32.

Sundaralingam, S. (1986). Biological, morphological and morphometric analysis of populations of *Microctonus aethiopoides* Loan (Hymenoptera: Braconidae). M.Sc. thesis, Pennsylvania State University, USA.

Tisdell, C. A. (1991). Economic impact of biological control of weeds and insects. In *Critical Issues in Biological Control* (ed. M. Mackauer, L. E. Ehler and J. Roland), pp. 301–316. Intercept, Andover.

8

Host specificity screening of insect biological weed control agents as part of an environmental risk assessment

Bernd Blossey

Introduction

Host specificity screening is the most important step that each potential weed biological control agent (whether insect or pathogen) has to pass prior to its introduction (Schroeder, 1983). Regardless of its control potential, only the demonstrated safety of plants of economic or ecological importance in the release area, will lead to approval for release of the control agent. The use of screening-protocols for weed biological control agents (Wapshere, 1975; Schroeder, 1983; CAB, 1986) are now widely accepted. Even less rigorous tests in the first part of this century have never led to the release of a 'new pest' (Crawley, 1989; Howarth, 1991).

The safety record of weed biocontrol using insects is excellent, nonetheless, the growing awareness of the public towards environmental interference by man, has led to legislative regulations such as the 'Biological Control Act of 1984' in Australia. Today, information on the environmental impact, alternatives, the relationship between local short-term uses and long-term productivity, and a risk analysis of the proposed action are required. Conflicts of interest over the implementation or safety of a specific biological control programme or agents, may still arise (Delfosse, 1990).

During the last decades especially, the critical objections of nature conservationists who fear for already stricken and endangered plant species

related to the target weed, have considerably delayed the progress of programmes (Schroeder and Goeden, 1986). Despite the excellent safety record of weed biocontrol, we have to assess the risks of introducing a foreign species into a new environment.

Insect foraging

A herbivorous insect has to overcome several difficulties prior to multiplication. The adult has to locate a host plant and eat to survive. Only a minority e.g. certain gall midges, are solely dependent on the reserves accumulated by the larvae. Most insect species can fly quite long distances. Butterflies, e.g. the monarch in North America, migrate thousands of kilometres. These long distance foragers locate promising patches of food plants largely by habitat characteristics, plant architecture, olfactorial cues or phenological development of their host plants (Bell, 1990).

Plant architecture, olfactorial stimuli, surface texture (i.e. trichomes, waxes etc.), taste, phenological stage (e.g. flowering), microsite characteristics, 'quality' of the plant as measured by the insect, and competitor or predator avoidance are important for acceptance or rejection by foraging adults and larvae (Bell, 1990). Regular feeding will only occur if essential requirements are met. Migrations might take place during the whole life cycle and it is of vital importance to locate new host plants. Most species also use host plants as courtship arenas. This increases the probability of mate finding.

Screening protocols

The aim of host specificity screening is to minimize risks of potential damage to fauna and flora within the future release area (see Hopper, Chapter 6; and Andow *et al.*, Chapter 10). The demand is that 'plants of economic, ecological, aesthetic, or other values must be safeguarded' (Klingman and Coulson, 1982). If this statement is taken seriously it means that every plant has to be safeguarded.

Investigations of long distance foraging during host specificity screening are impossible. Placing test plant species into the field and then waiting to see whether they will be colonized is an unacceptable experimental design. Field tests are carried out in areas with high host and insect populations, or insects are released in test areas. Thus, we do not obtain any information on host finding during migration. Screening tests focus on plant chemistry, leaf texture and plant architecture of test plant species.

Obviously it is impossible to test all crops and endemic plants prior to introduction of an agent. The task of host specificity screening is to include a series of those plant species that might serve as a potential host of the organism in question. These include:

1. Plant species related to the target weed.
2. Other host plants recorded in the literature.
3. Host plants of insect species related to the possible control agent.
4. Plant species with similar morphological or biochemical characteristics.
5. Plants growing in the same habitats as the target weed in the release area.
6. Crop and ornamental plants including those of potential economic value.

Generally a list of test plants is proposed by a researcher and reviewed by an advisory group. The make-up of this group varies from country to country, but people involved in all aspects of environmetal safety will comment on the proposal (see Lima, 1990 for details in the US). After revision, a final test plant list is prepared.

According to Schroeder (1983) the number of test plant species depends on

1. The taxonomic position, whether it is an isolated family or has many relatives.
2. The number of closely related wild or cultivated plants.
3. The geographical and/or ecological isolation of the release area.
4. The taxonomic position of the control organism; whether the taxon is known to be restricted to a small group of closely related plants.

In general the number of plants investigated varies between 40 and 50 but in large families, like thistles and knapweeds, up to 80 species have been exposed to potential control agents (Schroeder, 1983).

Test methods

Even though screening-protocols are now widely standardized, there is considerable debate concerning the validity of the results. A recent critical review of test procedures is given by Cullen (1990).

No-choice and multiple-choice tests

At present, no-choice or multiple-choice in the presence or absence of the normal host plant are being used to test feeding and oviposition of adults and feeding of larvae. Tests are carried out in confinement in the laboratory, or in quarantine in small tubes, Petri dishes or cages of considerable size depending on type of test and particular candidate. Gall-making insects, or internally feeding larvae, require much more elaborate tests than free-foraging larvae. More reliable results are expected to be obtained when insects are tested on potted plants instead of plant parts (Schroeder, 1983).

No-choice tests can exclude the majority of test plant species; more elaborate tests are required where initial feeding occurs. Multiple-choice tests, which include the host plant overcome some of the shortcomings of no-choice tests and they are generally more reliable. Data interpretation may still be problematical because insects in confinement show a much broader host range than in the field. This is of importance if members of Technical Advisory Groups are unfamiliar with the interpretation of test results in confinement, or foraging theory.

Field cage tests

Biocontrol practitioners used large walk-in cages, the size of cages being of overriding importance (Cullen, 1990). They allow plants to grow to their regular size and normal habitat characteristics can be matched. Problems may still arise with very mobile species such as Diptera and Lepidoptera or those species that depend on fertilized flowers. In addition, the species tested and their feeding damage has to be well known, because other insects present in large field cages may attack test plants and confound the results.

Open field tests

The results of field cage tests may still be of less value than those obtained from open field experiments (Cullen, 1990). The demand for reliable data has forced researchers to investigate the host specificity in the field (Groppe et al., 1990). However, plant quarantine regulations in many countries do not allow open field tests with alien plants and these therefore have to be tested in confinement.

Case study: evaluation of test results with biocontrol agents of purple loosestrife (*Lythrum salicaria* L.)

The following results summarize 5 years of investigation on host specificity of biological control agents of *L. salicaria*. The plant has invaded North American wetlands and is a threat to endemic flora and fauna (Thompson et al., 1987; Malecki et al., 1993). The screening involved no-choice and multiple-choice tests with leaves, cut stems and potted plants, as well as experiments in field cages and open field tests with six insect species. Laboratory and field experiments were carried out in northern Germany, quarantine testing at the Virginia Polytechnic Institute (VPI) at Blacksburg, Virginia.

A total of 44 plant species belonging to 16 families have been tested; results for four species will be discussed here in more detail. Plants and insects were reared either in the greenhouse, in the field, or in quarantine under various light and temperature regimes. Details of test procedures can be found in Kok et al. (1992) and Blossey et al. (1994).

Results of different test methods for the chrysomelid beetle *Galerucella calmariensis* L. are summarized in Table 8.1. The broadest host range was found when plant parts were offered in no-choice tests. Sometimes the feeding or oviposition rate was higher on test plants than on the control *L. salicaria*. Plants grown in the greenhouse or in quarantine develop different cell structures as well as physiological anomalies (i.e. changes in water content and secondary metabolites), which are important in host selection and acceptance by insect herbivores. Feeding and oviposition were reduced (sometimes non-existent) in multiple-choice tests or when potted plants were offered in the tests (Table 8.1).

Under open field conditions the least amount of feeding occurred. At Meggerdorf only newly emerged F_1 beetles attacked *Lythrum alatum* and *Decodon verticillatus*. However, the unexpected feeding of new generation beetles demonstrated the difficulties in the interpretation of test results.

Discussion

The host specificity of potential biological weed control agents is investigated to reduce the risk of potential negative environmental impacts. In weed biocontrol we are concerned with a biological control host (Cullen, 1990), thus an agent that severely damages a test plant species has to be excluded even though it is often argued in terms of physiological hosts (an agent that can successfully complete its entire development on a test plant species). A recent critical evaluation of the history of biological control by Howarth (1991) did not find evidence for serious negative impacts of insects used in biological weed control programmes. The review focused mainly on predatory organisms, which were introduced without any host range testing currently essential in biological weed control programmes. The current screening-procedures are rigorous enough to prevent any serious negative impacts to the environment.

However, several promising candidates failed to pass the tests, because they showed a broader host range during screening (Schroeder, 1983; Cullen,

Table 8.1. Comparison of host specificity screening results with *Galerucella calmariensis*

Test[a]	Plant parts	Lythrum salicaria	Lythrum alatum	Decodon verticillatus	Lagerstroemia indica
Laboratory					
No-choice					
Larval feeding	Leaves	+++	++	(+)	—
Larval development	Leaves	+++	+	—	—
Multiple-choice					
Adult feeding	Potted plants	+++	++	(+)	—
Oviposition	Potted plants	+++	+	—	—
Field					
Garden					
Adult feeding/oviposition/larval development	Potted plants	+++	—	—	—
Fehmarn					
Adult feeding/oviposition/larval development	Potted plants	+++	—	—	Not tested
Meggerdorf (F₁ adults)					
Adult feeding/oviposition/larval development	Potted plants	+++	++	+	Not tested
Quarantine					
adapted from Kok et al. *(1992)*					
No choice					
Larval development	Leaves	+++	+	—	—
Adult feeding	Cut shoots	+++	+++	++++	++++
Adult feeding	Potted plants	+++	++	++	—
Oviposition	Cut shoots	+++	++	+++	++++
Multiple choice					
Adult feeding	Cut shoots	+++	++	+	—
Oviposition	Cut shoots	+++	—	—	—

[a]+++ indicates normal feeding, oviposition or larval development; ++ indicates moderate feeding, oviposition or larval development; + indicates slight feeding, oviposition or larval development; (+) indicates occasional nibbling by larvae or adults; ++++ indicates that adult feeding and oviposition rate were higher on a test plant than on the control *L. salicaria* and — indicates that no feeding, oviposition nor larval development occurred.

1990; Shepherd, 1990). Often it is necessary to introduce several species to achieve control. To reject potential candidates because host specificity results were obtained from tests with questionable predictive power might exclude the most promising agents. Some researchers argue for the quantification of test results that should allow for estimation of a probability of host shift or host use (Cullen, 1990). Harris (1991) expects that a test plant species that supports less than 75% of the feeding on a control is not at risk in the field.

The results of screening with *G. calmariensis* illustrate some of the difficulties with host-range screening. We have to investigate different developmental stages, and the experimental design can alter the results in an unexpected and unknown way. The results we collect during the screening procedure are some of the best sources of information on insect behaviour, but they are poorly used. Currently we are even unable to predict the failure or establishment of a well-known insect in a well-known environment on its particular host plant (Crawley, 1989).

The reliability of test procedures has never been investigated after the release of an insect or pathogen. Therefore, we do not know whether any of the laboratory no-choice tests have predictive power, or are purely artefacts due to test conditions. Harris (1991) states that 'researchers lack enthusiasm for doing such studies as they perceive the reviewers to be unwilling to use any criteria except the survival or lack of it in laboratory feeding tests'. However, follow-up studies on impact on target weed populations as well as the realized host range in the field should be encouraged and adequately funded to improve recent procedures and to increase the success rate of weed biocontrol. The results of such studies should be discussed by Technical Advisory Groups so that their decisions can be based on scientifically sound test methods. Until such criteria become available tests should be designed that are as similar as possible to open field conditions, and in cases where different tests produce contradictory results the more natural test should be given greater weight.

At present we have to accept that ecology is still far from being predictive, especially if more than just a 'simple' plant–insect relationship is involved. Even in biological weed control, we are unable to predict success or failure of establishment of a well-known insect in a well-known environment on a particular host plant (Crawley, 1989). This is especially important when the release of genetically manipulated organisms into the environment is under consideration. We have to weigh with care the possible benefits against the risks. Once established, chances for removal do not exist.

Conclusions

Host specificity screening is the most important step that each potential biological weed control agent has to pass prior to introduction. Regardless of its control potential, only the demonstrated safety of non-target plants will lead to an approval for release. Test protocols and methods are widely standardized but the results of laboratory research are difficult to relate to field conditions. Insects in confinement often have broader host ranges than in the field. Tests in large walk-in cages or in the field solve some but not all of the problems of laboratory studies. Advisory groups that review petitions for field release often insist on starvation tests, and thus delay the development of test protocols that provide more reliable data. Investigations on the host specificity of potential biological control agents of *Lythrum salicaria* showed that results differed between no-choice or multiple-choice laboratory tests and field experiments; between tests with cut stems *versus* potted

plants, overwintered or newly emerged beetles and between different developmental stages. Studies that provide data on the actual host range of insect weed control agents after field release do not exist. Such data would provide the most promising means for the improvement of current screening protocols.

Acknowledgments

I thank the organizers of the Saariselka workshop for the invitation and creation of a stimulating atmosphere for discussion. D., Schroeder commented on an earlier draft of this paper. Work in Europe has been funded by the USDA-ARS, Beneficial Insects Introduction Laboratory, Beltsville MD, the United States Department of the Interior Fish and Wildlife Service, through Cornell University, Ithaca, NY, and the Departments of Agriculture and Wildlife, State of Washington.

References

Bell, W. J. (1990). Searching behaviour patterns in insects. *Annual Review of Entomology* **35**, 447–467.

Blossey, B., Schroeder, D., Hight, S. D. and Malecki, R. A. (1994). Host specificity and environmental impact of two leaf beetles for the biological control of purple loosestrife (*Lythrum salicaria*). *Weed Science* **42**: 134–140.

CAB (1986). *Screening organisms for biological control of weeds*. CAB, International Institute for Biological Control, Farnham Royal.

Crawley, M. J. (1989). The successes and failures of weed biocontrol using insects. *Biocontrol News and Information* **10**, 213–223.

Cullen, J. M. (1990). Current problems in host specificity screening. *Proceedings of the VII International Symposium on Biological Control of Weeds*, 6–11 March 1988, Rome, Italy (ed. E. S. Delfosse), pp. 27–36. Ministro dell'Agriculture e delle Foreste, Rome/CSIRO, Melbourne.

Delfosse, E. S. (1990). *Echium* in Australia: the conflict continues. *Proceedings of the VII International Symposium on Biological Control of Weeds*, 6–11 March 1988, Rome, Italy (ed. E. S. Delfosse), p. 117. Ministro dell'Agriculture e delle Foreste, Rome/CSIRO, Melbourne.

Groppe, K., Sobhian, R. and Kashefi, J. (1990). A field experiment to determine host specificity of *Larinus curtus* (Col., Curculionidae) and *Urophora sirunaseva* Hg. (Dipt., Thephritidae), candidates for the biological control of *Centaurea soltitialis* L. (Asteraceae) and *Larinus minutus* Gyllenhal, a candidate for biological control of *C. maculosa* Lam. and *C. diffusa* Lam. *Journal of Applied Entomology* **110**, 300–306.

Harris, P. (1991). Screening classical weed biocontrol projects and agents. In *Biological Control Quarantine: Needs and Procedures* (ed. J. R. Coulson, R. S. Soper and D. W. Williams), pp. 61–68. Proceedings of a workshop sponsored by USDA-ARS. US Department of Agriculture, Agriculture Research Service, ARS–99.

Howarth, F. G. (1991). Environmental impacts of classical biological control. *Annual Review of Entomology* **36**, 483–509.

Klingman, D. L. and Coulson, J. R. (1982). Guidelines for introducing foreign organisms into the United States for the biological control of weeds. *Weed Science* **20**, 661–667.

Kok, L. T., McAvoy, T. J., Malecki, R. A., Hight, S. D., Drea, J. J. Jr. and Coulson, J. R. (1992). Host specificity tests of *Galerucella calmariensis* (L.) and *G. pusilla* (Duft.) (Coleoptera: Chrysomelidae), potential biological control agents of purple loosestrife, *Lythrum salicaria* L. (Lythraceae). *Biological Control* **2**, 282–290.

Lima, P. J. (1990). United States Department of Agriculture (USDA) safeguards for introducing natural enemies for biological control of weeds.

Proceedings of the VII International Symposium on Biological Control of Weeds 6–11 March 1988, Rome, Italy (ed. E. S. Delfosse), pp. 109–115. Ministro dell'Agriculture e delle Foreste, Rome/CSIRO, Melbourne.

Malecki, R. A., Blossey, B., Hight, S. D., Schroeder, D., Kok, L. T. and Coulson, J. R. (1993). Biological control of purple loosestrife. *Bioscience* **43**, 680–686.

Schroeder, D. (1983). Biological control of weeds. In *Recent Advances in Weed Research* (ed. W. E. Fletcher), pp. 41–78. Commonwealth Agricultural Bureaux, Farnham Royal, UK.

Schroeder, D. and Goeden, R. D. (1986). The search for arthropod natural enemies of introduced weeds for biological control – in theory and practise. *Biocontrol News and Information* **7**, 147–154.

Shepherd, R. C. H. (1990). Problems which arise with host-specificity testing of insects. *Proceedings of the VII International Symposium on Biological Control of Weeds*, 6–11 March 1988, Rome, Italy (ed. E.S. Delfosse), pp. 85–92. Ministro dell'Agriculture e delle Foreste, Rome/CSIRO, Melbourne.

Thompson, D. Q., Stuckey, R. L. and Thompson, E. B. (1987). Spread, impact, and control of purple loosestrife (*Lythrum salicaria*) in North American Wetlands. US Fish and Wildl Serv, *Fish and Wildlife Research* **2**.

Wapshere, A. J. (1975). A protocol for programmes for biological control of weeds. *PANS* **21**, 295–303.

Part III AUGMENTATIVE BIOCONTROL

9

The use of exotic organisms as biopesticides: some issues

Jeff Waage

Introduction

There are currently two ways in which an exotic pathogen may be introduced into a country for biological control of a pest. It may enter as a 'classical' introduction, to be released and established for continuing control of a (usually exotic) pest, or it may enter as a formulated pesticide, either registered or in an experimental form. Either way, when released in the field, there is a likelihood of its establishment. There is a need, therefore, in both cases, to ensure that such introductions are safe to the environment, and pose acceptable risks to non-target organisms.

Introducing exotic pathogens – the classical precedent

Over the past century, the classical biological control of insect pests and weeds has been dominated by the introduction of insect control agents (see Hopper, Chapter 6; and Blossey, Chapter 8). The use of pathogens in classical biological control has been limited, but it is increasing as we come to understand more about the biology of pathogens and see the positive results from a handful of successful programmes.

Against insects pests, the success of the baculovirus of *Oryctes rhinoceros* on Pacific and Indian Ocean islands is well-known (Bedford, 1986). More recently, a number of exotic introductions have been made against other insect pests, including some 'new associations'. For instance, the fungus *Entomophaga praxibuli* has recently been

introduced into the US from Australia for grasshopper control (Anon, 1989) and an isolate of *Entomophaga grylli* has been introduced from Australia to the US against the indigenous grasshopper, *Phaulacridium vittatum* (Milner, 1985). For weeds, the past few decades have seen spectacular success with the introduction of rusts against *Chondrilla juncea* in Australia and *Rubus fructicosa* in Chile, and a smut against *Ageratina riparia* in Hawaii. Today, there are about 20 ongoing programmes in classical biological control of weeds with pathogens (Watson, 1991).

There is no inherent difference between the introduction and establishment of exotic pathogens for biological control and the introduction of exotic insect control agents. Procedures for classical biological control, described by Greathead (see Chapter 5), apply easily to both. In applying these procedures to pathogens of insects and weeds, the greatest concern is the risk to non-target organisms, particularly beneficial species such as crop plants or bees. More recently, concern has been directed as well to the possible effects of exotic biological control agents on indigenous flora and fauna, particularly where rare or endangered species are involved.

In general, the risk of negative effects on non-target organisms is usually minimized by the selection of highly specific agents. Specificity may be confirmed by exposure of the candidate agent to species of differing degrees of relatedness to the target pest, in order to establish the agent's host range (i.e. centrifugal screening; Wapshere, 1975), as well as by exposure to species of particular agricultural or environmental value which could, conceivably, be hosts, such as crops or rare butterflies.

With the introduction of exotic pathogens, specificity has been an important criterion for agent selection. However, as with programmes using insect agents, there has been a substantial difference in the degree of screening for specificity between insect and weed programmes. For exotic insect pathogens, host specificity testing has been limited. Where it is done, tests are usually made on species closely related to the pest, in order to confirm the host range. Alternatively, specificity may be argued on the basis of the high specificity of the group of organisms to which the agent belongs and/or the observation that the pathogen itself has not been recorded from non-target hosts of economic or conservation value in its area of origin. For weed pathogens, because of the risk to crop plants, the requirements for screening have been more extensive. In contrast to an insect agent for weed control, the host-plant list used for testing a rust or smut fungus may be 1.5 to 2 times as long, and the procedures far more complex (e.g. testing on varieties and growth stages of particular crops under different environmental conditions) (Evans, 1992).

Despite dramatic success over the past few decades, acceptance of pathogens as classical control agents for weeds has been slow, due to concern about infection of, or adaptation to, crop plants by exotic pathogens. This has caused some pathologists to claim that:

'despite the lack of documented serious conflicts, there is an air of pathophobia that has brought to a virtual standstill the application of the classical approach in the use of plant pathogens for weed control' (Freeman and Charudattan, 1985).

While this situation is changing gradually, pathophobia still lingers, and the detailed protocols for pathogen testing, and their cost constrain the development of promising agents. Nonetheless, there is likely to be an increase in programmes of introduction of exotic pathogens for classical biological control of insects and weeds. A high degree of specificity, e.g. at the species or genus level, will probably be required for these.

Pathogens as formulated biopesticides

Pathogens are finding their greatest utilization in biological control today as formulated biological pesticides. Here, mass production and application of a pathogen can overcome ecological problems of transmissibility and lead to a potent biopesticide which has the added benefits of being safe and compatible with the use of resident natural enemies in an integrated pest management (IPM) system. Furthermore, the technology of biopesticide production is such that it allows a tremendous versatility of supplier, this ranges from the large multinational corporation, to the small local industry to the farming cooperative. For developing countries in particular, the prospect of a local biopesticide industry would have many virtues, including reduction in use of foreign exchange to import chemicals, greater safety for the poorly trained farmer, and creation of employment.

In contrast to pathogens used in classical biological control, many of the commonly occurring pathogens that we use for biopesticides are not highly specific, for instance they may be specific at a genus or family level, rather than a species or genus level. Examples include *Bacillus thuringiensis*, *Beauveria bassiana* and *Metarhizium anisopliae* for insect pests. For weeds, mycoherbicidal candidates have usually been highly host-specific pathotypes of pathogens like *Colletotrichum gloeosporioides* and *Phytophthora palmivora* which, however, at the species level, are common pathogens of crops and other plants.

The environmental safety of biopesticides is generally addressed during the process of registration. Registration requirements for biopesticides have been reviewed by Greathead and Prior (1990) (see Klingauf, Chapter 28). The subject has been addressed by government regulatory authorities and by international organizations, including WHO, FAO and IOBC. The greatest concern in biopesticide registration, as for chemical pesticide registration, is safety to vertebrates. For biopesticides, this involves not only infectivity, but allergenicity (particularly with dry-spored pathogens) and toxicity of formulants and formulations. Because of their generally low potential for adverse effects, most authorities have recommended a tiered system, with the first tier, usually a laboratory screen, giving the maximum opportunity to see negative effects.

With respect to effects on non-target organisms, some registration requirements for bioinsecticides have included testing on bees and insect predators and parasitoids. For mycoherbicidal candidates, some tests on non-targets are usually done. These are focused on crop plants and, possibly, weeds in the habitats in which the pathogen will be used, because most candidate pathogens are not likely to move outside the crop. Thus, the technical details prepared for the product Collego, which is based on a pathotype of *Colletotrichum gloeosporioides* for the control of northern joint vetch, involved inoculations on 31 different field and vegetable crops, and some weeds.

Overall, requirements for registration of biopesticides have been less stringent in the area of host-specificity testing than procedures for safety testing in classical introductions. Further, there has been relatively less concern about the long-term fate of released pathogens in the environment. Indeed, biopesticides are often treated like chemical pesticides, as short-acting products with no residual effect. Because most biopesticides are used in areas where they are thought or known to be indigenous, there is no presumed risk associated with their persistence.

Why use biopesticides in an exotic context?

The development of pathogens as biopesticides will often result in the use of exotic pathogens. There are a number of reasons for this. First, the most virulent isolate of a pathogen may not necessarily come from the country where control is desired or from the pest species to be controlled. Thus, in a recent exploratory programme to find virulent isolates of *Metarhizium flavoviride* for the desert locust in Africa, the most virulent isolate found to date has come from a grasshopper in Australia (Prior, 1992). Several examples now exist for the related *M. anisopliae* in which high virulence to a particular insect pest is associated with isolates from quite unrelated insect hosts. These include weevil isolates virulent to Homoptera, and homopteran isolates virulent to moths and termites (Prior, 1990).

As a result of this complex relationship between geography, host-association and virulence, public and private organizations are now amassing enor-mous collections of pathogen genotypes to study as potential biopesticides, and are searching around the world to find new ones. But for this effort over the past two decades, we would not have such exciting products as *B.t. israelensis* for the control of dipteran disease vectors and *B.t. tenebrionis* for beetle pests.

The second reason that we find pathogens being used exotically as biopesticides is the simple commercial need for the producer to find markets, and for the user to find already formulated products. *Bacillus thuringiensis* (*Bt*) strains for a particular pest could be found in most countries, and there is evidence that they may be superior against that local pest compared to imported, generalized genotypes. But the latter comes in a package and can be imported tomorrow, whereas a local strain may require several years of development, without any guarantee that its local production is economically viable or competitive with internationally marketed strains.

Some case studies

Through its international programme, International Institute of Biological Control (IIBC) has encountered many and varied responses to the introduction of exotic pathogens as biopesticides, either commercially or for trials. There is a long tradition of research institutes importing exotic pathogens for tests on insects under laboratory and field conditions without any quarantine or safety testing procedures. In some countries formulated biopesticide products, e.g. *Bt*, are commonly sold over the counter even though they have not passed through any procedure to determine their safety as exotic micro-organisms. On the other hand, some countries have required that exotic biopesticides should be subject to the same regulations that apply to the introduction of exotic organisms. Thus, when a *Bt* isolate from UK was introduced to Kenya in 1990 for testing against *Helicoverpa armigera*, the introduction was considered by the same procedures as those that applied to classical agents. In the end, host range and other data supplied from the UK were sufficient to clear the isolate for introduction. In recent years, IIBC has found itself in the unusual position of developing the biopesticides whose primary value may be in their use in exotic contexts.

In this way, it has encountered some of the major issues in the movement of pathogens between countries. I will now briefly illustrate two such projects.

Development of *Metarhizium flavoviride* as a biopesticide for control of desert locust

In 1990, IIBC began a programme to develop a biological pesticide for the desert locust, in the wake of extensive and largely ineffective use of chemical pesticides in the last locust plague of the mid-1980s and the reaction of countries and donors alike that another, safe means of locust control was needed. The programme, which involves IIBC, the International Institute for Tropical Agriculture, the Département de Formation en Protection des Végétaux, Niger and a number of African countries, has focused on fungi as biopesticidal agents, and has involved extensive exploration in Africa and the Near East and bioassay to identify suitable pathotypes. Over 150 isolates of *Metarhizium* and *Beauveria* spp. have been screened. Nearly all of the 40 highly virulent isolates that have been found so far have been *Metarhizium* spp. and the great majority have come from acridoids (Prior *et al.*, 1994). However, some have come from regions where the desert locust does not occur, such as Australia. The chosen standard strain of *M. flavoviride* comes from the grasshopper, *Ornithacris cavroisi*, in Niger and has been shown to be highly virulent to four other genera of acridids.

The ultimate aim of the programme is to develop local production technologies for an effective mycoinsecticide that can be applied against locusts and grasshoppers throughout the hot, semi-arid regions of Africa and Asia. However, it is anticipated that such a formulation may be of value in other parts of the developing and developed world, and preliminary plans exist for studies in southern Africa, Australia, Latin America and the Caribbean. Field testing of *M. flavoviride* in novel oil formulations is now underway, however it is necessary to be flexible in the choice of field site, because outbreaks of locusts and grasshoppers are sporadic.

Testing has been undertaken in Niger, where the standard fungal strain is indigenous. Testing in Benin has also been done, after having first satisfied the Benin authorities that the organism was probably distributed naturally in the country, given its proximity to Niger and the similarity of their acridid faunas. However, in the development of trials in Madagascar, the government has required that only strains from Madagascar may be used, and these require preliminary safety testing on beneficial insects under glasshouse conditions. Thus, different countries have different requirements for the introduction and testing of this fungus. It is likely that the organism is ubiquitous, particularly given the dispersive nature of its hosts. Indeed, an essential element of the project has been biochemical work to find markers to distinguish the introduced pathotype from the 'background' of local strains where it is used. These markers are not only necessary for the interpretation of field effects, but they will be of value in planned ecological studies on the pathogen's effects on non-target organisms in the field.

Development of *Colletotrichum* sp. as a biopesticide for control of itch grass

Itch grass, *Rottboellia cochinchinensis*, is a serious, pan-tropical grassy weed of sugar cane, maize and upland rice. It is thought to be indigenous to Asia, but is a problem in Africa and particularly in Latin America, where it is exotic. In 1989, IIBC initiated a programme to find virulent pathogens of this weed to develop as a mycoherbicide for use in tropical countries. Extensive exploration was undertaken in Africa and Asia, and 800 isolates of necrotrophic fungi were made, 200 of which were subjected to a primary screen of the weed and key cereal crops. Thus, from the very beginning, the host specificity of the pathogen was an element in its selection. *Rottboellia* is a special case in this regard, because it is very closely related to maize. Indeed, it is the lack of specificity of available chemical herbicides that has stimulated the search for a more specific mycoherbicide.

Of all pathogens studies, only isolates of *Colletotrichum* sp. and the head smut *Sphacelotheca ophiuri* proved sufficiently specific and virulent (Ellison and Evans, 1992). The head smut is a possible candidate for exotic introduction (e.g. into Latin America), but *Colletotrichum* has potential as a mycoherbicide. Extensive screening of three virulent *Colletotrichum* isolates from Thailand and Sri Lanka was carried out on about 30 species and varieties of cereal crops. Although the *Colletotrichum* isolates were highly specific to *Rottboellia* (and, it proved subsequently, to particular biotypes of this weed), it was decided that field trials would be undertaken only in the country of origin of the particular isolate. In the event this was Thailand. A deliberate exotic introduction would have been a precedent for a *Colletotrichum*. Given the serious disease status of the species on many crops, and concern by some pathologists for the adaptability of pathotypes to new hosts, it was felt that such an introduction would constitute an unnecessary risk, however small that was.

The problem that now faces IIBC, after the satisfactory field trails in Thailand, is whether or how to introduce the fungus into Latin America, where *Rottboellia* is particularly serious and local *Colletotrichum* species are unknown. Would *Colletotrichum*, as a biopesticide, require clearance as a classical biological control agent in every Latin American (and Asian) country in which it might be used?

Some issues in the use of exotic pathogens as biopesticides

Are the risks associated with the introduction of exotic pathogens as biopesticides similar to those for a classical pathogen agent? The answer to this question may give us a starting point for the development of a consistent view on introduction procedures. It is important to recognize that the risks we are discussing are primarily those to non-target organisms. Environmental and human health risks are usually negligible, and will not be considered here. Further, let us concentrate on pathogenicity to non-target organisms, and set aside the subjective, subsequent question of what value we place on their preservation. In other words, let us concentrate on the subject of specificity of the biopesticide. Milner (1985) identifies three important questions with regard to the specificity of exotic pathogens: will the introduced pathogen show specificity at normal levels of inoculum; will this be retained at higher levels; and is it a stable trait after release?

Most pathogens used as biopesticides come from taxa that are opportunistic, they possess a capacity for saprophytic growth and a broad host-range at the species level, which suggests an evolutionary capacity for adaptation to new hosts, although particular pathotypes may be highly specific. These pathogens have, generally, lower specificity than those we commonly introduce in classical biological control, and perhaps a greater propensity for adaptation to new hosts. Hence, they may represent a higher risk to non-target organisms than most classically introduced species. This risk may be amplified at higher levels of inoculum, when non-targets are challenged more intensely. Further, such situations would be more frequent with biopesticides than with exotic introductions, due to their pattern of use.

Finally, the broad host-range at the species level suggests that pathogens used as biopesticides may be more likely to shift the host range than highly specific and co-evolved pathogens used in classical programmes. Such a shift may involve mutation or recombination, and the latter is made more likely by the presence of closely related species in the exotic environment.

On the basis of these criteria, exotic pathogens used as biopesticides may pose a generally greater risk to non-target species than those used classically, by virtue of their true or potential host range and the way in which they are used. However, this argument must be balanced against two other factors which reduce this risk. Most pathogens used as biopesticides disperse poorly, such that challenges to non-target organisms may be more limited than with classical agents, which by contrast, are selected for their dispersive ability. Most pathogens used as biopesticides may already be present in the area of introduction, at least at the species level. Thus, with reference to the case histories above, *Metarhizium flavoviride* in one form or other is likely to be found on acridids everywhere, as is *Colletotrichum sp.* on *Rottboellia*. The existence in the field of closely related species or pathotypes of the pathogen used in the biopesticide must reduce concern that the sprayed pathogen will cause environmental problems.

Bringing it all together

The fact that we now introduce exotic natural enemies into countries in different ways, as classical agents and as registered biopesticides, and use different procedures to do this owes more to history than to science. Biopesticides have, logically, been developed like chemical pesticides, usually by the same companies. As such, the organisms used have been selected for their high virulence and efficient production, formulation and application. Persistence in the environment has not been a priority, indeed it could be considered a disincentive from a marketing perspective. In the commercial, agrochemical world, these new biopesticides were simply not seen as organisms that would undergo substantial reproduction, and much less that they would spread out of the target crops and affect natural environments. Indeed, over 90% of today's bioinsecticide market involves products based on two kinds of organisms, *Bacillus thuringiensis* and heterorhabditid nematodes, which have no natural association with insect pests on plants, and hence persist very poorly there and have to be reapplied regularly. By contrast, other potential biopesticides, including fungi and viruses, have much greater persistence and are even capable of initiating epidemics in pest populations. In future, as IPM becomes a more common approach to pest management, it is likely that we will see these more persistent organisms used extensively as biocontrol products, as many arthropod predators and parasitoids are now, and the contrast between agents and biopesticides will become less distinct (Waage, 1995).

On the basis of the arguments above, I feel there is a case for considering the introduction of an exotic pathogen as a biopesticide in the same manner as a classical introduction, *if* it can be argued to be truly exotic, and with additional consideration given to its dispersal abilities and persistence. If we were to take this approach, existing, widely used biopesticides would be easily cleared on their safety record, while new products would be thoroughly tested before use in an exotic context. This would probably mean the inclusion of some centrifugal, host-range testing in addition to the specificity tests at present included in biopesticide registration requirements, as has been recommended by Greathead and Prior (1990). New countries may accept the data from the original registration and add to this any tests that they feel necessary, e.g. on desirable, local insects or plants, which would occur within the new area where the biopesticide would be used.

To make this approach more effective, we need to have a better understanding of what constitutes an exotic form of some taxonomically complex pathogens, and how likely they are to spread in the new environment and interact with indigenous forms there. This means, respectively, studies on the nature of virulence and specificity, and studies on the population dynamics and spatial ecology of pathogens. Further, to undertake both of these, we need to improve our methods for the detection and identification of particular pathotypes. It now appears with insect pathogens like *Bt*, *Beauveria* and *Metarhizium*, that effective biochemical techniques can be developed for this, and with these we can distinguish introduced pathotypes from those already present. These will be the tools with which we explore future aspects of specificity and environmental fate.

I have suggested that a responsible approach to the use of exotic pathogens as biopesticides may require greater consideration than is currently given to host range in biopesticide development or registration for use in a foreign country. However, there is the danger that overly restrictive regulatory procedures will block the development and use of products that are inherently safe.

From a commercial perspective, adding the costs of safety testing for each new country where the product may be sold might make the search for markets uneconomical. This, in turn, might reduce the viability of initial development of new biopesticides, because, easily accessible markets might be smaller. A case in point is the exclusion, until recently of *Bt* formulations from the Indian pest-control market, because of concern for effects on the silk moth industry. The inaccessibility of such an enormous market cannot but have delayed the development of *Bt*, as a commercial product in the subcontinent. Of course, as an alternative to safety testing for foreign markets local isolates of the desired pathogen could be found and developed for each country or region. This would, however, replace problems of safety testing with problems of fragmented research and development programmes and, ultimately, fragmented markets, yet another disincentive to biopesticide development.

With respect to all these problems, developing countries could be the primary losers. Multinationals may not see an economic benefit in the clearance of products for use in their small markets, and local biopesticide initiatives may not have the resources to find the necessary solutions.

Assuming that the present variation in approaches to the introduction of exotic pathogens can be rationalized to be both responsible and supportive of biopesticide development, what future developments can we anticipate? For bioinsecticides, most of the taxa likely to be of value are already in use, and information is accumulating in support of their safety in exotic environments. There should be few problems in ensuring the safety of their further development for international markets, but this process would be enhanced through a better understanding of the nature of specificity and pathogen dynamics in the field. For mycoherbicides, prospects for exotic use are much more limited in the short-term. To the best of my knowledge, no mycoherbicide has yet been used in an exotic context. The first use may occur in North America, with exchange of products between Canada and the US, but this is hardly a case of exotic introduction.

For introductions into new continents, the present level of concern for the risk of introducing a pathogen that will affect crops will probably make a fully classical procedure necessary for the foreseeable future. Indeed, most existing mycoherbicides would probably fail such a procedure were they to be considered for use abroad. Thus, the *Colletotrichum gloeosporioides*-based herbicide, Collego, which has been produced for control of northern joint vetch (*Aeschynomene virginica*) infects a range of legumes that includes crop species, and the *Phytophthora palmivora*-based herbicide, Devine, is pathogenic to melons. These host-ranges alone may block their introduction unless a very convincing case could be made for the restricted dispersal capacity of slime-spored mycoherbicides.

It is unfortunate but understandable that problems such as these will limit mycoherbicide development for weeds of global significance, and particularly for tropical weeds. On a case-by-case basis, the stringency of mycoherbicide screening will depend on the relatedness of the weed to plants of agricultural or conservation value. In this context, the instances cited above – Collego, Devine and the itchgrass mycoherbicide – present difficult cases. Far easier targets from a regulatory viewpoint might be sedges (e.g. *Cyperus rotundus*)

or water weeds, because of their relative taxonomic isolation from crops. However, difficult cases must be pursued if this important technology is to have global application. With sufficient precedent for safe, effective use, bioherbicides, like bioinsecticides, will be able to play an important role in IPM systems.

References

Anonymous (1989). Fungus for grasshopper control in US. *Agrow* **101**, 120.

Beford, G. O. (1986). Biological control of the rhinoceros beetle (*Oryctes rhinoceros*) in the South Pacific by a baculovirus. *Agricultural Ecosystems and the Environment* **15**, 141–147.

Ellison, C. A. and Evans, H. C. (1995). Present status of the biological control programme for the graminaceous weed *Rottboellia cochinchinensis*. *Proceedings of the Eighth International Symposium on Biological Control of Weeds* (ed. E. S. Delfosse and R. R. Scott). (In press.)

Evans, H. C. (1995). Pathogen – weed relationships: the practice and problems of host range screening. *Proceeding of the Eighth International Symposium on Biological Control of Weeds* (ed. E. S. Delfosse and R. R. Scott). (In press.)

Freeman, T. E. and Charudattan, R. (1985). Conflicts in the use of plant pathogens as biocontrol agents for weeds. In *Proceedings of the VI International Symposium on Biological Control of Weeds* (ed. E. S. Delfosse), pp. 351–357. Agriculture Canada: Ottawa, pp. 351–357.

Greathead, D. J. and Prior, C. (1990). The regulation of pathogens for biological control, with special reference to locust control. In *NORAGRIC Occasional Paper Series C*, pp. 67–80.

Milner, R. J. (1985). Field tests of a strain of *Entomophaga grylli* from the USA for biocontrol of the Australian wingless grasshopper, *Phaulacridium vittatum*. *Proceedings of the 4th Australasian Conference on Grassland Invertebrate Ecology*, Lincoln College, Canterbury, 13–15 May 1985, pp. 255–261.

Prior, C. (1990). The biological basis for regulating the release of micro-organisms, with particular reference to the use of fungi for pest control. *Aspects of Applied Biology* **24**, 231–238.

Prior, C. (1992). Discovery and characterization of fungal pathogens for locust and grasshopper control. In *Biological Control of Locusts and Grasshoppers* (ed. C. J. Lomer and C. Prior), pp. 159–180. CAB International, Ascot.

Prior, C., Lomer, C. J., Bateman, R. P., Codonou, I., Carey, M., Moore, D., Kooyman, C. and Sagbohan,

J. (1994). Selection of *Metarhizium* spp. isolates for field testing against locusts and grasshoppers. *Proceedings of the Sixth International Coloquium on Invertebrate Pathology and Microbial Control*, Montpellier, France, 28 August–2 September 1994. Published by the Society of Invertebrate Pathology (in press).

Waage, J. K. (1995). Integrated pest management and biotechnology: an analysis of the potential for integration. In *Biotechnology and Integrated Pest Management* (ed. G. J. Persley). CAB International, Ascot.

Wapshere, A. J. (1975). A protocol for programmes for biological control of weeds. *PANS* 21, 295–303.

Watson, A. K. (1991). The classical approach with plant pathogens. In Microbial Control of Weeds (ed. D. O. TeBeest), pp. 3–23, Chapman & Hall: London.

Postscript

Since the preparation of this paper in 1992, a draft Code of Conduct for the Import and Release of Biological Control Agents has been prepared by FAO, with assistance from IIBC, IOBC and various national biological control programmes. It describes a set of voluntary standards of conduct for public and private entities involved in the use and distribution of biological control agents. In this draft Code, no distinction is made between living, exotic biological control agents introduced in 'classical' programmes or as biopesticides. Both would require dossiers that include an analysis of host range of the agent and any potential hazards posed to non-target organisms, and both would require quarantine. In the case of repeated importation, as would be likely for some biopesticides, the draft Code recommends that product certification systems ensure that only imports of equivalent standard to the original import are released. The draft Code has been distributed to FAO member countries and regional plant protection organizations for comment, and will be put forward to the FAO Conference for endorsement in 1995. At the time of publication of this Symposium volume, FAO and IIBC are preparing guidelines on host-specificity testing to accompany the draft Code.

10

Use of *Trichogramma* in maize – estimating environmental risks

D. A. Andow, C. P. Lane and D. M. Olson

Introduction

Biological control of arthropod pests using arthropod natural enemies has been practised for centuries (Flint and van den Bosch, 1981), and although it has been widely appreciated that these natural enemies can have adverse effects on the environment, it has been largely assumed that they are either absolutely small or small relative to the benefits of biological control. For example, Samways (1988) suggested that there are no quantified cases where the introduction of an arthropod agent has been shown to have harmed a specific conservation programme or has been conclusively damaging to native fauna. This suggestion has proved difficult to test under natural conditions. For example, many natural enemies have been released to control forest Lepidoptera pests. Some of these enemies kill non-target, non-pest insects, and it has been suggested that such unintended mortality might destabilize these non-pest populations thus causing them to become sporadic pests. Recently, Pimentel *et al.* (1984) reviewed the evidence that biological control agents can cause adverse environmental effects. Their review clearly documented that biological control can entail environmental risk and that categorical dismissal of this concern is unwarranted. More recently, Howarth (1991) argued that the introduction of biological control agents into Hawaii and New Zealand is one of the major causes of extinctions of the native, insular, endemic arthropod faunae associated with these two islands. In most of those cases, vertebrate natural enemies were implicated, but Howarth

(1991) developed a reasonable argument that arthropod natural enemies could also affect the native faunae. While there is considerable debate over the nature, likelihood, and magnitude of the potential adverse effects of biological control agents on the environment, it is indisputable that such effects can occur (see Hopper, Chapter 6).

Biological control has been classified into three major strategies: (1) classical or inoculative biological control, in which parasitoids are released, expected to establish viable populations, increase in number and thus control pest populations; (2) inundative and augmentative biological control, in which large numbers of parasitoids are released against a pest and the large numbers are expected to suppress the pest population; and (3) conservation of parasitoids, in which the native parasitoids are managed to enhance their effects on the pest population. In this paper we will refer mainly to parasitoids, but this classification holds for any natural enemy, including predators and pathogens. From the perspective of environmental risk analysis, the classical strategy is likely to be riskier than the inundative strategy, which in turn, may be riskier than the conservation strategy. This is because the classical strategy involves the release of a parasitoid that does not normally occur in the habitat and it is then expected to reproduce. Exposure of threatened and endangered species to classical biological control agents will be determined by the population dynamics of the agents, which will be difficult to predict in sufficient detail *a priori*. Analysis of these risks might benefit from examination of realistic worst case scenarios. The inundative strategy entails fewer risks because the

101

introduced parasitoid is not expected to survive and establish viable populations. In those cases where the released organism might survive, naturally occurring populations of the species are controlled at low densities in or near the release site, so unless augmenting populations of the biological control agent destabilizes the natural controls on the native population of the agent, mass releases will be expected to die back to naturally occurring population densities. Inundative release does entail some risk, however, because the parasitoid is either new to the habitat or present in such high numbers that its influence on non-target host populations is magnified. In addition, there is a possibility that introgression of the released agent into the native population could change the population in ways that may increase environmental risks. Finally, the environmental risks associated with conservation strategies are likely to be related to the potential indirect effects of the habitat modifications that are intended to enhance parasitoid effectiveness. These ideas have not yet been empirically verified, but they may provide a framework to relate the nature of environmental risk to broad categories of biological control activity.

The likelihood of occurrence and the magnitude of the environmental risks associated with a particular biological control agent are probably dependent on local conditions. This implies that decisions related to the relative risks of biological control will require a case-by-case analysis until a conceptual framework that predicts occurrence and magnitude can be developed and verified. In this chapter, we describe a procedure for the evaluation of some of the environmental risks associated with inundative releases of parasitoids, and we apply the procedure to a specific illustrative case. Minnesota is one of the largest producers of sweet and field maize in the United States, and one of the key arthropod pests of maize in this region is the European corn borer, *Ostrinia nubilalis* (Lepidoptera: Pyralidae). We have evaluated the risks to populations of threatened and endangered species of Lepidoptera from potential inundative releases of *Trichogramma nubilale* (Hymenoptera: Trichogrammatidae) for the control of *O. nubilalis* in sweet and field maize in Minnesota.

Rare arthropods of Minnesota and risk analysis

There are several species of Lepidoptera in Minnesota that are classified as 'of special concern', 'threatened' or 'endangered' by the State of Minnesota (Dana and Huber, 1988). One species, the Karner blue (*Lycaeides melissa samuelis* Nabakov (Lycaenidae)) is also listed as a Federally endangered species because of its declining numbers throughout its geographic range. Many State listed species, including the Karner blue, are known to occur in counties of Minnesota where maize is widely grown. All of the historical and contemporary sites for the State endangered species are shown in Fig. 10.1, in addition to those of the Karner blue. The species are Uhler's arctic (*Oeneis uhleri varuna*), the Assiniboia skipper (*Hesperia assiniboia*), and Uncas skipper (*Hesperia uncas*). Karner blue, Assiniboia skipper, Dakota skipper (*Hesperia dacotae*, State threatened), Ottoe skipper (*Hesperia ottoe*, State threatened) and Poweshiek skipper (*Oarisma poweshiek*, State special concern) occur in sites near agricultural fields (R. Dana, personal communication). The distributions of these additional species are shown in Fig. 10.2 (both figures are from the Minnesota DNR Heritage Program database). Figs. 10.1 and 10.2 show both recent historical and contemporary distribution records, – some changes that have occurred in the distribution and abundance of these species are not completely represented. For example, the Karner blue is reported from two localities, one in east central Minnesota (probably extinct) and one in southeast Minnesota, in the Whitewater Wildlife Management Area in Winona County. In this paper we will focus on the issues related to the use of biological control agents near the population of Karner blue in southeastern Minnesota. The Karner blue is a specialist feeder on wild lupine (*Lupinus perennis* Fabaceae) in sandy soils intercalated in the oak savanna habitat near the Mississippi River in Minnesota. This area is surrounded by agricultural lands.

Because risk is always relative, alternatives to the action under consideration must be identified so that relative risks can be evaluated and compared. In our case the action under consideration is release of *T. nubilale* for inundative biological control of *O. nubilalis* in either sweet maize or field

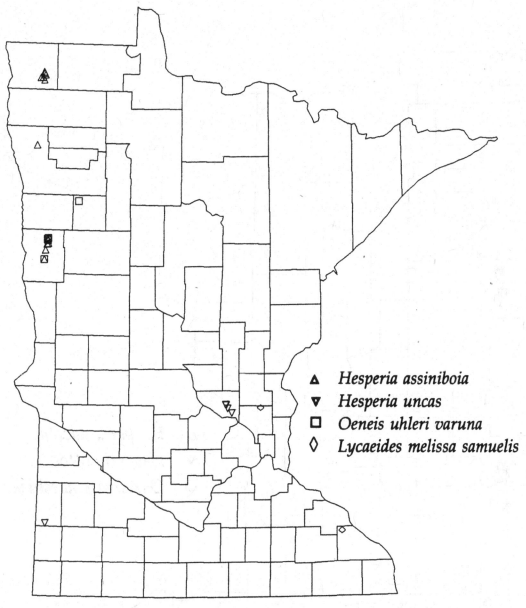

Source: MN Dept. of Natural Resources, Natural Heritage/Nongame Wildlife Programs

Fig. 10.1 Historical and recent distributional records of the three State endangered lepidopteran species, *Hesperia assiniboia* (Assiniboia skipper), *Hesperia uncas* (Uncas skipper), and *Oeneis uhleri varuna* (Uhler's arctic), and the Federally endangered Lepidopteran species, *Lycaeides melissa samuelis* (Karner blue) in Minnesota.

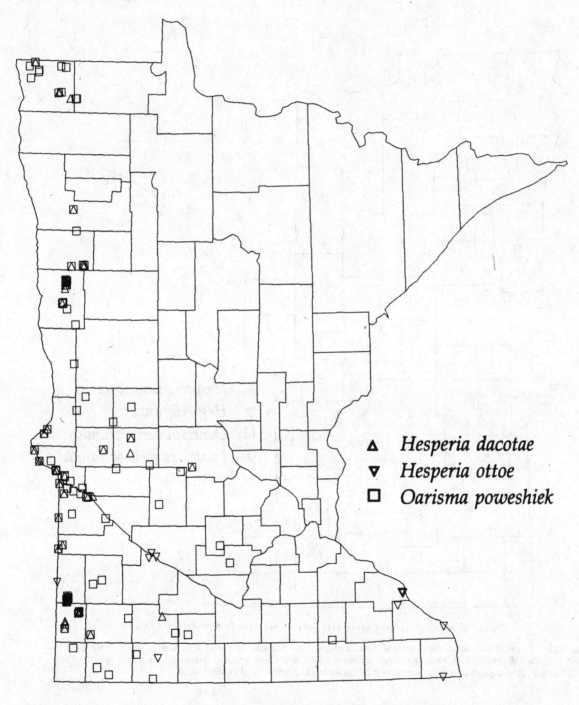

△ *Hesperia dacotae*
▽ *Hesperia ottoe*
☐ *Oarisma poweshiek*

Source: MN Dept. of Natural Resources, Natural Heritage/Nongame Wildlife Programs

Fig. 10.2 Historical and recent distributional records of the State threatened and special concern lepidopteran species *Hesperia dacotae* (Dakota skipper), *Hesperia ottoe* (Ottoe skipper), and *Oarisma powashiek* (Powashiek skipper) in Minnesota.

maize. Alternative actions include: (1) do nothing and tolerate the damage to the maize crop; and (2) use insecticides to control European corn borer. In Minnesota, it is nearly always economically feasible to use insecticides on sweet maize although usually it is better to do nothing than to apply insecticides in field maize. Thus, our analysis should focus on comparisons between no action for field maize and insecticide application for sweet maize, although all alternatives should be considered for both crops. This would be a rather lengthy process and so, in this paper, the more limited objective of quantifying the risk from inundative releases of *T. nubilale* will be pursued.

Environmental risk for any particular action is estimated from the degree of exposure of the environment to the potential risk-causing agent and the magnitude of the effects that the agent might have on the environment. This simple framework, risk = exposure × effects, will guide our analysis here. For biological control agents, exposure can be evaluated as the ability of the agent to disperse to the affected environment and survive and possibly reproduce, i.e. the fate of released organisms will indicate the level of environmental exposure to the biological control agent. Environmental effects analysis is more difficult to evaluate. In our case, effects analysis can be divided into at least two processes, the parasitism rate on non-target arthropod species that are rare or endangered, and the effect of the parasitism on the population dynamics of these species.

The distribution of field maize in Minnesota during 1986 is shown in Fig. 10.3. Minnesota is one of the two largest producers of sweet maize and the fourth largest producer of field maize in the United States (rank varies from year to year). Field maize is grown throughout the southern and central parts of the State, and cropping-area in the northwest has increased considerably during the past 5 years. The distribution of sweet-maize production is restricted to the Minnesota River valley and surrounding areas. There is little geographic overlap between sweet-maize and habitats of the threatened and endangered species listed above. Indeed, the major sweet-maize production areas are at least 100 km away from any sites where threatened or endangered species have been found. Risk analysis involving sweet maize will focus on relatively small-scale releases in isolated fields, and analysis involving field maize will consider larger scale releases.

It should be emphasized that this is an hypothetical analysis. Currently, *T. nubilale* is not a commercial product, and it will be several years before it can be determined if it will become one. Potential risks are estimated from data gathered in small-scale experiments. It seems appropriate, however, to provide a framework for analysis of environmental risk because *Trichogramma* spp. are used world-wide for the control of various arthropod pests.

Biology of *Trichogramma nubilale*

Trichogramma nubilale is an egg parasitoid of Lepidoptera and occurs naturally in Delaware, USA. Curl and Burbutis (1978) suggest that this species is a specialist parasitoid of *O. nubilalis* egg masses, and others have echoed this suggestion. Their evidence is based on observations that the species has been recovered only from *O. nubilalis* egg masses. In contrast, we believe that *T. nubilale* is an opportunistic generalist parasitoid of eggs of many species of Lepidoptera. Because all other species of *Trichogramma* are believed to be opportunistic generalists of insect eggs, there are no phylogenetic reasons for the belief that *T. nubilale* is a specialist. Furthermore, we have reared *T. nubilale* on eggs of *Manduca sexta* (Sphingidae) and *Mamestra brassicae* (Noctuidae), which demonstrates that other species are suitable hosts. Most critically, at constant temperature (24–29°C), the life cycle of *T. nubilale* is 10–12 days (Volden, 1983), while the life cycle of *O. nubilalis* is approximately 35 days at similar temperatures. Thus, in nature, parasitoids would be synchronized poorly with their hosts and there would be periods of perhaps 20 days when hosts would be relatively unavailable between the oviposition periods of *O. nubilalis*. Finally, it is frequently observed in Delaware that parasitism rates of *O. nubilalis* in maize are low throughout the spring and summer, but in late summer to early fall, parasitism rates rise rapidly. This seems to suggest that parasitoids transfer from some other host. Although our evidence that *T. nubilale* is an opportunistic generalist parasitoid is indirect, examination of more species of Lepidoptera in habitats other than maize fields would probably

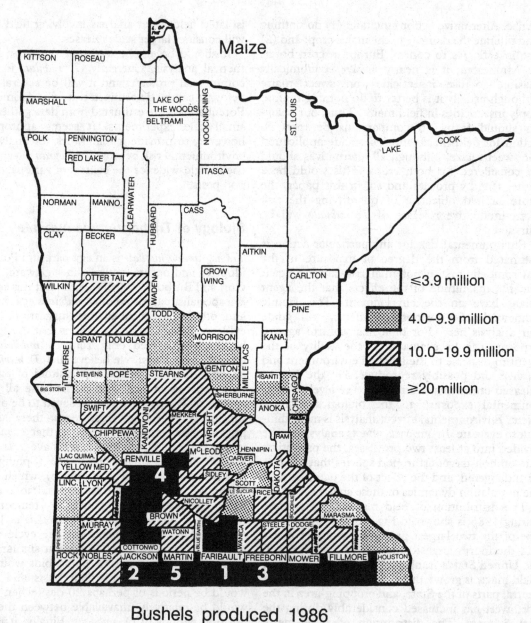

Bushels produced 1986

Fig. 10.3 Production of maize (all types) in Minnesota (from the Minnesota Department of Agriculture database).

reveal additional hosts. At present it is reasonable to assume that introductions of this species could affect populations of threatened or endangered Lepidoptera species.

Other species of *Trichogramma* can parasitize eggs of species in the Lycaenidae and Hesperidae. For example, eggs of the orange palm dart, *Cephrenes augiades sperthias* (Hesperidae), collected from wild and cultivated palms in Australia were parasitized by *Trichogramma* sp. (Oatman, 1986), and *Trichogramma* spp. were recoved from eggs of purple hairstreak, *Quercusia quercus cerris* (Lycaenidae), in England. *T. chilonis* parasitized eggs of *Deudorix epijarbas* (Lycaenidae) in laboratory and field studies in India (Rawat and Pawar, 1991), and this same species parasitizes many other host species, including a pyralid species related to *O. nubilalis* (Hirai, 1987). This brief review suggests that there are no apparent phylogenetic reasons to expect that *T. nubilale* could not parasitize eggs of the rare or endangered species of Lepidoptera in Minnesota when encountered.

We have exposed eggs of the Karner blue to *T. nubilale* in the laboratory and observed drilling and oviposition. *T. nubilale* killed the egg, but we do not know if it then developed in these eggs. It is not known if *T. nubilale* can parasitize any of the other species of rare or endangered species of Lepidoptera in Minnesota. Karner blue eggs are small (0.7 mm in diameter) and turban shaped and weigh approximately 0.06 mg. These eggs are smaller than those of many other species of butterfly, but are similar in size to the egg of *O. nubilalis* (0.05–0.07 mg, Leahy & Andow, 1994), from which one or two adult *T. nubilale* can emerge. We conclude that it is reasonable to assume that *T. nubilale* would parasitize the eggs of the Karner blue in Minnesota when encountered.

T. nubilale overwinters in egg masses of *O. nubilalis* at Delaware (Burbutis *et al.*, 1976), but is not known to overwinter successfully in Minnesota (Volden, 1983). If it does not overwinter, then the effects of releases of these parasitoids will be limited to the year of release, and only the effects of the released parasitoids and the subsequent 2–4 generations in the year of release need be considered.

Exposure to released *T. nubilale*

The primary issue in the evaluation of the potential exposure of rare or endangered Lepidoptera to *T. nubilale* involves determining whether *T. nubilale* will occur at the same time and in the same place as eggs of the rare or endangered Lepidoptera.

Phenology

T. nubilale will be released predominantly during the oviposition periods of *O. nubilalis*. In southern Minnesota *O. nubilalis* is bivoltine, with the first oviposition period occurring usually during early to mid-June and the second during mid-July to late-August. Peak oviposition during the second oviposition period normally occurs in late-July to early-August in field maize. Sweet maize is planted and matures over a longer period than field maize in Minnesota. Because *O. nubilalis* oviposit more eggs in maize at anthesis and immediately post-anthesis, late-planted sweet maize will be favoured for oviposition during mid- to late-August. As a consequence, releases of *T. nubilale* during mid- to late-August are likely to be concentrated in sweet maize. In northern Minnesota *O. nubilalis* is facultatively univoltine (some individuals in some years are bivoltine). The oviposition period occurs usually during mid-June to early-July. In northern Minnesota very little sweet maize is grown so *T. nubilale* would be released in field maize.

The phenology of oviposition by the rare Lepidoptera in Minnesota depends on the species (R. Dana, personal communication). The Karner blue is bivoltine and overwinters as eggs. These overwintering eggs are laid from early-August and persist until early-spring (April–May). The other oviposition period occurs during late-May to early-June. The skippers are all univoltine. The Dakota, Ottoe, and Poweshiek skippers oviposit from late-June through July, and the Assiniboia skipper oviposits during mid- to late-August.

Nearly all of the rare Lepidoptera oviposit during times that *T. nubilale* might be released. The Karner blue and northern populations of Dakota and Poweshiek skippers, in particular, appear to be at risk because their oviposition periods overlap considerably with that of *O. nubilalis*. Ottoe skippers and southern populations of Dakota and

Poweshiek skippers may be at periodic risk. The Assiniboia skipper does not appear to be at great risk because its oviposition period occurs considerably later than that of *O. nubilalis* and its distribution does not overlap with major sweet-maize production areas.

Spatial co-occurrence

Whether *T. nubilale* and eggs of any of the rare Lepidoptera would co-occur will depend on the number of *T. nubilale* released, their mortality rate subsequent to release, and their dispersal distance relative to the distance from the release site to the habitats of the rare Lepidoptera. These qualitative ideas can be made more quantitative as follows: let $n(x,t)$ be the number of *T. nubilale* at a location x meters from the release site at t days after release, N_0 be the number of *T. nubilale* released, m be the daily mortality rate (assumed to be constant over time), and d be the distance from the release site to the habitat with the rare Lepidoptera. If $N(t)$ is the number of *T. nubilale* that pass through the habitat with the rare Lepidoptera, then $N(t) =$ the number alive at time t multiplied by the proportion of those alive that have passed through the habitat of interest:

$$N(t) = (N_0 \exp[-mt]) \frac{\int\limits_{d}^{\infty} n(x,t)dx}{\int\limits_{0}^{\infty} n(x,t)dx}. \qquad (10.1)$$

To estimate $N(t)$, three terms must be estimated, N_0, m, and $n(x,t)$. The estimation processes are described below.

Numbers of released *T. nubilale*

The numbers of *T. nubilale* that will be released, N_o, for control of *O. nubilalis* is unknown at this time. Current research suggests release rates ranging from 175 000–1 870 000 females/ha will be efficacious (Andow and Prokrym, 1991; Prokrym *et al.*, 1992), although release rates this high are clearly uneconomical. Economical values for use in sweet maize may be about 100 000 females/ha (Prokrym *et al.*, 1992), which is similar to the release numbers of other *Trichogramma* spp. used against *Ostrinia* spp. in other parts of the world. For illustrative purposes, we shall assume that about 100 000 females/ha will be released in sweet maize and 30 000 females/ha in field maize.

Field size is likely to be quite variable. For sweet maize that is grown outside of the major production areas, field sizes as large as 1 ha may be planted near habitats with rare Lepidoptera species. Field sizes for field maize in Minnesota are in the order of 200–300 ha, but land near habitats with rare or endangered Lepidoptera tends to be of poorer agricultural quality, so field sizes are likely to be considerably smaller, perhaps in the order of 100 ha. Using these values, we calculate that for sweet maize, approximately 100 000 females may be released, and for field maize, approximately 3 000 000 females may be released near habitats with rare Lepidoptera.

Dispersal

Estimation of $n(x,t)$ was done experimentally and will be described in greater detail in other publications (Andow, unpublished data). Two rectangular fields, which were about 30×45 m and separated by at least 300 m, were planted to maize in a 0.5×0.75 m lattice. A fixed number of *T. nubilale* was released at the centre of each field, and sentinel egg masses of *O. nubilalis* were placed at various distances from the release point and replaced every 1 to 3 days for about 1 week after the release. These experiments were repeated for 3 years from 1985 to 1987. Egg masses were evaluated to determine if they were parasitized and the proportion of egg masses parasitized at each distance from the release point was calculated for each field and each time after the release. A Weibull function was fitted to these data by weighted non-linear least-squares for each field and time (weighted by the inverse of the standard error of the mean proportion parasitized at each distance). The Weibull function is:

$$a \exp(-(x^c/b)), \qquad (10.2)$$

where x is distance, a is a normalization parameter, b is a parameter related to variance, and c is the shape parameter. The least-squares estimation procedure may give biased estimates of the parameters of this function. The Weibull function was chosen because although it is an empirically derived function, it contains the normal function ($c = 2$) and the exponential function ($c = 1$), which are widely used in the dispersal literature as special cases.

Of the six fields, *T. nubilale* dispersed the farthest in the 1985 East field. We focus our analysis on the results from this field so that we can model a worst-case scenario. Results from this field will be used to predict the greatest amount of dispersal of *T. nubilale* and thereby the greatest exposure to endangered species in nearby habitats. With our methods, $n(x,t)$ could not be estimated directly because we measured parasitism rates. Assuming that parasitism rates of egg masses are directly proportional to the density of *T. nubilale* that are present to parasitize them, $n(x,t_F)$ for fixed t_F will be estimated as a proportionality constant with the Weibull functions. The estimated function for the 1985 East field, 6 days after release, is shown in Fig. 10.4. To provide continuous estimates of $n(x,t)$ from $n(x,t_F)$ for fixed t_F, we fitted regression models to the three Weibull parameters (*a*, *b* and *c*), with time as the independent variable. The estimated values and the fitted equations for the three Weibull parameters based on the 1985 East data are shown in Fig. 10.5. The parameter $a(t)$ was fitted using linear regression, $b(t)$ was fitted using a logarithmic model, and $c(t)$ was fitted using an exponential model. Simulations using these estimated functions suggest that the model is most sensitive to the estimation of *c*. In particular, when *c* becomes small, more organisms disperse longer distances. The fitted equation for $c(t)$ becomes smaller than any of the estimated values (Fig. 10.5), so this particular model is consistant with a worst-case scenario, i.e. the model will predict greater amounts of dispersal than the data. These considerations lead to the

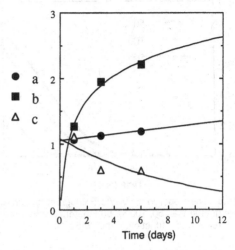

Fig. 10.5 Parameter values used in equation 10.3 to estimate dispersal of *Trichogramma nubilale* from a release site at different times after release. The three parameters, *a*, *b*, and *c*, are those from the Weibull function. Points indicate parameter estimates from experimental data, and the lines are the extrapolations from the data. The extrapolations are consistent with a worst-case scenario and probably overestimate the dispersal capacity of *T. nubilale*.

following estimate of $n(x,t)$:

$$n(x,t) = a(t) \exp\left(-\frac{x^{c(t)}}{b(t)}\right), \quad (10.3)$$

where,

$$a(t) = 1.049 + 0.024\ t;$$
$$b(t) = 1.291 + 1.240 \log t;$$
$$c(t) = 1.079 \exp[-0.117\ t].$$

Using equation 10.3 and integrating the area under the curve for different distances versus time, we can calculate the proportion of viable individuals that will disperse farther than 50, 100, and 500 m from the release site as a function of time after release (Fig. 10.6). This suggests that 10 days after release, most viable individuals (70%) will be found more than 100 m away from the release point.

Mortality
The rate of mortality of released *T. nubilale* was estimated from field data. Andow and Prokrym (1991) conducted experiments on plots 8.6× 16 m and estimated the disappearance rate of *T. nubilale* to be 0.52 day^{-1}. This disappearance rate includes both mortality and dispersal from the

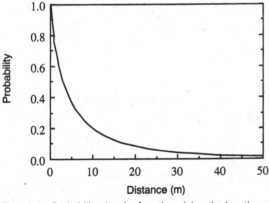

Fig. 10.4 Probability density function giving the location of *Trichogramma nubilale* 6 days after a release at the origin of the abscissa, showing the shape of the Weibull function with parameter values as follows: $a = 1.189$, $b = 2.216$, and $c = 0.594$. These parameter values are the same as those used in equation 10.4.

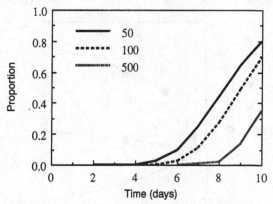

Fig. 10.6 The proportion of viable *Trichogramma nubilale* that are greater than 50, 100, or 500 meters away from the point of release as a function of the time since release.

plots. To estimate the potential effect of dispersal on disappearance rate, we used results from the experiments described above to estimate dispersal, and then subtracted this estimate from 0.52. Again, to model a worst-case scenario, we used data collected from the 1985 East field because estimates of mortality based on these data would give the lowest estimates that our data could support. By using the lowest estimated mortality, more *T. nubilale* would survive for longer periods, and more would ultimately enter habitats that contained rare or endangered species of Lepidoptera. Fitted parameters of the Weibull function from data obtained 6 days after release (Fig. 10.5) were used to estimate dispersal rates, not those estimated by regression. For ease of calculation, we assume here that the plots were circular, of the same area and with a diameter of 13.24 m. With these assumptions, the following equation was used to estimate a daily emigration rate from the plots:

$$\frac{\int_{6.62}^{\infty} 1.189 \exp\left\{-\frac{x^{0.594}}{2.216}\right\}dx}{\int_{0}^{\infty} 1.189 \exp\left\{-\frac{x^{0.594}}{2.216}\right\}dx} = 1 - (1 - e)^6$$

(10.4)

where, e is the daily emigration rate from the circular plot. The left-hand side of this equation is emigration from the plot after 6 days estimated from the dispersal equation 10.3, and the right-hand side is emigration estimated from an assumption of constant daily emigration. Despite the theoretical difficulties underlying equation 10.4, we can work through the calculations, and estimate $e = 0.107$ day^{-1}. When this is combined with the disappearance rate, mortality is estimated to be 0.413 day^{-1}.

Estimated exposure

The expected number of *T. nubilale* passing through habitats with rare Lepidoptera will depend on all of the above considerations and the distance from the release site to the habitats. Entering all of the estimated values into equation 10.1, we can estimate the number of *T. nubilale* that reach various distances from the release site. For releases of *T. nubilale* in sweet maize (100 000 females), the expected number of parasitoids that reach habitats greater than 100, 500, and 1000 m away from the release site are shown in Fig. 10.7. By day 9, nearly 1300 parasitoids may have reached the habitats only 100 m away, and by day 11, nearly 600 parasitoids may have reached habitats about 1 km away. About 30 times these numbers (39 000 and 18 000) might reach the habitats from a release in field maize. Keeping in mind that these are worst-case estimates, these results suggest that there is a modest but significant potential risk involved in the release of *T. nubilale* near habitats with rare Lepidoptera.

These estimates represent average numbers of

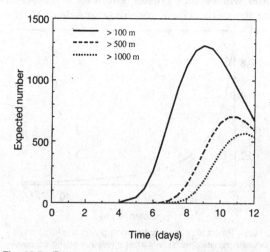

Fig. 10.7 The expected number of viable *Trichogramma nubilale* that are greater than 100, 500, or 1000 meters away from the point of release as a function of the time since release.

parasitoids that might reach the habitats. Further work on this problem could use a stochastic model to calculate the expected variation in this average number to determine the utility of the averages. If the variation is large, then the averages might not be the best numbers to use for minimizing short-term risk. In addition, we have not yet taken into account the potential effects due to the offspring of the released parasitoids. If the parasitoids have a high reproductive rate on *O. nubilalis* then many more might reach habitats that contain rare or endangered species of Lepidoptera.

Actual distances to Karner blue habitats

We have conducted intensive surveys of the distribution of Karner blue in Winona county. The five Karner blue habitats in the area are only 0.5–0.9 km (mean 0.66 km) from the nearest agricultural field. A more vivid picture of the proximity of Karner blue habitat to agricultural land is illustrated in Fig. 10.8, which shows one of the larger populations of Karner blue in Minnesota (each square indicates the location of at least one Karner blue adult) in its typical habitat of oak

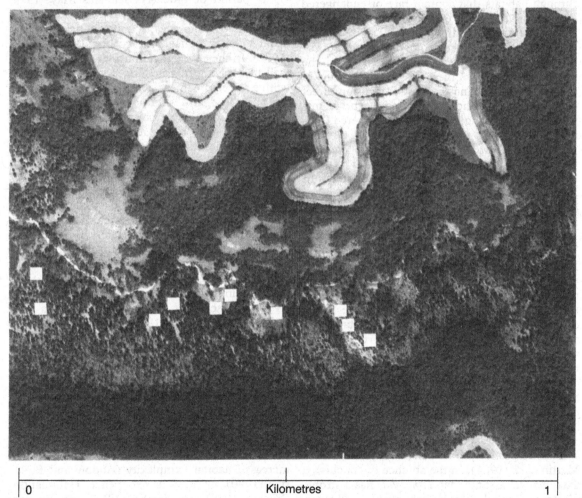

0	Kilometres	1

□ = Areas with *Lycaeides melissa samuelis*

Fig. 10.8 Aerial photograph of the Cuthrell South site in the Whitewater Wildlife Management Area in Winona County, Minnesota, USA, showing the distribution of *Lycaeides melissa samuelis* (Karner blue) adults in relation to nearby agricultural fields.

savanna woodlands. This population is only 0.6 km from the nearest agricultural field, which has been planted with maize. These data provide further evidence that the potential risk from releases of *T. nubilale* is not negligible.

Effects of released *T. nubilale*

Evaluation of the potential effects of the released parasitoids on rare Lepidoptera is considerably more difficult than the evaluation of exposure. Given that 600 to 39000 females may pass through habitats that contain rare or endangered Lepidoptera, to what extent will they enter and stay in the habitat and find and parasitize eggs of these rare species? If they do parasitize eggs of the rare species, what effect might this have on the population biology of these rare species? Many of these questions cannot be answered using direct experimental methods because experiments would entail placing the rare or endangered Lepidoptera at risk in the experiment itself. A more reasonable alternative would be to conduct research on *T. nubilale* in habitats similar to those harbouring rare or endangered Lepidoptera. In addition, we can use data on *T. nubilale* in other habitats, data on other parasitoids, and theoretical investigations.

Finding the habitat

Little work has been done on how parasitoids find appropriate habitats, and no information exists that would aid prediction as to whether or not *T. nubilale* will find the habitats with rare Lepidoptera. There are a number of anecdotal statements about habitat preferences of *Trichogramma* spp., but no studies have examined this question rigorously. The evidence on habitat preferences is based on collections of hosts from various habitats. Because some species emerged from hosts collected in woodlands and others emerged from hosts collected in fields, several researchers have referred to these species as woodland and field species, respectively (Flanders, 1937; Kot, 1964; Martin *et al.*, 1981). In the absence of conclusive data, we will assume that *T. nubilale* has a strong preference for oak savanna habitats and will enter and forage in Karner blue habitats. Because oak savanna habitat is only a part of the total habitat area that surrounds a maize field, this assumption is consistent with our worst-case scenario.

Leaving the habitat

Foraging theory predicts that parasitoids will forage in a way that maximizes the rate of encounter with unparasitized hosts. For parasitoids that are not egg limited and that forage in patchy environments, one way in which this would be accomplished is to forage in habitats that have high densities of unparasitized hosts. This can be accomplished by using a giving-up-time (GUT) strategy, in which a parasitoid leaves a habitat if it has not encountered unparasitized hosts after some period of time (Hassell and May, 1974; Murdoch and Oaten, 1975). Parasitoids that are egg limited or forage in uniform habitats may exhibit other behaviours (Gibb, 1962; Krebs, 1973). In the case under consideration *T. nubilale* will be foraging in a heterogeneous environment with a full complement of eggs upon eclosion, and the ability to mature additional eggs later in life, so the GUT model might be applicable. Mechanistically, GUT-type behaviour can occur if a parasitoid modifies its searching behaviour upon host or host cue encounter by increasing its turning rates or decreasing its searching speed, an activity described as area-restricted search (Curio, 1976).

Several laboratory and field studies of parasitoids demonstrate area-restricted search (e.g. Laing, 1937; Wylie, 1958; Hafez, 1961; Doutt, 1965; Spradberry, 1970; Richardson and Borden, 1972; Murdie and Hassell, 1973), and others demonstrate that parasitoids spend most time foraging in high density host patches (Hassell, 1971; Hirose *et al.*, 1976; van Lenteren and Bakker, 1978; Waage, 1979, 1983; Galis and van Alphen, 1981; van Alphen and Galis, 1983; Summy *et al.*, 1985; Morrison, 1986), or in patches with greater concentrations of host kairomones (Waage, 1979; Galis and van Alphen, 1981; Morrison and Lewis, 1981, Dicke *et al.*, 1984). Factors that reduce the GUT include encounters with parasitized hosts (Morrison and Lewis, 1981; Bakker and van Alphen 1988), encounters with other parasitoids (Hassell, 1971), increased habitat complexity (Andow and Prokrym, 1990), or previous experience. Habituation (Waage, 1979) may decrease GUT and operant conditioning (van Alphen and van Harsel, 1982) may increase GUT.

Host density, host kairomone concentration, habitat complexity, and previous experience may influence GUT for *T. nubilale* in habitats with rare or endangered Lepidoptera. Intraspecific encounters with parasitized hosts or other parasitoids are not likely at the densities *T. nubilale* are likely to attain in these habitats. Interspecific encounters, if they occur, will probably reduce GUT, thus causing *T. nubilale* to leave faster, and this would mitigate potential adverse effects on rare or endangered Lepidoptera. The effect of host kairomone concentration, habitat complexity, and previous experience on GUT is difficult to predict without further research, and these factors will be disregarded here. Although the rare Lepidoptera have very low densities, *T. nubilale* is probably polyphagous, so the density of all potential hosts in these habitats must be estimated before it is possible to predict that parasitoids would remain for a long or short time period in these habitats.

Estimation of total host density for *T. nubilale* has not yet been done in any of the Karner blue habitats. *Erynnis persius* is another lupine-feeding butterfly that occurs in the same Whitewater Wildlife Management Area (R. Dana, personal communication). There are more than 10 other species of butterfly in these Karner blue sites (Lane, 1992), Cuthrell (1990) reported 43 other species of butterfly in Karner blue habitats (Table 10.1), and there is an unknown number of moth species in these sites. Thus, it is possible that a Karner blue habitat would be a potentially excellent habitat for *T. nubilale*. For illustrative purposes, and to remain consistent with a worst-case scenario, we will assume that no *T. nubilale* leave once they find a Karner blue habitat.

Host finding

Assuming that *T. nubilale* will find and stay in habitats with rare or endangered Lepidoptera, it may or may not be able to find the eggs of these rare species. On 5 August 1992, we conducted an experiment to determine if *T. nubilale* could find eggs of the Karner blue. Lupine Valley is in the Whitewater Wildlife Management Area and is typical oak savanna habitat, supporting a large population of wild lupine, but is 3 km from the nearest population of Karner blue. The Karner blue lays its eggs on wild lupine, on nearby forbs and grasses, and on lupine litter. We placed 100 egg masses of *O. nubilalis* at the common oviposition sites of Karner blue, and released 1000

female *T. nubilale* no more than 6 m from the egg masses to determine if the parasitoids could find hosts located where Karner blue normally oviposits. It would have been difficult to use actual Karner blue eggs because they are hard to find, and it would have been ethically questionable to expose large numbers of an endangered species to the parasitoid. *O. nubilalis* egg masses are appropriate because *T. nubilale* readily accepts them, and the experiment did not confound variability in host acceptance with potential host finding. Egg masses were exposed for 2 days, collected and their fate was determined in the laboratory as described in Andow (1990). One egg mass was parasitized by *Trichogramma* sp. (*T. minutum* complex, J. Pinto, personal communication), none were parasitized by *T. nubilale*, but 60% were preyed upon. Although this experiment was conducted under cool wet conditions, the data suggest that *T. nubilale* is not particularly effective at finding eggs of the Karner blue.

These results are further supported by the following theoretical argument. The amount of surface area that *Trichogramma* must search to find hosts has a significant effect on the rate of host location (Need and Burbutis, 1979; Ables *et al.*, 1980; Burbutis and Koepke, 1981). Indeed, there was an inverse relationship between plant surface area and parasitism rates for *T. nubilale* (Need and Burbutis, 1979; Burbutis and Koepke, 1981). It is unlikely that *Trichogramma* spp. search plants in a random manner. In maize plants near anthesis, *T. nubilale* searches preferentially at heights nearer to the ear than the ground, or in the canopy (unpublished data). Therefore, depending on where the parasitoids prefer to search, parasitism rates may be higher or lower than predicted from uniform random search. Although it is not well supported, we assume that search by *T. nubilale* in an oak savanna habitat is a uniform random process.

Oak savanna habitats have a far more complex vegetation structure and a far greater plant surface area than maize fields. This implies that parasitism efficiency would be lower in Karner blue habitats than in maize fields. We estimate that the plant-surface area in oak savanna habitats is twice that in a mature maize field (which is approximately 18 m^2 plant surface/m^2, Prokrym *et al.* 1992), and we assume that the searching efficiency would be 0.0014% parasitism/female per day (searching efficiency estimated in Andow and

Table 10.1. Butterfly species found in the habitat of the Karner blue. Species are as listed in Cuthrell (1990) and plant associations are as in Macy and Shepard (1941), Opler and Krizek (1984) and Pyle (1981)

Species name	Preferred food plants
Amblyscirtes vialis	Poaceae
Ancyloxypha numitor	Poaceae
Asterocampa celti	*Celtis occidentalis*
Asterocampa clyton	*Celtis occidentalis*
Atrytone delaware	—
Basilarchia arthemis	*Salix, Populus, Betula, Crategus*
Basilarchia archippus	*Salix, Populus*
Celastrina ladon	*Prunus, Cornus, Vaccinium*
Cercyonis pegala	Poaceae
Charidryas gorone	*Helianthus, Ambrosia,* Asteraceae
Colias eurytheme	*Trifolium*
Colias philodice	Fabaceae
Danaus plexippus	*Asclepias*
Enodia anthedon	Poaceae
Epargyreus clarus	Fabaceae
Erynnis baptisiae	*Baptisia tinctoria*
Erynnis juvenalis	*Quercus,* Leguminosae
Erynnis persius	*Lupinus, Salix, Populus*
Euphyes ruricola	Cyperaceae
Everes comyntas	*Trifolium,* Fabaceae
Harkenclenus titus	*Prunus, Aronia*
Heraclides cresophontes	*Zanthoxylem, Ptelea*
Hesperia leonardus	*Panicum, Agrostis, Andropogon*
Lycaena phlaeas	*Rumex*
Megisto cymela	Poaceae
Nymphalis antiopa	*Salix, Ulmus, Populus*
Papilio polyxenes	Apiaceae
Phyciodes tharos/morpheus	*Aster*
Pieris rapae	Brassicaceae, Tropaeolaceae
Ponanes hobomok	Poaceae
Polites origines	Poaceae
Polygonia comma	*Ulmus, Urtica, Humulus*
Polygonia progne	*Ribes, Ulmus*
Pontia protodice	*Lepidium*
Pterourus glaucus	*Salix, Betula, Fraxinus, Prunus*
Pyrisitia lisa	
Satyrium edwardsii	*Quercus*
Speyeria cybele	*Viola*
Strymon melinus	*Astragalus, Quercus*
Thorybes bathyllus	Fabaceae
Vanessa atalanta	*Urtica, Boehmeria, Humulus*
Vanessa cardui	*Urtica, Althaea*
Wallengrenia egeremet	*Panicum,* Poaceae

Prokrym, 1991). If all of the parasitoids are searching in 1 ha of oak savanna habitat (this is the size of the Karner blue habitat shown in Fig. 10.8), then we calculate that 600 female *T. nubilale* can cause 0.0001% parasitism, 1300 can cause 0.0002% parasitism, 18000 can cause 0.0025% parasitism, and 39000 can cause as much as 0.0054% parasitism per day, if they are uniformly distributed throughout the habitat. If the habitat was only 0.1 ha and the plant surface area was 9 m^2 plant surface/m^2, then maximum daily parasitism would be 0.22%. Because the Karner blue sometimes lays eggs on complex surfaces, the locating efficiency of *T. nubilale* will probably be even less than that which we have calculated (Andow and Prokrym, 1990).

The maximum estimated parasitism rate for a worst-case scenario is 0.22% day^{-1}. Although this value is extremely small, it is based on extrapolation of field data well beyond the reasonable limits of the data, and at least some of the assumptions should be verified experimentally before this value is accepted. We developed this estimate to illustrate how the effects of introduced parasitoids can be estimated.

Effects of parasitism

If it is assumed that *T. nubilale* reaches Karner blue habitats and parasitizes Karner blue eggs, we then consider whether the mortality exerted by *T. nubilale* on Karner blue populations is merely proportional to the parasitism rates, or whether it is magnified or diminished. These effects will depend on the biology of the species parasitized. Considerable experience with inundative releases suggest that greater numbers of released parasitoids exert greater levels of mortality on the target pest (reviewed in part in Stinner, 1977; Hassan *et al.*, 1988). For example, on *O. nubilalis*, *T. nubilale* induced egg mortality is directly related to population suppression of later larval stages (Prokrym *et al.*, 1992; Andow and Olson, 1992).

The effects of *Trichogramma* induced mortality could be magnified if the parasitoids could regulate their host population. The results from several field studies provide evidence for density-dependent parasitism on the larval stages of the host population (Embree, 1966; Munster-Svendsen, 1980; Stamp, 1981; Heads and Lawton, 1983; Roland, 1986), but none of these report the presence of egg parasitoids. Other studies suggest that population regulation of Lepidoptera in the larval or pupal stages can also occur by a delayed density-dependent mechanism (Courtney, 1986) or by intraspecific competition within the host population (Dempster, 1983). The only population of insects that appears to be regulated by egg parasitoids is *Papilio xuthus*. The egg parasitoids *T. papilionis*, *T. australicum*, and *T. dendrolimi* probably regulate *P. xuthus* populations because the host is multivoltine, very abundant, and has overlapping generations. These conditions provide the egg parasitoids with an abundant and temporally stable supply of hosts, which allows them to build up the large populations that are capable of regulating a host population (Hirose *et al.*, 1980). Because the rare and endangered

Lepidoptera in our region are not multivoltine, and have discrete generations, it seems unlikely that their populations will be regulated by introductions of *T. nubilale*.

Based on our observations on the fate of Lepidoptera eggs placed where the Karner blue typically oviposits, we believe that its populations are unlikely to be reduced by introductions of *T. nubilale*. Only 1% of the egg masses was parasitized by *Trichogramma* sp., none was parasitized by *T. nubilale*, but 60% were preyed upon by unidentified predators. It appears that egg predation may be very significant for the Karner blue, but that small, periodic 1–2% increases in egg parasitism would be less likely to have a large effect on Karner blue populations.

The effects of *Trichogramma* induced mortality could be lessened if there were compensatory mortality in later developmental stages of the rare or endangered Lepidoptera. Such compensatory mortality occurs in larval stalkborer populations, *Chilo partellus* (van Hamburg, 1980), and is probably caused by density-dependent dispersal and 'contest' competition among larvae in stalks. Van Hamburg and Hassell (1984) suggested that if egg parasitism did not reduce populations to less than 10 larvae/plant, density-dependent mortality during the later larval stages would compensate completely for the egg mortality. As mentioned above, density-dependent larval mortality occurs commonly in arthropods, so that some compensation for egg mortality is likely. There are few population data on rare or endangered species of Lepidoptera, and what exists cannot even support speculation on the occurrence of compensatory mortality. For example, although many studies have been conducted on the rare nymphalid, *Euphydryas editha bayensis*, the role of density-dependent mortality in the population dynamics of the species is not known. There are no data on the Karner blue related to compensatory mortality. While compensatory mortality is likely to occur, it is possible that rare or endangered species may have different population dynamics with little or no compensatory mortality.

Conclusions

The nature, likelihood of occurrence, and risks associated with a particular biological control agent are difficult to estimate. Using *T. nubilale* as an example, we have illustrated a method to evaluate the magnitude of risk to populations of the endangered Karner blue. We considered the following points:

1. Host acceptance of eggs of Karner blue by *T. nubilale*, assumed to be possible
2. Overwintering ability of *T. nubilale* in Minnesota, assumed not to be possible
3. Timing of releases relative to the time of oviposition by the rare Lepidoptera, determined that there would be significant overlap for some of the species, including Karner blue, in some parts of the state
4. Dispersal from release sites to habitats with rare Lepidoptera, making the following estimations

| | Numbers moving more than | |
	100 m	500 m
From sweet maize	1300	600
From field maize	39 000	18 000

a. Numbers of released *T. nubilale* were assumed to be 100 000 in sweet maize and 3 000 000 in field maize
b. Dispersal was estimated using a Weibull function fitted to field data
c. Mortality was assumed to be constant in space and time, and was estimated from field data to be -0.413 day^{-1}
d. Distance to Karner blue habitat was 500–1000 m, but in subsequent analysis, we assumed this habitat was only 100 m from release sites

5. Habitat finding by *T. nubilale* was assumed to be a passive process, but that all dispersing females would find Karner blue habitats
6. Habitat leaving was assumed not to occur, so all *T. nubilale* stayed in a Karner blue habitat once they found it
7. Host finding was assumed to be similar to that in maize fields, and was estimated from field data
8. Parasitism rate on the rare species was estimated for a 0.1 ha habitat with 9 m^2 plant surface area/m^2, 100 m away from a release in field maize, and was estimated to be 0.22% day^{-1}
9. Effects of this level of parasitism on Karner blue populations is uncertain, but likely to be small

It should be emphasized that this is an hypothetical analysis. It is based on numerous untested but testable assumptions and extrapolations from field data that would seem to be well beyond their reasonable limits. It is the method of analysis that should be examined, and we believe that this approach to the problem of estimating the magnitude of environmental risk is sound.

References

Ables, J. R., McCommas, D. W., Jones, S. L. and Morrison, R. K. (1980). Effect of cotton plant size, host egg location, and location of parasite release on parasitism by *Trichogramma pretiosum*. *Southwestern Entomological* 5, 261–265.

Andow, D. A. (1990). Characterization of predation on egg massess of *Ostrinia nubilalis* (Lepidoptera: Pyralidae). *Annals of the Entomological Society of America* 83, 482–486.

Andow, D. A. and Olson, D. (1992). Biological control of European corn borer with *Trichogramma*. Illinois Agricultural Pesticides conference 1992, Summaries of Presentations. University of Illinois at Urbana-Champaign.

Andow, D. A. and Prokrym, D. R. (1990). Plant structural complexity and host-finding by a parasitoid. *Oecologia* 82, 162–165.

Andow, D. A. and Prokrym, D. R. (1991). Release density, efficiency and disappearance of *Trichogramma nubilale* for control of European corn borer. *Entomophaga* 36, 105–113.

Bakker, K. and van Alphen, J. J. (1988). The influence of encounters with parasitized hosts on the time spent searching on a patch by *Leptopilina heterotoma* (Thompson) (Hymenoptera, Cynipidae). In *Parasitoid Insects, Proceedings European Workshop, Lyon* (ed. M. Bouletream and G. Bonnet), pp. 39–40. Institut National de la Recherche Agronomique, Paris.

Burbutis, P. P. and Koepke, C. H. (1981). European corn borer control in peppers by *Trichogramma nubilale*. *Journal of Economic Entomology* 74, 246–247.

Burbutis, P. P., Curl, G. D. and Davis, C. P. (1976). Overwintering of *Trichogramma nubilale* Ertle and Davis in Delaware. *Environmental Entomology* 5, 888–890.

Courtney, S. P. (1986). The ecology of pierid butterflies: dynamics and interactions. *Advances in Ecological Research* 15, 51–131.

Curio, E. (1976). The Ethiology of Predation. Springer-Verlag, Berlin.

Curl, G. D. and Burbutis, P. P. (1978). Host-preference studies with *Trichogramma nubilale*. *Environmental Entomology* 7, 541–543.

Cuthrell, D. (1990). Status of the Karner blue

butterfly, *Lycaeides mellissa samuelis* Nabokov, in Minnesota. Report for the Minnesota Nongame Wildlife Program.

Dana, R. and Huber, R. L. (1988). Butterflies. In *Minnesota's Endangered Flora and Fauna* (ed. B. Coffin and L. Pflannmuller), pp. 380–395. University of Minnesota Press, Minneapolis, Minnesota.

Dempster, J. P. (1983). The natural control of populations of butterflies and moths. *Biological Reviews* 58, 461–481.

Dicke, M., van Lenteren, J. C., Boskamp, G. J. and van Dongen-van Leeuwen, E. (1984). Chemical stimuli in host-habitat location by *Leptopilina heterotoma* (Thomson) (Hymenoptera: Eucoilidae), a parasite of *Drosophila*. *Journal of Chemical Ecology* 10, 695–712.

Doutt (1965). Biological Characteristics of Entomophogous Adults. In *Biological Control of Insect Pests and Weeds* (ed. P. Debach), pp. 145–167. Rienhold Pub. Co., New York.

Embree, D. G. (1966). The role of introduced parasites in the control of the winter moth in Nova Scotia. *Canadian Entomologist* 98, 1159–1168.

Flanders, S. E. (1937). Notes on the life history and anatomy of *Trichogramma*. *Annals of the Entomological Society of America* 30, 304–308.

Flint M. L. and van den Bosch, R. (1981). *Introduction of Integrated Pest Management*. Plenum Press, New York.

Galis, F. and van Alphen, J. J. (1981). Patch time allocation and search intensity of *Asobara tabida* Nees (Hym: Braconidae). *Netherlands Journal of Zoology* 31, 701–712.

Gibb, J. A. (1962). Tinbergen's hypothesis of the role of specific searching images. *Ibis* 104, 106–111.

Hafez, M. (1961). Seasonal fluctuations of population density of the cabbage aphid, *Brevicoryne brassicae* (L.), in the Netherlands, and the role of its parasitoid *Aphidius (Diaeretiella) rapae* (Curtis). *Tijdschrift over Plantziekten* 67, 445–548.

Hassan, S. A., Kohler, E. and Rost, W. M. (1988). Mass production and utilization of *Trichogramma*: 10. Control of the codling moth *Cydia pomonella* and the summer fruit tortix moth *Adoxophyes orana* (Lep.: Tortricidae). *Entomophaga* 33, 413–420.

Hassell, M. P. (1971). Mutual interference between searching insect parasites. *Journal of Animal Ecology* 40, 473–486.

Hassell, M. P. and May, R. M. (1974). Aggregation in predators and insect parasites and its effect on stability. *Journal of Animal Ecology* 43, 567–594.

Heads, P. A. and Lawton, J. H. (1983). Studies on the natural enemy complex of the holly leafminer: The effects of scale on the detection of aggregation

responses and the implications for biological control. *Oikos* 40, 267–276.

Hirai, K. (1987). Biology and parasitism of *Trichogramma chilonis* Ishii (Hymenoptera, Trichogrammatidae), as an egg parasite of the soybean podborer, *Leguminivora glycinivorella* Matsumura (Lepidoptera, Tortricidae). *Bulletin of the Tohoku National Agricultural Experiment Station* 75, 41–64.

Hirose, Y., Kimoto, H. and Hiehata K. (1976). The effect of host aggregation on parasitism by *Trichogramma papilionis* Nagarkatti (Hymenoptera: Trichogrammatidae), an egg parasitoid of *Papilio xuthus* Linne. (Lepidoptera: Papilionidae). *Applied Entomology and Zoology* 11, 116–125.

Hirose, Y., Suzuki, Y., Takagi, M., Hiehata, K., Yamasaki, M., Kimoto, H., Yamanaka, M., Iga, M. and Yamaguchi, K. (1980). Population dynamics of the citrus swallowtail, *Papilio xuthus* Linne (Lepidoptera: Papilionidae): mechanisms stabilizing its numbers. *Researches on Population Ecology* 21, 260–285.

Howarth, F. G. (1991). Environmental impacts of classical biological control. *Annual Review of Entomology* 36, 485–509.

Kot, J. (1964). Experiments in the biology and ecology of species of the genus *Trichogramma* Westwood and their use in plant protection. *Ekologia Polska* 12(A), 243–303.

Krebs, J. R. (1973). Behavioral aspects of predation. In *Perspectives in Ethology* (ed. P. P. G. Bateson and P. H. Klopfer), pp. 73–111. Plenum Press, New York.

Laing, M. (1937). Host finding by insect parasites. I. Observations on the finding of hosts by *Alysia manducator*, *Mormoniella vitripennis* and *Trichogramma evanescens*. *Journal of Animal Ecology* 6, 298–317.

Lane, C. P. (1992). The Minnesota 1991 status of the Karner Blue butterfly (*Lycaeides melissa samuelis*: Lycaenidae) and its associated plant resources. Report to the US Fish and Wildlife Service, Office of Endangered Species, Washington, DC.

Leahy, T. C. and Andow, D. A. (1994). Egg weight, fecundity and longevity are increased by adult feeding in *Ostrinia nubilalis* Hübner (Lepidoptera: Pyralidae). *Environmental Entomology* 87, 342–349.

Macy, R. W. and Shepard, H. H. (1941). *Butterflies*. University of Minnesota Press, Minneapolis.

Martin, P. B., Lindgren, P. D., Greane, G. L. and Grissell, E. E. (1981). The parasitoid complex of three noctuids (Lepidoptera) in a Northern Florida cropping system: seasonal occurrence, parasitization, alternate hosts, and influence of host habitat. *Entomophaga* 26, 401–419.

Morrison, G. (1986). Searching time aggregation and

density dependent parasitism in a laboratory host-parasitoid interaction. *Oecologia* **68**, 298–303.

Morrison, R. K. and Lewis, W. J. (1981). The allocation of searching time by *Trichogramma* spp.: an experimental analysis. *Environmental Entomology* **9**, 79–85.

Munster-Svendsen, M. (1980). The distribution in time and space of parasitism in *Epinotia tedella* (Cl.) (Lepidoptera: Tortricidae). *Ecological Entomology* **5**, 373–383.

Murdie, G. and Hassell, M. P. (1973). Food distribution, searching success and predator-prey models. *The Mathematical Theory of the Dynamics of Biological Populations* (ed. M. S. Bartlett and R. W. Hiorns), pp. 87–101. Academic Press, London.

Murdoch, W. W. and Oaten, A. (1975). Predation and population stability. *Advances in Ecological Research* **9**, 2–132.

Need, J. T. and Burbutis, P. P. (1979). Searching efficiency of *Trichogramma nubilale*. *Environmental Entomology* **8**, 224–227.

Oatman, E. R. (1986). Parasites reared from eggs of the Orange Palm Dart, *Cephrenes augiades sperthias* (Lepidoptera: Hesperiidae), in Australia. *Proceedings of the Entomological Society of Washington* **88**, 393–394.

Opler, P. A. and Krizek, G. (1984). *Butterflies East of the Great Plains*. Johns Hopkins University Press, Baltimore, MD.

Pimentel, D., Glenister, C., Fast, S. and Gallahan, D. (1984). Environmental risks of biological pest controls. *Oikos* **42**, 283–290.

Prokrym, D. R., Andow, D. A., Ciborowski, J. D. and Sreenivasam, D. D. (1992). Suppression of *Ostrinia nubilalis* by *Trichogramma nubilale* in sweet corn. *Entomologia Experimentalis et Applicata* **64**, 73–85.

Pyle, R. M. (1981). *The Audubon Society Field Guide to North American Butterflies*. Knopf, New York.

Rawat, U. S. and Pawar, A. D. (1991). Field recovery of *Trichogramma chilonis* Ishii (Hymenoptera: Trichogrammatidae) from *Deudorix epijarbas* Moore (Lepidoptera: Lycaenidae) in Himachal Pradesh, India. *Entomon* **16**, 49–52.

Richardson, J. V. and Borden, J. H. (1972). Host-finding behavior of *Coeloides brunneri* (Hymenoptera: Braconidae). *Canadian Entomologist* **104**, 1235–1250.

Roland, J. (1986). Parasitism of winter moth in British Columbia during buildup of its parasitoid *Cyzenis albicans*: attack rate on oak *v.* apple. *Journal of Animal Ecology* **55**, 215–234.

Samways, M. J. (1988). Classical biological control and insect conservation: are they compatible? *Environmental Conservation* **15**, 349–354.

Spradberry, J. P. (1970). Host finding by *Rhyssa*

persuasoria (L.) an ichneumonid parasite of sircid wood wasp. *Animal Behaviour* **18**, 103–114.

Stamp, N. E. (1981). Effect of group size on parasitism in a natural population of the Baltimore checkerspot *Euphydryas phaeton*. *Oecologia Berlin* **49**, 201–206.

Stinner, R. E. (1977). Efficacy of inundative releases. *Annual Review of Entomology* **22**, 515–531.

Summy, K. R., Gilstrap, F. E. and Hurt, W. G. (1985). *Aleurorcanthus woglum* (Hom.: Aleyrodidae) and *Encarsia opulenta* (Hym.: Encrytridae): density-dependent relationship between adult parasite aggregation and mortality of the host. *Entomophaga* **30**, 107–112.

van Alphen, J. J. and Galis, F. (1983). Patch time allocation and parasitization efficiency of *Asobara tabida*, a larval parasitoid of *Drosophila*. *Journal of Animal Ecology* **52**, 937–952.

van Alphen, J. J. and van Harsel, H. H. (1982). Host selection by *Asobara tabida* Nees (Braconidae; Alysiinae), a larval parasitoid of fruit inhabiting *Drosophila* species. III. Host species selection and functional response. In *Foraging Behavior of Asobara tabida, a Larval Parasitoid of Drosophilidae* (ed. J. J. van Alphen), pp. 61–93. Ph.D. thesis, University of Leiden.

van Hamburg, H. (1980). The grain-sorghum stalkborer, *Chilo partellus* (Swinhoe) (Lepidoptera: Pyralidae): survival and location of larvae at different infestation levels in plants of different ages. *Journal of the Entomological Society of South Africa* **43**, 71–76.

van Hamburg, H. and Hassell, M. P. (1984). Density dependence and the augmentative release of egg parasitoids against graminacious stalkborers. *Ecological Entomology* **9**, 101–108.

van Lenteren, J. C. and Bakker, K. (1978). Behavioral aspects of the functional response of a parasite (*Pseudocoila bochei* Weld) to its host (*Drosophila melanogaster*). *Netherlands Journal of Zoology* **28**, 213–233.

Volden, C. S. (1983). Effects of temperature, feeding and host density on the reproductive biology of *Trichogramma nubilale* and *Trichogramma ostrinae*, M.S. thesis, University of Minnesota.

Waage, J. K. (1979). Foraging for patchily-distributed hosts by the parasitoid, *Nemeritis canescens*. *Journal of Animal Ecology* **48**, 353–371.

Waage, J. K. (1983). Aggregation in field parasitoid populations: foraging time allocation by a population of *Diadegma* (Hymenoptera: Ichneumonidae). *Ecological Entomology* **8**, 447–453.

Wylie, H. G. (1958). Factors that effect host finding by *Nasonia vitripennis* (Walk.) (Hymenoptera: Pteromalidae). *Canadian Entomologist* **90**, 597–608.

11

Entomopathogenic nematodes in biological control: feasibility, perspectives and possible risks

Ralf-Udo Ehlers and Arne Peters

Introduction

Although many mermithid, tylenchid, aphelenchid and rhabditid nematodes are known to be important antagonists of insects, very few nematode species have so far been used for biological control. The mermithid *Romanomermis culicivorax* was used in field trials to control malaria-transmitting mosquitos (Petersen and Cupello, 1981). However, with the discovery of the larvicide *Bacillus thuringiensis* var. *israelensis* this nematode lost its significance. In Australia, the neotylenchid parasite *Deladenus siricidicola* is successfully released to control the introduced European sawfly *Sirex noctilio* (Bedding, 1984). Other species have never gone beyond the scientific research level, with the exception of rhabditid nematodes of the genus *Steinernema* and *Heterorhabditis* and their bacterial symbionts *Xenorhabdus* spp., which are widely used for biological control of insect larvae in cryptic environments.

Biology

The nematode's life cycle consists of a free-living and a propagative phase. The only stage occurring outside the host insect is the infective third stage dauer juvenile (Fig. 11.1), which is developmentally arrested and morphologically and physiologically adapted for long-term survival under detrimental environmental conditions in the soil.

The dauer juveniles (DJs) do not take up nourishment and carry cells of their bacterial symbiont *Xenorhabdus* sp. in the intestinal lumen (Fig. 11.2) (Bird and Akhurst, 1983; Endo and Nickle, 1991). They actively seek out suitable host insects and penetrate the haemocoel where they find favourable conditions for propagation. The DJs start feeding, release the symbiotic bacteria into the haemolymph and enter into the propagative phase of the life cycle. Provided that the insect's humoral and cellular defence mechanisms do not succeed in the elimination of the nematode–bacterium complex, the host will die within 2 days after infection. The bacteria proliferate, the nematodes feed on the bacteria cells and the host tissue and develop into adults. *Steinernema* spp. are amphimictic, thus, they can only propagate if at least one male and one female nematode is present in the insect cadaver. In contrast, the heterorhabditid DJs develop into self-fertilizing hermaphrodites and only the subsequent adult generations that develop from eggs are either amphimictic (Poinar, 1990) or automictic. The females or hermaphrodites lay eggs from which the first junveniles (J1) hatch within about 2 days (Lunau *et al.*, 1993). The life cycle of both, steinernematid and heterorhabditid species, consists of four juvenile stages (J1–J4). Under permissive nutritional conditions another propagative cycle will take place. When the nematode population has reached a certain density threshold an alternative pathway leads to the formation of the

119

Fig. 11.1 Infective dauer juveniles (approximately 800 μm long) of the entomopathogenic rhabditid nematode *Steinernema feltiae.*

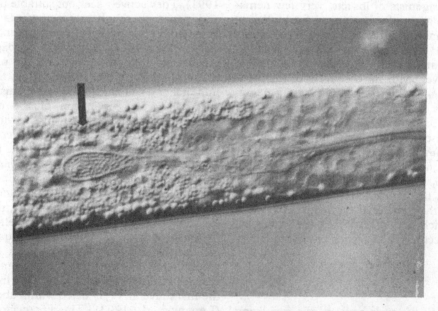

Fig. 11.2 Nematode vesicle containing cells of the bacterial symbiont *Xenorhabdus bovienii* (arrow) located in the anterior part of the dauer juvenile's intestine of the entomopathogenic nematode *Steinernema feltiae.*

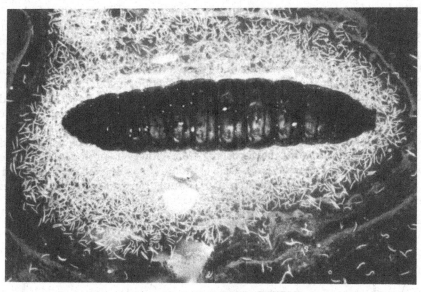

Fig. 11.3 Dauer juveniles of *Heterorhabditis* sp. emigrate from a last instar *Galleria mellonella* 18 days after infection and subsequent propagation within the dead host insect. The red colour of the cadaver is caused by pigment production of the bacterial symbiont *X. luminescens*.

morphologically distinct J2Ds, which embed cells of the bacterial symbiont and further develop into DJs (Strauch *et al.*, 1994). They leave the insect cadaver (Fig. 11.3) and search for new hosts. The life cycle of *Heterorhabditis* spp. is shown in Fig. 11.4.

Symbiosis

Many factors characterize the symbiotic relationships of the nematode–bacterium complex: successful propagation of the nematode depends on the presence of *Xenorhabdus* spp. in the insect cadaver. Without its symbiont the nematode cannot reproduce (Poinar and Thomas, 1966). On the other hand, *Xenorhabdus* spp. are unable to survive without the nematode. The DJ protects the bacterium against detrimental environmental conditions in the soil (Fig. 11.2). *Xenorhabdus* spp. have never been isolated from the soil. They seem to lack efficient mechanisms to compete against other micro-organisms in the soil environment. The nematode carries the bacterium into a sterile surrounding where a successful growth of *Xenorhabdus* spp. is secured. *Xenorhabdus* spp. produce antibiotics (Akhurst, 1982; Paul *et al.*, 1981) that inhibit the colonization of the insect

cadaver by other micro-organisms, which could prevent the reproduction of the nematode. Both symbiosis partners contribute to overcome the insect's defence mechanisms: *S. carpocapsae* produces an inhibitor of the antibacterial insect-immunoprotein Cecropin (Götz *et al.*, 1981). In the supernatant fluid from *X. nematophilus* cultures a factor was found that inhibits the activation of the pro-phenoloxidase system (Bréhelin *et al.*, 1990). Phenoloxidase enhances the recognition of non-self particles in the haemolymph and contributes to the encapsulation of these bodies. This factor may be identical to a proteolytic protein excreted by *Xenorhabdus* spp. (Boemare and Akhurst, 1988; Schmidt *et al.*, 1988). However, Yokoo *et al.* (1992) found that both organisms (*S. carpocapsae* and *X. nematophilus*) inhibited the activation of the pro-phenoloxidase cascade.

Mass production and commercialization

In vivo mass production has been reported from Italy (Deseö *et al.*, 1990). To reduce production costs, *in vitro* propagation was already a topic of research when entomopathogenic nematodes were used for the first time in the history of biological control against larvae of the Japanese beetle *Popillia japonica* (Glaser, 1931; Glaser and Farrell, 1935). The method that is nowadays most commonly applied to commercially produce entomopathogenic nematodes was developed by Bedding (1981, 1984). The nematodes are propagated in monoxenic three-dimensional solid phase cultures. The process, like the *in vivo* production, lacks economies of scale as labour costs remain constant and significant. More advanced liquid-culture techniques have been developed to produce *S. carpocapsae* in large-scale bioreactors (Friedman, 1990) and research is underway to improve the liquid culture of *Hetero-*

rhabditis sp. (Ehlers, 1990; Ehlers *et al.*, 1992).

However, unsolved problems arise with storage and transport of entomopathogenic nematodes. Although the DJ is an arrested state that tolerates environmental extremes, it remains sensitive to desiccation, high temperature, and lack of oxygen. These factors have to be considered with regard to product development and distribution. Components used for formulation, e.g. clay and alginate can immobilize the DJs and retard their desiccation. However, prolonged shelf life (more than 6 months) has not yet been achieved and improved formulation methods need further development (Georgis, 1990).

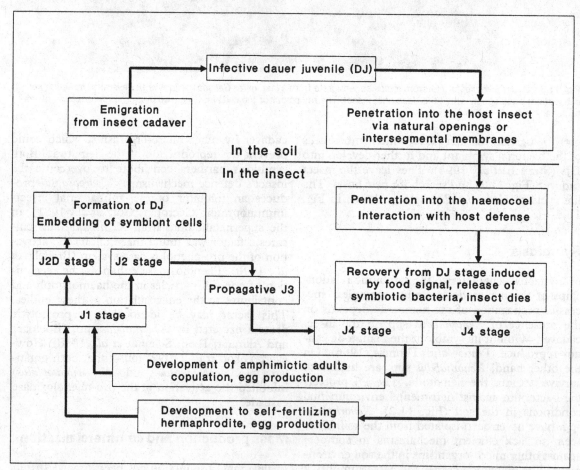

Fig. 11.4 Lifecycle of *Heterorhabditis* spp. The free-living infective dauer juvenile (DJ) penetrates into the host insect. Propagation takes place inside the host with 4 juvenile stages (J1 to J4) and the adult stages. The DJs always develop to hermaphrodites, while development without an intermediate DJ stage leads to amphimictic adults. The formation of the DJ is initiated in the J1 stage. This leads to the formation of a second stage pre-dauer juvenile (J2D), which already has some characters of the DJ.

Biotechnological production largely contributed to the reduction of application costs in inundative biological control. One million DJs of *Heterorhabditis* sp. (HSH) produced in solid state cultures cost about 1 DM. In comparison, the same amount produced in 5000 l liquid-culture bioreactors would reduce the costs to less than a tenth. These data were calculated from results obtained in laboratory-scale bioreactors (Ehlers *et al.*, 1992) and were extrapolated to an industrial-scale plant (von Reibnitz and Ehlers, 1992). A redution of the product costs will certainly promote a more widespread use of entomopathogenic nematodes in biocontrol. However, high levels of investment are needed for the development and construction of an industrial-production plant. Besides such an enterprise cannot easily change the production process from one nematode species to another. Considering these constraints only a manifold increase of sales can justify the production on a cubic-meter scale and probably only a few different nematode species or strains will be commercially available in the future. Thus, an expansion to export markets is a prerequisite for a further scale-up going hand-in-hand with the reduction of the product price. This development can only be realized when government officials allow the import and application of non-endemic species in their countries. However, these decisions can only be made when further data on the possible impact of non-endemic nematodes on the environment and the geographical distribution of the different species are available.

Taxonomy and geographical distribution

The ability to produce entomopathogenic nematodes commercially on a large scale contributed to their worldwide use in biological control. If the release of non-indigenous nematode species into the environment has to be avoided, information on the occurrence of endemic species is needed. A prerequisite for obtaining these data is the possibility of assigning available nematode isolates to well described taxa. However, taxonomy of entomopathogenic nematodes is still a matter of discussion (Mráck and Bednarek, 1991). Steinernematids have been distinguished by morphological characters of males and DJs and several new species have been described (Poinar, 1990).

Unfortunately, cross-breeding of different isolates (Akhurst and Bedding, 1975) is not always used to ascertain the taxonomy at the species level. So far this method could not be used to differentiate heterorhabditid isolates (hermaphrodites) and DNA analysis has been recommended for species characterization in this genus (Curran and Webster, 1989; Smits *et al.*, 1991). Recently, Dix *et al.* (1992) developed a technique to cross-breed second generation amphimictics of *Heterorhabditis* spp. Their results indicate that strains with identical restriction fragment length polymorphism (RFLP) patterns do not necessarily cross-breed or produce fertile offspring. Thus, the development of reliable methods to identify species of the genus *Heterorhabditis* needs further research.

Entomopathogenic nematodes are geographically widespread and their occurrence has been reported from all continents except the Antarctic. *H. bacteriophora* and *S. carpocapsae* seem to be widely distributed (Poinar, 1990). *S. feltiae* was isolated in Europe and Asia and was probably introduced into Australia, where it occurs mainly in Tasmania, the island which was first colonized by European settlers (Akhurst and Bedding, 1986). Other species, however, are geographically restricted, like e.g. *S. glaseri* in America or *S. scapterisci* in South America (in Uruguay). Due to the increase of interest in entomopathogenic nematodes, new strains are currently isolated worldwide. Accepted and currently described or redescribed species are listed in Table 11.1.

Isolation techniques and estimation of population density

Most of the described species were isolated by a baiting technique. Final instar larvae of *Galleria mellonella* are placed into a soil sample. They are recovered after a few days and then examined for nematode infestation (Bedding and Akhurst, 1975). This method will only give qualitative results. To enhance the efficiency of an estimate for the prevalence of a nematode species Fan and Hominick (1991) propose a consecutive exposure of *Galleria* larvae to soil samples. They suggest that this method could also be used to estimate population dynamics. A similar method is proposed by Bednarek and Nowicki (1991). They estimated that up to 12 000 DJs/m^2 of *Steinernema*

Table 11.1. Described species and undescribed isolates which are currently described or could not be assigned to described species with references and redescriptions in brackets ([1] = nomen nudum)

	References of description and redescription
Steinernema spp.	
S. affinis	Bovien, 1937 (Poinar, 1988a)
S. anomali	Kozodoi, 1984
S. bicornutum	Tallòsi *et al.* (submitted)
S. carpocapsae	Weiser, 1955
S. caudatum	Xu, Z. *et al.*, 1991
S. feltiae	Filipjev , 1934 (Poinar and Lindhardt, 1971)
S. glaseri	Steiner, 1929
S. intermedia	Poinar, 1985
S. kraussei	Steiner, 1923 (Mracek, 1977; Mracek *et al.*, 1992)
S. kushidai	Mamiya, 1988
S. rara	Doucet 1986 (Poinar *et al.*, 1988)
S. ritteri	Doucet and Doucet, 1990
S. scapterisci	Nguyen and Smart, 1990
Heterorhabditis spp.	
H. bacteriophora	Poinar, 1975
H. indicus	Poinar *et al.*, 1992
H. megidis	Poinar *et al.*, 1987
H. zealandica	Wouts, 1979 (Poinar, 1990)
Undescribed isolates of Steinernema *spp.*	
S. longicaudum[1]	Shen, 1992
S. serratum[1]	Liu, 1992
strain M, N, W1, W2	Akhurst and Bedding, 1986
strain *riobravis*	Raulston *et al.*, 1992
strain NC 513	Curran, 1989
Undescribed isolates of Heterorhabditis *spp.*	
H. brevicaudis[1]	Liu, 1992
Irish strains	Smits *et al.*, 1991
NW-European strains	Smits *et al.*, 1991

occur in crop soil. Laboratory tests revealed that only 40–50% of inoculated nematodes could be recovered with the baiting method.

Bedding *et al.* (1983) tested various nematode species and strains against different insect larvae and *G. mellonella* was shown to be the most susceptible insect. Consequently, they recommend this insect for surveys to isolate entomopathogenic nematodes from soil samples. However, with the isolation of *S. scapterisci* from its original host *Scapteriscus* sp., a nematode was found that rarely infects *G. mellonella* and does not propagate in this insect. Therefore, it is unlikely that this species would have been isolated by the baiting method with *G. mellonella*. Surveys to detect entomopathogenic nematodes in German soils (Sturhan, 1990) used a centrifugation method to isolate DJs from soil samples. Up to 100 000 *S.*

affinis DJs/m^2 were isolated. Compared with the results obtained by the baiting technique, it seems obvious that the centrifugation method also extracts nematodes which may not be infective, or which possibly do not react towards *G. mellonella* larvae. With the exception of *S. affinis* and *S. feltiae*, which were also isolated by the baiting technique (Ehlers *et al.* 1991), another six *Steinernema* spp. were found. Three of them could be identified as *S. carpocapsae*, *S. intermedia* and *S. kraussei*. The remaining three species have so far not been described or isolated by the baiting method. *S. kraussei* was found as frequently as *S. feltiae*, when isolated with the centrifugation method (Sturhan, personal communication) but is rarely trapped by *Galleria* baits (Peters, unpublished). Thus, it is most probable that several species cannot be isolated by baiting with *G.*

mellonella and may, so far, have been overlooked. The results obtained with *Galleria* should, therefore, be interpreted with care and further data are needed to comment finally on the presence of different nematode species in different countries and certain ecosystems. Methods to estimate population dynamics of entomopathogenic nematodes have yet to be improved.

Host range and host specificity

When entomopathogenic nematodes are used for the control of particular insect pests it is necessary first to select the most virulent species or strain, and second to predict their possible impact on non-target insects. If permanent establishment is to be achieved, recycling in target or non-target hosts is a prerequisite. Thus, information on the host range and specificity are required.

Entomopathogenic nematodes are considered to have a broad host range (e.g. Poinar, 1986; Klein, 1990). *S. carpocapsae* proved to be infectious in 250 insect species from over 75 families in 11 orders (Poinar, 1979). These results mainly refer to laboratory bioassays on moist filter paper in Petri dishes with high DJ concentrations and they generate the image of nematodes being extreme generalists. However, even laboratory experiments produce mixed results due to the individual or stage dependent activity of insect defence reactions against the nematodes (Ehlers and Gerwien, 1993). Furthermore, the potential to kill an insect in a laboratory assay cannot be transferred to field conditions, particularly when high control is required (Gaugler, 1988).

The occurrence of a variety of species, even at one location (Akhurst and Bedding, 1986; Sturhan, personal communication) implies an evolutionary development that results in distinct characters of every species, including development of specific nematode–host associations. Considerable evidence indicates that nematodes possess a restricted host range, Bedding *et al.* (1983) found that no one species is the best control agent for all, or even most, insect species. Remarkable differences in LD_{50}s of different nematode species and strains against several insects of different orders were found. *Otiorhynchus sulcatus* was highly susceptible to *H. bacteriophora* ($LD_{50} = 4$ DJs), in contrast, *S. feltiae* was considered to be non-pathogenic ($LD_{50} = 2500$) (Bedding *et al.*, 1983). *S. glaseri* and *H. bacteriophora* are considered to be more effective for field control of scarab larvae than e.g. *S. carpocapsae* or *S. feltiae* (Klein, 1990). Larvae of the cranefly *Tipula paludosa* were highly susceptible to *S. anomali* and *S. feltiae* in moist sand (70–90% mortality), whereas *Heterorhabditis* sp. North-West European group (*sensu* Smits *et al.*, 1991), the species widely used for control of *O. sulcatus*, hardly affected these larvae (10% mortality in laboratory bioassay). Even the closer related species *S. affinis* caused only 4% mortality (Ehlers and Gerwien, 1993). The specific host–pathogen relation of *S. scapterisci* to Orthoptera (Nguyen and Smart, 1991) has already been mentioned.

Certainly the specificity would be better understood if insect populations were monitored during surveys to isolate entomopathogenic nematodes. Laboratory bioassays can contribute to understanding specific host–pathogen interactions related to the host defence mechanisms against the nematode–bacterium complex, and host recognition by the nematode. However, the host range should not be defined only by the ability of a certain strain to penetrate and kill an insect under laboratory conditions. Abiotic and biotic environmental factors have an important influence on the host range (Ehlers, 1992) and experiments assessed under outdoor conditions have provided more realistic data on the host range and the specificity of certain nematode species (Georgis and Gaugler, 1991).

Host specificity is related to an impact on non-target species. Our understanding of nematode host specificity is insufficient to judge if nematodes are safe for non-target invertebrates in all habitats. However, entomopathogenic nematodes would be better described as opportunistic rather than general antagonists.

Factors influencing the control potential and persistence

Biotic and abiotic factors that influence the efficacy of entomopathogenic nematodes and their persistence in soil environments have been extensively studied (e.g. Ishibashi and Kondo, 1990;

Kaya, 1990; Kung *et al.*, 1990*a,b*, 1991; Womersley, 1990) and a summary of our current knowledge would certainly be beyond the scope of this paper. Although much information has already been gathered, Gaugler (1988) stated: 'We lack definitive information on the fate of nematodes introduced into the soil, on factors regulating their population dynamics, on optimal conditions for epizootic initiation, and on the ecological barriers to infection'.

The control efficacy of a certain nematode–bacterium complex is defined by its phenotype interacting with the phenotype of the target insect within the limits of environmental conditions (Ehlers, 1992). Some biotic and abiotic factors that influence the control potential are listed in Fig. 11.5. To estimate the environmental impact of a nematode release, the persistence of the DJs in the soil is of major interest. Environmental conditions that favour the control efficacy do not principally correspond with conditions that increase the persistence of the DJs. However, a successful infestation and propagation of the nematodes, either in target or in non-target hosts,

is a prerequisite for prolonged survival and steinernematids and heterorhabditids have been reported to persist for several years in different soil types (Jackson and Wouts, 1987; Kaya, 1990; Rovesti *et al.*, 1991).

Use of entomopathogenic nematodes in integrated pest management

Increasing concerns about the use of chemical insecticides have enhanced further development of alternative control measures. The special conditions that limit the efficacy of chemical control against soil-inhabiting insects (microbial degradation, absorption, insect resistance) and possible environmental risks (i.e. ground-water pollution) favoured the search for biological agents that may be used in soil environments. The natural habitat of entomopathogenic nematodes is the soil and numerous field tests have proved their control potential (e.g. Klein, 1990). Several advantages are associated with the use of nematodes in integrated pest management (IPM): They can be mass-produced with *in vitro* methods in large-

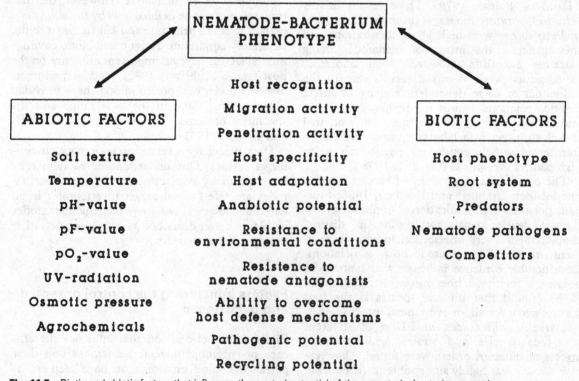

Fig. 11.5 Biotic and abiotic factors that influence the control potential of the nematode–bacterium complex.

scale bioreactors and stored for several months at low temperature. The application is done with available plant-protection equipment. After introduction into the soil they actively seek out their host insect and kill it within a few days. Their control potential is not limited by agrochemicals, thus they can be integrated easily into standard chemical-control practice (e.g. Rovesti *et al.*, 1988; Rovesti and Deseö, 1990; Backhaus, 1991). As for all biocontrol agents, certain environmental conditions have to be considered: a loss of pathogenicity due to ultra-violet (UV) radiation (Gaugler *et al.*, 1992) must be avoided. To guarantee nematode movement, they must be applied to a moist soil surface. Irrigation enhances the penetration of the DJs into the soil. Transport conditions and market distribution systems have yet to be adjusted to the special needs of the DJs (moisture and oxygen supply) to secure maximum survival.

Currently, only a few species are used in IPM. *S. carpocapsae*, *S. feltiae*, *H. bacteriophora* and *Heterorhabditis* sp. North West European group, are commercialized in the US, Australia, Europe and China. Mainly for the control of *O. sulcatus* and sciariid flies approximately 250 ha in Europe were treated with nematodes in 1992. Due to high product costs, the application in the industrialized countries is still limited to high-value crops (mainly ornamentals, tree nurseries, home gardens). This situation will probably change in the near future: with the improvement of solid-state culture techniques for production of *H. bacteriophora* in Australia, and the development of liquid-culture techniques for the production of *S. carpocapsae*, nematodes for treatment of approximately 20 000 ha will be available in 1993. In China, hundreds of hectares have already been treated with *S. carpocapsae*, produced in solid-state cultures, against the peach fruit moth *Carposina nipponensis* (Wang, 1990) and over 100 000 shade trees have been injected with the same nematode to control the cossid *Holcocercus insularis* (Bedding, 1990).

Almost all documented applications of entomopathogenic nematodes represent examples of inundative augmentation, probably due to a lack of information about the antagonistic potential of a naturally occurring population and because the inundative strategy has often been proved to be highly effective. However, there is evidence for the potential of alternative strategies e.g. inoculative release and augmentation by environmental manipulation: Jackson and Wouts (1987) observed an increase in infestation by grass grubs after a single application of *H. bacteriophora* in New Zealand. In the first year after application of 0.5 million DJ/m^2 a relatively low reduction of 9% was observed, whereas in the following year up to 56% mortality, caused by nematode infection, was recorded. Significant antagonistic potential of indigenious steinernematids has been documented. Recently, Raulston *et al.* (1992) found a new *Steinernema* species, parasitizing about 10% of *Spodoptera exigua* and *Helicoverpa zea* in cornfields in the lower Rio Grande Valley. In Austria (Fischer and Führer, 1990), Czechoslovakia (Mráček, 1986) and Southern Germany (Eichhorn, 1988) *S. feltiae* and *S. kraussei* are the most important factors in the mortality of spruce false webworm *Cephalcia abietis* (Hymenoptera, Pamphiliidae). Extreme soil acidification due to acid-air contamination significantly reduced the amount of nematode infested nymphs. Nematode density and infestation rate could be increased by raising the pH above 4.0 after application of lime or magnesium fertilizers (Führer and Fischer, 1991). A successful example of using environmental manipulation to increase the nematode density is reported by Brust (1991). No tillage and the limited presence of weeds in maize reduced the damage caused by southern corn rootworm *Diabrotica undecimpunctata howardi* along with an increase of the population density of *H. bacteriophora*. Examples for a classical biological control approach (permanent establishment) have so far not been reported, except for the release of *S. scapterisci*, isolated in Uruguay, to control mole crickets in Florida. A successful establishment was reported. Five years after release, mean nematode infection of *Scapteriscus borellii* was 35.8% and 7.8% for *S. vicinus* (Parkman and Frank, 1992).

Perspectives and genetic improvement

The overall potential of entomopathogenic nematodes in IPM may still be underestimated. Currently, inundative augmentation is economically feasible in high value crops only. Production and formulation techniques are currently in progress, and will certainly result in a further reduction of the application costs, thus enhancing the spectrum of combatable pest insects. Improvement of formulation techniques is of major concern as application densities of *Heterorhabditis* sp. (HSH) could be reduced to about 200 000 DJs/m^2 only when the majority of DJs survive transport conditions. To avoid loss of energy reserves and guarantee maximum survival of the DJs an immobilization and induction of a quiescent state (anhydrobiosis or cryobiosis) is required. Selection of more virulent nematodes and bacteria along with the isolation of other species will also be a subject for future research.

The commercial use of entomopathogenic nematodes is also limited because of their low tolerance of environmental extremes. Genetic improvements could help to overcome these limitations and have often been applied for biocontrol agents (e.g. Hoy, 1985; and see van Lenteren, Chapter 3). However, improved traits can easily be lost during mass production or after field release (Roush, 1990). In this respect heterorhabditid nematodes possess a major advantage: considering the hermaphroditic propagation of *Heterorhabditis* spp., they are ideal targets for genetic improvement. Selfing over several generations resulted in highly homozygous in-bred lines, which were used to calculate the genetic variability of beneficial traits in the wild-type population of *H. bacteriophora*. High heritability values for several beneficial traits, e.g. host-finding, UV and heat tolerance were found. This suggests that improvement by selection and cross-breeding is possible. Unfortunately, the heritability of the desiccation tolerance was low, thus, mutagenisis may be a better strategy for improvement of this trait (Glazer *et al.*, 1991). Methods for genetic research of *Caenorhabditis elegans* have been successfully tested with *H. bacteriophora* and some mutants are available (Zioni *et al.*, 1992). Entomopathogenic nematodes are closely related to *Caenorhabditis elegans*, the most completely understood eukaryotic animal in terms of molecular and classical genetics, development and anatomy, and it is most probable that many genes already identified in the *C. elegans* genome are shared by nematodes of the genus *Heterorhabditis*. The spectacular results achieved in *C. elegans* genetics can rapidly promote scientific progress in the field of genetic improvement of entomopathogenic nematodes (Fodor *et al.*, 1990).

The same is true for the bacterial symbionts of the nematodes, considering the close phylogenetic relationship between the genus *Xenorhabdus* and *Escherichia coli* (Ehlers *et al.*, 1988). Several *Xenorhabdus* genes have already been cloned in *E. coli* (Frackman and Nealson, 1990). Transformation systems, e.g. the plasmid vector pHK 17, used for genetic studies of *E. coli*, have been adapted in order to study *Xenorhabdus* (Xu *et al.*, 1989). Transposon insertion mutants of *X. nematophilus* were produced and characterized (Xu *et al.*, 1991). These techniques can be used to answer fundamental questions about the biology, physiology, pathogenicity and the symbiotic relationship with the nematode.

Serious problems arise from the instability of *X. luminescens*, which may convert from the wild-type into a so-called secondary form that inhibits the growth of *Heterorhabditis* spp. in *in vitro* cultures (Akhurst, 1980; Ehlers *et al.*, 1990). The mechanisms of this phase shift and the function of the secondary form are poorly understood (Akhurst *et al.*, 1992). Molecular techniques could help to reveal the genetic background of the instability and genetic engineering is a possible approach in order to overcome the adverse influence of secondary-form *Xenorhabdus* spp. on nematode reproduction.

Griffin and Downes (1991) observed a decline in mortality of insect larvae caused by heterorhabditid strains correlated with a decrease in temperature. Although nematodes may still be active at low temperatures and penetrate the insect, mortality is often delayed due to reduced growth of *X. luminescens* in the haemolymph (Ehlers, 1991). Selection or genetic manipulation can help to overcome this limitation and allow field applications in temperate climates early, or late, in the year. Several other biotechnological approaches to improve beneficial traits of the nematode–bacterium complex are possible and will probably be worked on in the near future.

Possible environmental impacts and risks

Negative effects of entomopathogenic nematodes and their symbionts on vertebrates have not been reported (e.g. Gaugler and Boush, 1979; Poinar *et al.*, 1982; Obendorf *et al.*, 1983; Poinar, 1990).

Young tadpoles only were vulnerable when covered with 10000 DJs in a Petri dish (Poinar, 1989). These results could not be repeated when the nematodes were tested under seminatural conditions in an aquarium (Jung, personal communication).

Available information to predict the environmental impact of a release of entomopathogenic nematodes in the field is poor. Consequently, a risk assessment can only be based on the current knowledge of the biology and ecology of the nematode–bacterium complex in combination with previously published studies.

Nematodes of both genera, *Steinernema* and *Heterorhabditis*, are restricted to soil environments and their host spectrum is limited to arthropods, particularly insects. Due to their susceptibility to desiccation and UV radiation a long-term impact on above-ground insects can be discounted and can only occur within a short period after application on the foliage. An impact on aquatic arthropods is unlikely, as penetration of an insect demands a solid support. The DJs rapidly sediment to the ground, and thus cannot infest filter-feeding arthropods, e.g. mosquito larvae (Molta and Hominick, 1989). In water saturated soil, nematode movement is severely restricted due to the lack of surface-tension forces that are necessary for locomotion and penetration (Womersley, 1990). Hence entomopathogenic nematodes of the genera *Steinernema* and *Heterorhabditis* will only have a direct environmental impact on soil-inhabiting arthropods.

A limited mortality of beneficial or non-target insects cannot be excluded since the nematodes are capable of infecting many different host species. Effects can probably be observed when nematodes are used inundatively. An impact of entomopathogenic nematodes on the weed-controlling agent *Hylobius transversovittatus* was assessed under simulated field conditions. Due to the protected habitat of the weevil larvae inside the roots of *Lythrum salicaria*, nematode-caused mortality was of minor importance and would certainly not hamper the establishment of this weevil in North America (Blossey and Ehlers, 1991).

Data about the impact on other non-target insects have been obtained from laboratory bioassays: *H. bacteriophora* (strain NC) and *S. carpocapsae* (strain ALL) had little or no effects on carabid, staphylinid and cicindelid beetles (Georgis & Wojcik, 1987). Pölking and Heimbach (1992) recorded susceptibility of larvae of the carabid *Poecilus cupreus* to *Heterorhabditis* sp. (strain HSH). *S. carpocapsae* adversely affected parasitoids (Diptera and Hymenoptera) of *Pseudaletia unipuncta* (Lepidoptera) either by direct infection of the parasitoid larvae or by causing premature death of the insect host (Kaya, 1978, 1984; Kaya and Hotchkin, 1981).

Changes in non-target insect population density after field application were studied by Rethmeyer (1991). *S. feltiae*, *Heterorhabditis* sp. (HL'81) and *Heterorhabditis* sp. (HDO1) were applied on up to 100 m^2 plots in wheat, an orchard, a forest and a forest edge. Insect and arachnid populations were collected on the treated and control plots with pitfall traps and photoeclectors. No significant impact on the population density of major groups and on the diversity of the species was observed. Special emphasis was given to the changes in the density of coleopteran species: of over 100 different species and genera recorded, the density of only two carabid (*Amara ovata*, *Trechus quadristriatus*), two chrysomelid (*Longitarsus* sp., *Phyllotreta* sp.), one curculionid (*Strophosoma melanogrammum*) and one staphylinid (Aleocharinae) beetle species was significantly lower on the treated plots compared with the control. However, in the orchard a significantly higher number of individuals of the species *Amara ovata*, *A. similata*, *Barypeithes araneiformis*, and *B. trichopterus* was recorded on the plots treated with *S. feltiae* and of *Platynus dorsalis*, *Longitarsus* sp. and *Sitona lineatus* on plots with *Heterorhabditis* sp. (HL'81) when compared with the untreated plots.

These data emphasize the insignificant impact of entomopathogenic nematodes on non-target hosts. The commonly held opinion that entomopathogenic nematodes have a wide host-range is based on the observation of their general insect pathogenicity. However, differences in virulence of specific nematode species to certain hosts are obvious, thus mortality among non-target insects is certainly restricted to a few species and will only affect part of the population.

Competitive interactions with other entomopathogens, insect parasitoids and predators that represent an antagonistic potential to the target insect or other noxious pests must also be taken into consideration. Soil organisms with a significant impact on the abundance of insects can, with only a few exceptions, be assigned to the entomopathogens, e.g. *Bacillus* spp., fungi, viruses and other nematodes. The majority of arthropod parasitoids and predators are mainly active above the soil surface. Competitive interactions between *Beauveria bassiana* (Barbercheck and Kaya, 1991), *Bacillus thuringiensis* and the *Lymantria dispar* nuclear polyhydrosis virus (Bednarek, 1986) with entomopathogenic nematodes were examined. A simultaneous infestation and propagation of both pathogens was seldom observed and a successful reproduction of the different pathogens seems to be a matter of which one comes first. A reduction of the antagonistic potential of other pathogens by entomopathogenic nematodes can certainly be neglected. In contrast, increase in control by synergistic interactions between nematodes and other pathogens was observed. For example, *S. carpocapsae* emigrating from nuclear polyhydrosis virus (NPV) infested *Spodoptera exigua* larvae were able to transmit this virus (Kaya and Burlando, 1989).

An extirpation of a target or non-target insect has not yet been documented and is most unlikely to occur after nematode application. Up to 100% mortality was sometimes observed when 0.5 million *Heterorhabditis* sp. (HSH)/m² were used to control *O. sulcatus* in steamed peat-moss in the glasshouse. In aged potting media the control potential was already reduced (Arndt, personal communication). Hence it is essential to distinguish between results obtained under controlled and field conditions. In natural environments the nematode density is quickly reduced to a naturally occurring level. Ishibashi and Kondo (1986) applied nematodes to untreated bark compost and observed a 90% reduction within one month after application along with an increased density of rhabditid and mononchid nematodes, the latter group resembling nematode predators. Considering the rapid decline of the inundatively augmented nematode density, dramatic effects on target and non-target populations will be of brief duration. Besides, pathogenicity mechanisms of all described nematode species are quite similar due to their symbiotic association with *Xenorhabdus* spp. As these nematodes have been isolated from many different soil habitats and have been reported from all over the world, it can be sup-

posed that every soil inhabiting insect has had contact with entomopathogenic nematodes and has developed defence mechanisms to survive nematode attack. Success of the haemolymph defence mechanisms depends on the number of penetrated DJs (Peters and Ehlers, unpublished). After inundative release the persisting nematodes will therefore probably be tolerated, especially by non-target insects when a preliminary virulence selection has evolved target specificity.

If climatic and physical factors permit the survival of a certain nematode–bacterium complex, a single application (inundative or inoculative) could theoretically result in a successful establishment of a non-endemic or genetically improved strain or species. Generally speaking, negative impacts by introduction of beneficial organisms are either caused by an unusually high susceptibility of a potential host because of absent coevolution or by the absence of antagonists of the introduced species. A dispersal of the beneficial organism beyond the inoculation site is a prerequisite for such effects. Compared to beneficial arthropods, nematodes appear to possess a limited potential for vertical dispersal due to their relatively low motility. However, the possibility of dispersal by infected hosts should not be underestimated. In laboratory studies, *S. carpocapsae* were dispersed up to 11 m by infected adult *Spodoptera exigua* (Timper *et al.*, 1988). *S. scapterisci* was dispersed over distances of about 25 km from the inoculation site 5 years after release (Parkman and Frank, 1992). Hence, an uncontrolled distribution cannot be excluded. But this does not necessarily involve an uncontrolled augmentation in the new environment. Nematode population dynamics are influenced by antagonists, e.g. nematophagous fungi. *Drechmeria coniospora*, *Monacrosporium ellipsosporum*, *Arthrobotrys oligospora* and *Hirsutella rhossiliensis* are known to be antagonists of all nematode groups (Stirling, 1991) and their effect on *Steinernema* and *Heterorhabditis* spp. has also been reported (Poinar and Jansson, 1986*a,b*; Timper and Kaya, 1992). Predators, like soil inhabiting mites (Epsky *et al.*, 1988), collembols, tardigrades, mononchid and dorylaimid nematodes (Kaya, 1990) are known to prey upon entomopathogenic nematodes. All these antagonists are non-specific nematode antagonists and are ubiquitous. When an exotic nematode is released its uncontrolled propagation is not possible as it encounters many antagonists that will limit its population density. The only specific antagonist so far reported is a microsporidian parasite found in *S. glaseri* (Poinar, 1988*b*). Infection with *Pasteuria penetrans*, an important specific antagonist of nematodes, has so far not been reported from entomopathogenic species.

Although inundative release is the major approach to realize the potential of entomopathogenic nematodes, inoculative release may be a useful approach for biological control of insects in low-value crops. In these crops inundative control is excluded for economic reasons. Control rates above 80% can only be achieved with high nematode densities, which cannot be maintained over a longer period by inoculative control. Consequently, this method can only be applied when high pest population densities cause economic losses, and control results of 50% or lower would be sufficient to reduce plant damage. Another advantage of a high economic threshold density is the possible support of the necessary nematode recycling. The inoculative method may be highly effective if the 'new association' strategy (Hokkanen and Pimentel, 1989) is applied. However, this strategy would require that serious impacts on endemic insect populations can be neglected. Entomopathogenic nematodes introduced into a habitat that contains susceptible non-target hosts will presumably have an environmental impact. However, environmental impact and environmental risks have to be distinguished. Concerning the long coevolutionary development of nematode–insect associations and the ubiquitous prevalence of non-specific nematode antagonists, the risk of an introduction of a non-endemic nematode strain is certainly of minor significance compared with the benefits that result from successful biocontrol and substitution of hazardous agrochemicals.

References

Akhurst, R. J. (1980). Morphological and functional dimorphism in *Xenorhabdus* spp., bacteria symbiotically associated with insect pathogenic nematodes *Neoaplectana* and *Heterorhabditis*. *Journal of General Microbiology* 121, 303–309.

Akhurst, R. J. (1982). Antibiotic activity of *Xenorhabdus* spp., bacteria symbiotically associated with insect pathogenic nematodes of the families Heterorhabditidae and Steinernematidae. *Journal of General Microbiology* 128, 3061–3066.

Akhurst, R. J. and Bedding, R. A. (1975). A simple cross-breeding technique to facilitate species determination in the genus *Neoaplectana*. *Nematologica* 24, 328–330.

Akhurst, R. J. and Bedding, R. A. (1986). Natural occurrence of insect pathogenic nematodes (Steinernematidae and Heterorhabditidae) in soil in Australia. *Journal of the Australian Entomological Society* 25, 241–244.

Akhurst, R. J., Smigielski, A. J., Mari, J., Boemare, N. and Mourant, R.G. (1992). Restriction analysis of phase variation in *Xenorhabdus* spp. (Enterobacteriaceae), entomopathogenic bacteria associated with nematodes. *Systematic and Applied Microbiology* 15, 469–473.

Backhaus, G. (1991). Insektenparasitäre Nematoden gegen den Gefurchten Dickmaulrüßler. *Gartenbau und Gartenwesen* 2, 80–85.

Barbercheck, M. E. and Kaya, H. K. (1991). Competitive interactions between entomopathogenic nematodes and *Beauveria bassiana* (Deuteromycotina, Hyphomycetes) in soilborne larvae of *Spodoptera exigua* (Lepidoptera, Noctuidae). *Environmental Entomology* 20, 707–712.

Bedding, R. A. (1981). Low cost in-vitro mass production of *Neoaplectana* and *Heterorhabditis* species (Nematoda) for field control of insect pests. *Nematologica* 27, 109–114.

Bedding, R. A. (1984). Large scale production, storage and transport of the insect-parasitic nematodes *Neoaplectana* spp. and *Heterorhabditis* spp. *Annals of Applied Biology* 104, 117–120.

Bedding, R. A. (1990). Logistics and strategies for introducing entomopathogenic nematode technology into developing countries. In *Entomopathogenic nematodes in biological control* (ed. R. Gaugler and H. K. Kaya), pp. 233–248. CRC Press, Boca Raton, FL.

Bedding, R. A. and Akhurst, R. J. (1975). A simple technique for the detection of insect parasitic nematodes in soil. *Nematologica* 21, 109–110.

Bedding, R. A., Molyneux, A. S. and Akhurst, R. J. (1983). *Heterorhabditis* spp., *Neoaplectana* spp., and *Steinernema kraussei*: Interspecific and intraspecific differences in infectivity for insects. *Experimental Parasitology* 55, 249–257.

Bednarek, A. (1986). Development of the *Steinernema feltiae* (Filipjev) entomogenous nematode (Steinernematidae) in the conditions of occurrence in the insect's body cavity of other pathogens. *Annals of Warsaw Agricultural University – Animal Science* 20, 69–74.

Bednarek, A. and Nowicki, T. (1991). New estimation

method for the density of entomogenous nematodes (Rhabditida: Steinernematidae) in the soil. *Revue de Nématologie* **14**, 638–639.

Bird, A. F. and Akhurst, R. J. (1983). The nature of the intestinal vesicle in nematodes of the family Steinernematidae. *International Journal of Parasitology* **13**, 599–606.

Blossey, B. and Ehlers, R.-U. (1991). Entomopathogenic nematodes (*Heterorhabditis* spp. and *Steinernema anomali*) as potential antagonists of the biological weed control agent *Hylobius transversovittatus* (Coleoptera: Curculionidae). *Journal of Invertebrate Pathology* **58**, 453–454.

Boemare, N. and Akhurst, R. J. (1988). Biochemical and physiological characterization of colony form variants in *Xenorhabdus* spp. (Enterobacteriaceae). *Journal of General Microbiology* **134**, 751–761.

Bovien, P. (1937). Some types of association between nematodes and insects. *Videndskabelige Meddedelser fra Dansk Naturhistorik Forening, Copenhagen* **101**, 1–114.

Bréhelin, M., Drif, L. and Boemare, N. (1990). Depression of defence reactions by Steinernematidae and their associated bacteria. *Proceedings of the Vth International Colloquium on Invertebrate Pathology and Microbial Control, Adelaide, Australia, August 1990*, 213–217.

Brust, G. E. (1991). Augmentation of an endemic entomogenous nematode by agroecosystem manipulation for the control of a soil pest. *Agriculture Ecosystems and Environments* **36**, 175–184.

Curran, J. (1989). Chromosome number of *Steinernema* and *Heterorhabditis* species. *Revue de Nématologie* **12**, 145–148.

Curran, J. and Webster, J. M. (1989). Genotypic analysis of *Heterorhabditis* isolates from North Carolina. *Journal of Nematology* **21**, 140–145.

Deseö, K. V., Ruggeri, L. and Lazzari, G. (1990). Mass-production and quality control of entomopathogenic nematodes in *Galleria mellonella* L. larvae. *Proceedings of the Vth International Colloquium on Invertebrate Pathology and Microbial Control, Adelaide, Australia, August 1990*, p. 250.

Dix, I., Burnell, A. M., Griffin, C. T., Joyce, S. A. and Nugent, M. J. (1992). The identification of biological species in the genus *Heterorhabditis* (Nematoda: Heterorhabditidae) by crossbreeding second generation amphimictic adults. *Parasitology* **104**, 509–518.

Doucet, M. M. A. (1986). A new species of *Neoaplectana* Steiner 1929 (Nematoda: Steinernematidae) from Cordoba, Argentina. *Revue de Nématologie* **9**, 317–323.

Doucet, M. M. A. and Doucet, M. E. (1990). *Steinernema ritteri* n. sp. (Nematoda: Steinernematidae) with a key to the species of the genus. *Nematologica* **36**, 257–265.

Ehlers, R.-U. (1990). Einsatz der Biotechnologie zur Produktion insektenparasitärer Nematoden für die biologische Bekämpfung von Schadinsekten. *Vorträge zur Hochschultagung der Agrarwissenschaftlichen Fakultät der Christian-Albrechts-Universität zu Kiel* **72**, 207–215.

Ehlers, R.-U. (1991). Wie kalt darf's sein – Nematoden gegen Dickmaulrüssler. *Deutsche Baumschule, April 1991*, 158–160.

Ehlers, R.-U. (1992). Environmental and biotic factors influencing the control potential of entomopathogenic nematodes of the genus *Steinernema* and *Heterorhabditis*. In *Nematology from Molecule to Ecosystem* (ed. P. W. T. Maas and E. J. Gommers), pp. 201–212. European Society of Nematologists, Dundee.

Ehlers, R.-U. and Gerwien, A. (1993). Selection of entomopathogenic nematodes (Steinernematidae and Heterorhabditidae, Nematoda) for the biological control of cranefly larvae *Tipula paludosa* (Tipulidae, Diptera). *Zeitschrift für Pflanzenkrankheiten und Pflanzenschutz* **100**, 343–353.

Ehlers, R.-U., Wyss, U. and Stackebrandt, E. (1988). 16S rRNA cataloguing and the phylogenetic position of the genus *Xenorhabdus*. *Systematic and Applied Microbiology* **10**, 121–125.

Ehlers, R.-U., Stoessel, S. and Wyss, U. (1990). The influence of phase variants of *Xenorhabdus* spp. and *Escherichia coli* (Enterobacteriaceae) on the propagation of entomopathogenic nematodes of the genera *Steinernema* and *Heterorhabditis*. *Revue de Nématologie* **13**, 417–424.

Ehlers, R.-U., Deseö, K. V. and Stackebrandt, E. (1991). Identification of *Steinernema* spp. (Nematoda) and their symbiotic bacteria *Xenorhabdus* spp. from Italian and German soils. *Nematologica* **37**, 360–364.

Ehlers, R.-U., Osterfeld, K. H., Krasomil-Osterfeld, K. and Lunau, S. (1992). Propagation of entomopathogenic nematodes in bioreactors. *Mitteilungen der Biologischen Bundesanstalt* **283**, 376.

Eichhorn, O. (1988). Untersuchungen über die Fichtengespinstblattwespen *Cephalcia* spp. Panz. (Hymenoptera, Pamphiliidae) II. Die Larven- und Nymphenparasiten. *Journal of Applied Entomology* **105**, 105–140.

Endo, B. Y. and Nickle, W. R. (1991). Ultrastructure of the intestinal epithelium, lumen, and associated bacteria in *Heterorhabditis bacteriophora*. *Journal of the Helminthological Society of Washington* **58**, 202–212.

Epsky, N. D., Walter, D. E. and Capinera, J. L. (1988). Potential role of microarthropods as biotic mortality factors of entomogenous nematodes (Rhabditidae: Steinernematidae,

Heterorhabditidae). *Journal of Economical Entomology* 81, 821–825.

Fan, X. and Hominick, W. M. (1991). Efficiency of the *Galleria* (wax moth) baiting technique for recovering infective stages of entomopathogenic rhabditids (Steinernematidae and Heterorhabditidae) from sand and soil. *Revue de Nématologie* 14, 381–387.

Filipjev, I. N. (1934). Eine neue Art der Gattung *Neoaplectana* Steiner nebst Bemerkungen über die systematische Stellung der letzteren. Miscellanea Nematologica, *Magasin de Parasitologie del'Académie des Sciences de l'URSS* 4, 229–240.

Fischer, P. and Führer, E. (1990). Effect of soil acidity on the entomophilic nematode *Steinernema kraussei* Steiner. *Biology and Fertility of Soils* 9, 174–177.

Fodor, A., Vecseri, G. and Farkas, T. (1990). *Caenorhabditis elegans* as a model for the study of entomopathogenic nematodes. In *Entomopathogenic Nematodes in Biological Control* (ed. R. Gaugler and H. K. Kaya), pp. 249–270. CRC Press, Boca Raton, FL.

Frackman, S. and Nealson, K. H. (1990). The molecular genetics of *Xenorhabdus*. In *Entomopathogenic Nematodes in Biological Control* (ed. R. Gaugler, and H. K. Kaya), pp. 301–318. CRC Press, Boca Raton.

Friedman, M. J. (1990). Commercial Production and Development. In *Entomopathogenic Nematodes in Biological Control* (ed. R. Gaugler and H. K. Kaya), pp. 153–172. CRC Press, Boca Raton.

Führer, E. and Fischer, P. (1991). Towards integrated control of *Cephalcia abietis* L. as a defoliator of the Norway spruce in Central Europe. *Forest Ecology and Management* 39, 87–95.

Gaugler, R. (1988). Ecological considerations in the biological control of soil-inhabiting insects with entomopathogenic nematodes. *Agriculture Ecosystems and Environment* 24, 351–360.

Gaugler, R. and Boush, G. M. (1979). Nonsusceptibility of rats to the entomogenous nematode *Neoplectana carpocapsae*. *Environmental Entomology* 8, 658–660.

Gaugler, R., Bednarek, A. and Campbell, J. F. (1992). Ultraviolet inactivation of heterorhabditid and steinernematid nematodes. *Journal of Invertebrate Pathology* 59, 155–160.

Georgis, R. (1990). Formulation and application technology. In *Entomopathogenic Nematodes in Biological Control* (ed. R. Gaugler and H. K. Kaya), pp. 173–191. CRC Press, Boca Raton, FL.

Georgis, R. and Gaugler, R. (1991). Predictability in biological control using entomopathogenic nematodes. *Journal of Economic Entomology* 84, 713–720.

Georgis, R. and Wojcik, W. (1987). The effect of entomogenous nematodes *Heterorhabditis heliothidis* and *Steinernema feltiae* on selected predatory soil insects. *Journal of Nematology* 19, 524.

Glaser, R. W. (1931). The cultivation of a nematode parasite of an insect. *Science* 73, 614–615.

Glaser, R. W. and Farell, C. C. (1935). Field experiments with the Japanese beetle and its nematode parasite. *Journal of the New York Entomological Society* 43, 345–354.

Glazer, I., Gaugler, R. and Segal, D. (1991). Genetics of the entomopathogenic nematode *Heterorhabditis bacteriophora* (strain HP 88): The Diversity of beneficial traits. *Journal of Nematology*, 23, 324–333.

Götz, P., Boman, A. and Boman, H. G. (1981). Interactions between insect immunity and an insect-pathogenic nematode with symbiotic bacteria. *Proceedings of the Royal Society of London* 212, 333–350.

Griffin, C. T. and Downes, M. J. (1991). Low temperature activity in *Heterorhabditis*, sp. (Nematoda: Heterorhabditidae). *Nematologica* 37, 83–91.

Hokkanen, H. M. T. and Pimentel, D. (1989). New associations in biological control: theory and practice. *Canadian Entomology* 121, 829–840.

Hoy, M. A. (1985). The potential of genetic improvement of predators for pest management programs. In *Genetics in Relation to Insect Management* (ed. M. A. Hoy and J. J. McCelvey Jr.), pp. 106–115. Rockefeller Press, New York.

Ishibashi, N. and Kondo, E. (1986). *Steinernema feltiae* (DD-136) and *S. glaseri*: Persistence in soil and bark compost and their influence on native nematodes. *Journal of Nematology* 18, 310–316.

Ishibashi, N. and Kondo, E. (1990). Behavior of infective juveniles. In *Entomopathogenic Nematodes in Biological Control* (ed. R. Gaugler and H. K. Kaya), pp. 139–150. CRC Press, Boca Raton, FL.

Jackson, T. A. and Wouts, W. M. (1987). Delayed action of an entomophagous nematode *Heterorhabditis* sp. (V16) for grass grub control. *Proceedings of the 40th New Zealand Weed and Pest Control Conference*, 33–35.

Kaya, H. K. (1978). Interaction between *Neoaplectana carpocapsae* (Nematoda: Steinernematidae) and *Apanteles militaris* (Hymenoptera: Braconidae), a parasitoid of the armyworm, *Pseudaletia unipuncta*. *Journal of Invertebrate Pathology* 31, 358–364.

Kaya, H. K. (1984). Effect of the entomogenous nematode *Neoaplectana carpocapsae* on the tachinid parasite *Compsilura concinnata* (Diptera: Tachinidae). *Journal of Nematology* 16, 9–13.

Kaya, H. K. (1990). Soil ecology. In *Entomopathogeniic Nematodes in Biological Control* (ed. R. Gaugler and H. K. Kaya), pp. 93–115. CRC Press, Boca Raton, FL.

Kaya, H. K. and Burlando, T. M. (1989). Development of *Steinernema feltiae* (Rhabditida: Steinernematidae) in diseased insect hosts. *Journal of Invertebrate Pathology* **53**, 164–168.

Kaya, H. K. and Hotchkin, P. G. (1981). The nematode *Neoaplectana carpocapsae* Weiser and its effect on selected ichneumonid and braconid parasites. *Environmental Entomology* **10**, 474–478.

Klein, M. G. (1990). Efficacy against soil-inhabiting insect pests. In *Entomopathogenic Nematodes in Biological Control,* (ed. R. Gaugler and H. K. Kaya), pp. 195–214, CRC, Press, Boca Raton, FL.

Kozodoi, E. M. (1984). A new entomophagous nematode *Neoaplectana anomali* sp.n.(Rhabditida, Steinernematidae) and its biology. *Zoologicheskii Zhurnal* **63**, 1605–1909.

Kung, S. P., Gaugler, R. and Kaya, H. K. (1990*a*). Soil type and entomopathogenic nematode persistence. *Journal of Invertebrate Pathology* **55**, 401–406.

Kung, S. P., Gaugler, R. and Kaya, H. K. (1990*b*). Influence of soil pH and oxygen on persistence of *Steinernema* spp. *Journal of Nematology* **22**, 440–445.

Kung, S. P., Gaugler, R. and Kaya, H. K. (1991). Effect of soil temperature, moisture, and relative humidity on entomopathogenic nematode persistence. *Journal of Invertebrate Pathology* **57**, 242–249.

Liu, J. (1992). Taxonomic study of the genus: *Steinernema* Travassos and *Heterorhabditis* Poinar. *Proceedings of the XIX International Congress of Entomology, Beijing, China,* 28 June – 4 July 1992, p. 318.

Lunau, S., Stoessel, S., Schmidit-Peisker, A. J. and Ehlers, R.-U. (1993). Establishment of monoxenic inocula for scaling up *in vitro* cultures of the entomopathogenic nematodes *Steinernema* spp. and *Heterorhabditis* spp. *Nematologica* **39**, 385–399.

Mamiya, Y. (1988). *Steinernema kushidai* n.sp. (Nematoda: Steinernematidae) associated with scarabaeid beetle larvae from Shizuoka, Japan. *Applied Entomology and Zoology* **23**, 313–320.

Molta, N. B. and Hominick, W. M. (1989). Dose- and time- response assessments of *Heterorhabditis heliothidis* and *Steinernema feltiae* (Nem.: Rhabditida) against *Aedes aegypti* larvae. *Entomophaga* **34**, 485–493.

Mráček, Z. (1977). *Steinernema kraussei,* a parasite of the body cavity of the sawfly, *Cephaleia abietes,* in Czechoslovakia. *Journal of Invertebrate Pathology* **30**, 87–94.

Mráček, Z. (1986). Nematodes and other factors controlling the sawfly *Cephaleia abietes* (Pamphiliidae, Hymenoptera) in Czechoslovakia. *Forest Ecology and Management* **15**, 75–79.

Mráček, Z. and Bednarek, A. (1991). The morphology of lateral fields of infective juveniles of entomogenous nematodes of the family Steinernematidae (Rhabditida). *Nematologica* **37**, 63–71.

Mráček, Z., Weiser, J., Bures, M. and Kahounova, L. (1992). *Steinernema kraussei* (Steiner, 1923) (Nematoda, Rhabditida) – Rediscovery of its type locality in Germany. *Folia Parasitologica* **39**, 181–182.

Nguyen, K. B. and Smart, G. C. Jr (1990). *Steinernema scapterisci* n. sp. (Rhabditida: Steinernematidae). *Journal of Nematology* **23**, 187–199.

Nguyen, K. B. and Smart, G. C. (1991). Pathogenicity of *Steinernema scapterisci* to selected invertebrates. *Journal of Nematology* **23**, 7–11.

Obendorf, D. L., Peel, B., Akhurst, R. J. and Miller, L. A. (1983). Non-susceptibility of mammals to the entomopathogenic bacterium *Xenorhabdus nematophilus.* *Environmental Entomology* **12**, 368–370.

Parkman, J. P. and Frank, J. H. (1992). Infection of sound-trapped mole crickets, *Scaptericus* spp. by *Steinernema scapterisci.* *Florida Entomologist* **75**, 163–165.

Paul, V. J., Frautschy, S., Fenical, W. and Nealson, K. H. (1981). Antibiotics in microbial ecology. Isolation and structure assign ment of several new antibacterial compounds from the insect-symbiotic bacteria *Xenorhabdus* spp. *Journal of Chemical Ecology* **7**, 589–597.

Petersen, J. J. and Cupello, J. M. (1981). Commercial development and future prospects for entomogenous nematodes. *Journal of Nematology* **13**, 280–284.

Poinar, G. O. Jr (1975). Description and biology of a new parasitic rhabditoid *Heterorhabditis bacteriophora* n.gen., n.sp. *Nematologica* **21**, 463–470.

Poinar, G. O. Jr (1979). *Nematodes for Biological Control of Insects.* CRC Press, Boca Raton, FL.

Poinar, G. O. Jr (1985). *Neoaplectana intermedia* n.sp. (Steinernematidae: Nematoda) from South Carolina. *Revue de Nématologie* **8**, 321–327.

Poinar, G. O. Jr (1986). Entomophagous nematodes. In *Biological Plant and Health Protection* (ed. J. M. Franz), pp. 95–121. G. Fischer Verlag, Stuttgart.

Poinar, G. O. Jr (1988*a*). Redescription of *Neoaplectana affinis* Bovien (Rhabditida: Steinernematidae). *Revue de Nématologie* **11**, 143–147.

Poinar, G. O. Jr (1988*b*). A microsporidian parasite of *Neoaplectana glaseri* (Steinernematidae: Rhabditida). *Revue de Nématologie* **11**, 359–361.

Poinar, G. O. Jr (1989). Non-insect hosts for the entomogenous rhabditoid nematodes *Neoaplectana* (Steinernematidae) and *Heterorhabditis* (Heterorhabditidae). *Revue de Nématologie* **12**, 423–428.

Poinar, G. O. Jr (1990). Biology and taxonomy of Steinernematidae and Heterorhabditidae. In *Entomopathogenic Nematodes in Biological Control* (ed. R. Gaugler and H. K. Kaya), pp. 23–61. CRC Press, Boca Raton, FL.

Poinar, G. O. Jr and Jansson, H. B. (1986a). Susceptibility of *Neoaplectana* spp. and *Heterorhabditis heliothidis* to the endoparasitic fungus *Drechmeria coniospora*. *Journal of Nematology* 18, 225–230.

Poinar, G. O. Jr and Jansson, H. B. (1986b). Infection of *Neoaplectana* and *Heterorhabditis* (Rhabditida: Nematoda) with the predatory fungi, *Monacrosporium ellipsosporum* and *Arthrobotrys oligospora* (Monilialis: Deuteromycetes). *Revue de Nématologie* 9, 241–244.

Poinar, G. O. Jr and Lindhardt, K. (1971). The re-isolation of *Neoaplectana bibionis* Bovien (Nematodea) from Danish bibionids (Diptera) and their possible use as biological control agents. *Entomologica Scandinavica* 2, 301–303.

Poinar, G. O. Jr and Thomas, G. M. (1966). Significance of *Achromobacter nematophilus* Poinar and Thomas (Achromobacteriaceae: Eubacteriales) in the development of the nematode, DD-136 (*Neoaplectana* sp., Steinernematidae). *Parasitology* 56, 385–390.

Poinar, G. O. Jr, Thomas, G. M., Presser, S. B. and Hardy, J. L. (1982). Inoculation of entomogenous nematodes, *Neoaplectana* and *Heterorhabditis*, and their associated bacteria, *Xenorhabdus* spp., into chicks and mice. *Environmental Entomology* 11, 137–138.

Poinar, G. O. Jr, Jackson, T. and Klein, M. (1987). *Heterorhabditis megidis* sp.n. (Heterorhabditidae: Rhabditida), parasitic in the Japanese beetle, *Popillia japonica* (Scarabaeidae: Coleoptera), in Ohio. *Proceedings of the Helminthological Society of Washington* 54, 53–59.

Poinar, G. O. Jr, Mráček, Z. and Doucet, M. M. A. (1988). A re-examination of *Neoaplectana rara* Doucet, 1986 (Steinernematidae: Rhabditida). *Revue de Nematologie* 11, 447–449.

Poinar, G. O. Jr, Karunakar, G. K. and David, H. (1992). *Heterorhabditis indicus* n.sp. (Rhabditida : Nematoda) from India: separation of *Heterorhabditis* spp. by infective juveniles. *Fundamental and Applied Nematology* 15, 467–472.

Pölking, A. and Heimbach, U. (1992). Effect of some biological control agents on two predatory beetles in laboratory experiments. *Mitteilungen der Biologischen Bundesanstalt* 283, 378.

Raulston, J. R., Pair, S. D., Loera, J. and Cabanillas, H. E. (1992). Prepupal and pupal parasitism of *Helicoverpa zea* and *Spodoptera frugiperda* (Lepidoptera: Noctuidae) by *Steinernema* sp. in cornfields in the lower Rio Grande Valley. *Journal of Economic Entomology* 85, 1666–1670.

Rethmeyer, U. (1991). Auswirkungen eines Einsatzes entomopathogener Nematoden auf die Bodenfauna verschiedener Biotope. Ph.D. thesis, Technische Hochschule, Darmstadt, Germany.

Roush, R. T. (1990). Genetic variation in natural enemies: critical issues for colonization in biological control. In *Critical Issues in Biological Control* (ed. M. Mackauer, L. E. Ehler and J. Roland), pp. 263–288. Intercept, Andover, Hants.

Rovesti, L. and Deseö, K. V. (1990). Compability of chemical pesticides with the entomopathogenic nematode *Steinernema carpocapsae* Weiser and *S. feltiae* Filipjev (Nematoda: Steinernematidae). *Nematologica* 36, 237–245.

Rovesti, L., Heinzpeter, E. W., Tagliente, F. and Deseö, K. V. (1988). Compatibility of pesticides with the entomopathogenic nematode *Heterorhabditis bacteriophora* Poinar (Nematoda: Heterorhabditidae). *Nematologica* 34, 462–476.

Rovesti, L., Heinzpeter, E. W. and Deseö, K. V. (1991). Distribution and persistence of *Steinernema* spp. and *Heterorhabditis* spp. (Nematodes) under different field conditions. *Anzeiger für Schädlingskunde, Pflanzenschutz und Umweltschutz* 64, 18–22.

Schmidt, T. M., Bleakley, B. and Nealson, K. H. (1988). Characterization of an extracellular protease from the insect pathogen *Xenorhabdus luminescens*. *Applied and Environmental Microbiology* 54, 2793–2797.

Shen, C. P. (1992). Description of a entomopathogenic nematode, *Steinernema longicaudum* sp. nov. and its application. *Proceedings of the XIX International Congress of Entomology, Beijing, China, 28 June–4 July, 1992*, p. 318.

Smits, P. H., Groenen, J. T. M. and De Raay, G. (1991). Characterization of *Heterorhabditis* isolates using DNA restriction fragment length polymorphism. *Revue de Nématology* 14, 445–453.

Steiner, G. (1923). *Aplectana kraussei* n. sp., eine in der Blattwespe *Lyda sp.* parasitierende Nematodenform. *Zentralblatt für Bakteriologie, Abteilung II* 59, 14–18.

Steiner, G. (1929). *Neoaplectana glaseri*, n. g., n. sp. (Oxyuridae), a new nemic parasite of the Japanese beetle (*Popillia japonica*, NEWM.). *Journal of the Washington Academy of Science* 19, 436–440.

Stirling, G. R. (1991). *Biological Control of Plant Parasitic Nematodes*. CAB International, Wallingford.

Strauch, O., Stoessel, S. and Ehlers, R.-U. (1994). Culture conditions define automictic or amphimictic reproduction in entomopathogenic rhabditid nematodes of the genus *Heterorhabditis*. *Fundamental and Applied Nematology* 17, 575–582.

Sturhan, D. (1990). Studies on the distribution of entomoparasitic nematodes in the Federal Republic of Germany. Mitteilungen der Biologischen Bundesanstalt, 266, 453.

Tallòsi, B., Peters, A. and Ehlèrs, R.-U. (submitted). Steinernema bicornutum sp.n. (Rhabditida: Nematoda) from Vojvodina, Yugoslavia. Russian Journal of Nematology.

Timper, P. and Kaya, H. K. (1992). Impact of a nematode-parasitic fungus on the effectiveness of entomopathogenic nematodes. Journal of Nematology 24, 1–8.

Timper, P., Kaya, H. K. and Gaugler, R. (1988). Dispersal of the entomogenous nematode Steinernema feltiae (Rhabditida: Steinernematidae) by infected adult insects. Environmental Entomology 17, 546–550.

von Reibnitz, C. and Ehlers, R.-U. (1992). Cost estimation for the production of entomopathogenic nematodes applied in biological plant protection. Mitteilungen der Biologische Bundesanstalt 283, 377.

Wang, J. (1990). Use of nematode Steinernema carpocapsae to control the major apple pest Carposina nipponensis in China. Proceedings of the Vth International Colloquium on Invertebrate Pathology and Microbial Control, Adelaide, Australia, August 1990, p. 392.

Weiser, J. (1955). Neoaplectana carpocapsae, n.sp. (Anguillulata: Steinernematinae), novy cizopasnik housenek obalece jablecného, Carpocapsa pomonella

L. Vestnik Ceskoslovenske Zoologické Spolecnosti 19, 44–51.

Womersley, C. Z. (1990). Dehydration survival and anhydrobiotic potential. In Entomopathogenic Nematodes in Biological Control (ed. R. Gaugler and H. K. Kaya), pp. 117–137. CRC Press, Boca Raton, FL.

Wouts, W. M. (1979). The biology and life cycle of a New Zealand population of Heterorhabditis heliothidis (Heterorhabditidae). Nematologica 25, 192–202.

Xu, J. M., Lohrke, S., Hurlbert, I. M. and Hurlbert, R. E. (1989). Transformation of Xenorhabdus nematophilus. Applied and Environmental Microbiology 55, 806–812.

Xu, J. M., Olson, M. E., Khan, M. L. and Hurlbert, R. E. (1991). Characterization of Tn5-induced mutants of Xenorhabdus nematophilus ATCC 19061. Applied and Environmental Microbiology 57, 1173–1180.

Yokoo, S., Tojo, S. and Ishibashi, N. (1992). Suppression of the prophenoloxidase cascade in the larval haemolymph of the turnip moth, Agrotis segetum by an entomopathogenic nematode, Steinernema carpocapsae and its symbiotic bacterium. Journal of Insect Physiology 38, 915–924.

Zioni, S., Glazer, I. and Segal, D. (1992). Phenotypic and genetic analysis of a mutant of Heterorhabditis bacteriophora strain HP88. Journal of Nematology 24, 359–364.

12

Pseudomonads as biocontrol agents of diseases caused by soil-borne pathogens

Geneviève Défago and Christoph Keel

Introduction

Progress in agronomy over the last 30 years has allowed world food production to almost double. In part, this was possible through the control of diseases caused by air-borne pathogens. However, the damage caused by soil-borne pathogens has increased during the same period; this has resulted in important yield reduction in protected and unprotected crops. To date, satisfactory methods of control have not been developed. Few plant varieties are resistant to soil-borne pathogens. Unfortunately, crop rotation is often not possible for economic reasons. Pesticides are seldom effective in soil and growing concern for the environment restricts their use. Biological control of root diseases can be considered as an alternative for the future. The pseudomonads are currently receiving attention world-wide because some strains, added at a few grams or milligrams per hectare, effectively increase crop yield and improve plant health. These beneficial effects are well documented, however the mechanisms involved are only partially understood. A better understanding of these mechanisms is crucial to safety if pseudomonads are to be released on a large scale. Data on safety are scarce because the investigation of potential risks associated with the release of pseudomonads has just begun. This review will mainly focus on the results obtained with strain CHA0 of *Pseudomonas fluorescens*.

Beneficial effect of resident populations of pseudomonads

Resident populations of pseudomonads are part of the natural disease suppressiveness of soils and of suppressiveness induced by monoculture or solarization (Smiley, 1979; Scher and Baker, 1980; Rovira and Wildermuth, 1981; Cook and Weller, 1987; Défago and Haas, 1990; Gamliel and Katan, 1991). It is important to realize that natural suppressiveness occurs only in distinct areas. It is accepted that interactions between populations of resident pseudomonads and some abiotic or biotic factors account for this site specificity. For example, soils which are naturally suppressive to black root rot of tobacco, a disease caused by *Thielaviopsis basicola*, were found scattered in a region of 32 km² at Morens (Switzerland) and populations of fluorescent pseudomonads tightly bound to roots were shown to be involved in disease suppression (Gasser and Défago, 1981; Stutz *et al.*, 1986; Défago and Haas, 1990). The suppressive soils are derived from a weathered ground moraine that contains native vermiculitic clay minerals. In contrast, some neighbouring soils, which are derived from a weathered molasse and contain secondary illitic clay minerals, are conducive to black root rot (Stutz *et al.*, 1985, 1989). Various *Pseudomonas* strains were isolated from the suppressive soils. One of them, *P. fluorescens* strain CHA0, was studied in detail. This strain induces the suppression of black root rot of

137

tobacco when added to soil derived from a geological origin similar to that of the natural suppressive soil and fails to suppress disease in a
conducive soil adjacent to the natural suppressive
soil of Morens (Stutz et al., 1989). Two reasons
may explain this failure: (i) the poor survival of
strain CHA0 in the conducive soil of Morens
(Stutz et al., 1989) and (ii) the incapacity of strain
CHA0 to get enough iron out of clays of the
conducive soil to synthesize HCN (Keel et al.,
1989). Hydrogen cyanide plays an important role
in the suppression of black root rot of tobacco by
strain CHA0 (Voisard et al., 1989). Thus, the site
specific behaviour of strain CHA0 is in agreement
with the localization of the natural suppressiveness in distinct areas.

Beneficial effect of added pseudomonads

The application of plant-beneficial *Pseudomonas*
strains to seeds, plants or soil can significantly
improve crop yield in greenhouse and field experiments. This beneficial effect is amply documented
(e.g. Schroth and Hancock, 1982; Burr and Caesar, 1984; Schippers et al., 1987; Davison, 1988;
Kloepper et al., 1988, 1989; Schippers, 1988;
Weller, 1988; Défago and Haas, 1990; Rovira et
al., 1992; Weller et al., Chapter 13). This effect is
due to:

1. the suppression of diseases or deleterious effects
   caused by micro-organisms;
2. a beneficial effect on other biocontrol agents or on
   mycorrhiza;
3. an increased availability of mineral nutrients;
4. a direct growth promotion.

Suppression of diseases and of deleterious effects

The addition of pseudomonads to a crop can
increase the yield. This beneficial effect is either
due to the suppression of well-known diseases
caused by major soil-borne pathogens or to the
suppression of deleterious effects of minor pathogens (micro-organisms which reduce plant growth
but without any disease symptoms being obvious)
(see review Défago and Haas, 1990). One *Pseudomonas* strain may suppress more than one disease
and protect more than one crop. A few strains,

when added to the soil, give systemic protection to
the leaves and stems against pathogens (Alström,
1991; van Peer et al., 1991; Wei et al., 1991). For
example, strain CHA0 suppresses diseases caused
by *Gaeumannomyces graminis* on wheat, by *Thielaviopsis basicola* on tobacco, by *Fusarium oxysporum*
spp. on tomato and by *Pythium ultimum* on various
crops. Strain CHA0 is also able to protect systemically. While this strain colonizes only the root
system of tobacco plants, it reduces the number
and the diameter of necroses caused by the
tobacco necrosis virus (TNV) on leaves (Maurhofer, unpublished results).

Other beneficial effects

The beneficial effect of mycorrhiza and of fungi
used as biocontrol agents can be increased by the
addition of pseudomonads (e.g. Park et al., 1988;
Duponnois and Garbaye, 1991; Fuchs and
Défago, 1991; Lemanceau and Alabouvette,
1991; Duponnois, 1992). For example, tomatoes
growing in the soil of commercial greenhouses or
in hydroponic culture are better protected against
wilt disease with a mixture of non-pathogenic
Fusarium and pseudomonads than with the non-
pathogenic *Fusarium* alone (Lemanceau et al.,
1992; Fuchs and Défago, 1991).

In some cases pseudomonads promote plant
growth in the absence of pathogenic micro-
organisms. An increased availability of mineral
nutrients or the production of plant-growth regulators may be responsible for this increase
(Kloepper et al., 1981; Lifshitz et al., 1987; Pietr et
al., 1991).

Inconsistency of the beneficial effect

Added *Pseudomonas* strains may consistently
reduce disease severity and increase plant yields
under laboratory conditions and in experimental
greenhouses. However, in commercial greenhouses and under field conditions, i.e. in different
locations and/or different years, the performance
of the same biocontrol strain tends to be variable
(Weller, 1988; Kloepper et al., 1989). Many
reasons can account for this inconsistency. Two of
them are:

1. the genetic instability of the pseudomonads during

Table 12.1. Extracellular metabolites and enzymes produced by *Pseudomonas fluorescens* strain CHA0

Metabolites or enzymes	Properties	References
2,4-Diacetylphloroglucinol[a]	Antibiotic, herbicide	Keel *et al.* (1990, 1991, 1992); Maurhofer *et al.* (1992)
Indoleacetate	Growth hormone	Oberhänsli *et al.* (1991)
HCN	Biocide	Ahl *et al.* (1986); Voisard *et al.* (1989); Keel *et al.* (1989)
Lipase		Sacherer, unpublished
Monoacetylphloroglucinol	Antibiotic	Défago *et al.* (1990)
Proteases		Sacherer, unpublished
Pyoluteorin[a]	Antibiotic, herbicide	Défago *et al.* (1990)
Pyrrolnitrin	Antibiotic	Unpublished
Pyoverdine	Siderophore	Keel *et al.* (1989)
Salicylate	Inducer of resistance	Meyer *et al.* (1992)

[a]These metabolites were detected in the rhizosphere of wheat colonized by strain CHA0 (Maurhofer *et al.*, 1991; Keel *et al.*, 1992).

maintenance in the laboratory and during large-scale production;

2. the influence of environmental factors on the bacterial metabolites and the mechanisms involved in disease suppression.

A better understanding of the genetic instability and of the mechanisms involved in disease suppression might allow the consistency of the beneficial effect of added pseudomonads to be increased.

Genetic instability

Pseudomonads are maintained and mass produced on laboratory media. Like other soil microorganisms (e.g. *Streptomyces* spp.; Birch *et al.*, 1990), they are able to modify their genetic constitution so that they are better adapted for growth under such artificial conditions than in their natural habitat. Any mutation is likely to affect the saprophytic activity of the strain in the soil environment and the effectiveness to control disease under field conditions (Baker and Défago, 1988; Weller, 1988; Dutrecq *et al.*, 1991). Strain CHA0 is genetically unstable when stored in nutrient yeast broth for several days. The spontaneous mutants often have pleiotropic defects. They have lost the ability to produce tryptophan side chain

oxidase (TSO), several antibiotics and extracellular enzymes and to suppress black root rot of tobacco (Oberhänsli and Défago, 1991; Sacherer, unpublished results). Spontaneous mutants have not been isolated from plants growing in field trials (Oberhänsli and Défago, 1991). Correct handling of strain CHA0 during laboratory maintenance minimizes the danger of spontaneous mutations. However, these mutations can occur during mass production, e.g. for field application.

Bacterial metabolites and mechanisms involved in disease suppression

It is generally assumed that beneficial pseudomonads colonize plant roots and use root exudates to synthesize metabolites, which allow them to suppress disease. A *Pseudomonas* strain can produce a wide array of secondary metabolites, extracellular enzymes and other active substances on laboratory media (Table 12.1; Leisinger and Margraff, 1979; Kiprianova and Smirnov, 1981; Défago and Haas, 1990). These compounds can be chemically characterized and their role in disease suppression can be assessed by testing biocontrol activity of non-producing mutants. When a mutation results in a reduction of disease suppression, the relevant genomic fragment can be cloned to demonstrate complementation. However, it is necessary to check whether the loss of production of a metabolite allows the bacteria to over-produce another metabolite; this over-production will impede the genetic analysis.

Several genes that are part of the suppressive ability of the *Pseudomonas* were identified, some were cloned and sequenced (Table 12.2; Thomashow and Weller, 1988; Toyoda *et al.*, 1988; Voisard *et al.*, 1989; Vincent *et al.*, 1991; Fenton *et al.*, 1992; Keel *et al.*, 1992; Laville *et al.*, 1992; and see Dowling *et al.*, Chapter 14). It should be kept in mind that more than one metabolite may operate to suppress a given disease and that suppression of different plant diseases by one particular strain of *Pseudomonas* may involve different metabolites. These metabolites protect roots probably through the following mechanisms: root colonization; antibiosis; competition for iron; degradation of pathogenicity, and germination factors of the pathogen; plant growth-promotion; induction of plant defence mechanisms.

Root colonization

The ability of pseudomonads to colonize the root for an extended period is a prerequisite for biocontrol activity (Weller, 1988; Parke, 1990; Bull *et al.*, 1991; Lugtenberg *et al.*, 1991). It is not clear if an effective biocontrol agent needs to colonize the whole root system intensively, in the sense of being present at the sites where the infection takes place, or if it can be only present at some sites and trigger systemic resistance in the rest of the root system. Often, beneficial root-associated *Pseudomonas* increase in number in response to the diseases they suppress; if the suppression is ineffective, the number may be depressed (Weller, 1988; Mazzola and Cook, 1991). Strain CHA0 colonizes wheat roots up to a depth of 1 m and survives in soil for several months (Défago *et al.*, 1987, 1990). The strain was found not only on the root surface but also in the root cortex of tobacco near small wounds, as shown by indirect immuno-fluorescence staining (Défago *et al.*, 1990; Berling, 1991). Thus, it is not yet clear if strain CHA0 is truly saprophytic or if under certain conditions it behaves as an endophytic bacteria or as a weak pathogen able to induce resistance in the plant.

The molecules involved in root colonization and survival of pseudomonads in soil are not well-

known. Production of the antibiotic phenazine-1-carboxylic acid (PCA) by *P. fluorescens* strain 2–79 contributes to the bacterium's long-term survival in soil (Mazzola *et al.*, 1992). The production of a specific siderophore and the utilization of a broad spectrum of siderophores are part of the ability of a *P. putida* strain to grow on roots in a soil with low iron availability (Bakker *et al.*, 1986). Many other characteristics may influence root colonization (see review by Weller, 1988; and see Weller *et al.*, Chapter 13) however, their molecular basis is not yet understood.

The ability to colonize the root is not sufficient for a *Pseudomonas* strain to be an effective biocontrol agent.

Antibiosis

Many but not all *Pseudomonas* strains that suppress diseases, inhibit the growth of the target pathogen *in vitro*. Several antibiotic compounds were isolated from the growth media of the bacteria. The compounds of known chemical structure are listed in Défago and Haas (1990). The application of molecular genetics and biochemical techniques have proved the role of oomycin A, phenazines (PCA), HCN and 2,4-diacetyl-phloroglucinol (Phl) in disease suppression (Fravel, 1988; Thomashow and Weller, 1988; Homma and Suzui, 1989; Voisard *et al.*, 1989; Défago and Haas, 1990; Gutterson, 1990; Keel *et al.*, 1990; Haas *et al.*, 1991; Howie and Suslow, 1991; Shanahan *et al.*, 1992). This latter compound is very promising. Phl-producing *Pseudomonas* strains are among the best biocontrol strains collected in Europe and in America. Regions of chromosomal DNA involved in Phl synthesis have been cloned and DNA probes are available (Vincent *et al.*, 1991; Fenton *et al.*, 1992; Keel *et al.*, 1992). Furthermore, a gene, i.e. *gacA*, which regulates the production of several metabolites including Phl has been identified, cloned and sequenced (Laville *et al.*, 1992). In addition, a cosmid was isolated, which carried a 22 kb insert of CHA0 DNA and enhanced, in a CHA0 background, the production of Phl and Plt *in vitro* and in the rhizosphere of wheat (Haas *et al.*, 1991; Maurhofer *et al.*, 1991, 1992). The recombinant strain provides improved protection for cucumber against *Pythium* and *Fusarium* but is deleterious

Table 12.2 Effect of structural and regulatory genes on disease suppression by *Pseudomonas fluorescens* CHA0 in different plant – pathogen systems under gnotobiotic conditions

Gene(s)/Plasmid	Gene products and regulatory effects involved	Plant	Pathogen	Effect[a]	Reference
hcn	Hydrogen cyanide	Flax	*Fusarium oxysporum* f. sp. *lini*	None	Keel, unpublished
		Cress	*Pythium ultimum*	+	Keel, unpublished
		Corn	*P. ultimum*	None	Keel, unpublished
		Corn	*Rhizoctonia solani*	None	Keel, unpublished
		Cucumber	*P. ultimum*	+	Keel, unpublished
		Cucumber	*R. solani*	None	Keel, unpublished
		Tobacco	*Thielaviopsis basicola*	+	Haas *et al.* (1991), Voisard *et al.* (1989)
		Wheat	*Gaeumannomyces graminis* var. *tritici*	None	Haas *et al.* (1991), Wüthrich and Défago (1991)
		Wheat	*P. ultimum*	None	Wüthrich and Défago (1991)
		Wheat	*R. solani*	None	Keel, unpublished
phl	2,4-diacetylphloroglucinol	Tobacco	*T. basicola*	+	Keel *et al.* (1990), (1992)
		Wheat	*G. graminis* var. *tritici*	+	Keel *et al.* (1991), (1992)
plt	Pyoluteorin	Cress	*P. ultimum*	+	Maurhofer, unpublished
		Cucumber	*P. ultimum*	None	Maurhofer, unpublished
		Wheat	*P. ultimum*	None	Maurhofer, unpublished
pvd	Pyoverdine	Tobacco	*T. basicola*	None	Keel *et al.* (1989)
		Wheat	*G. graminis* var. *tritici*	None	Keel and Défago (1991)
		Wheat	*P. ultimum*	None	Keel and Défago (1991)
tso	Tryptophan side chain oxidase, indoleacetate	Tobacco	*T. basicola*	None	Oberhänsli *et al.* (1991)
		Wheat	*G. graminis* var. *tritici*	None	Oberhänsli *et al.* (1991)
gac	Global antibiotic and cyanide control	Tobacco	*T. basicola*	+	Laville *et al.* (1992)
		Cucumber	*P. ultimum*	+	Laville *et al.* (1992), Sacherer, unpublished
		Wheat	*G. graminis* var. *tritici*	None	Laville *et al.* (1992)
		Wheat	*P. ultimum*	None	Laville *et al.* (1992), Sacherer, unpublished
		Corn	*P. ultimum*	None	Sacherer, unpublished
anr	Anaerobic growth and cyanide control	Tobacco	*T. basicola*	+	Keel and Haas, unpublished
pME3090	Antibiotic overproduction	Corn	*P. ultimum*	Toxic	Maurhofer *et al.* (1992)
		Corn	*R. solani*	Toxic	Maurhofer *et al.* (1992)
		Cress	*P. ultimum*	Toxic	Maurhofer *et al.* (1992)
		Cucumber	*P. ultimum*	Improved	Maurhofer *et al.* (1992)
		Cucumber	*Phomopsis sclerotioides*	Improved	Maurhofer, unpublished
		Cucumber	*F. o.* f. sp. *cucurbitaceae*	Improved	Maurhofer, unpublished
		Tobacco	*T. basicola*	Toxic	Maurhofer *et al.* (1991)
		Wheat	*G. graminis* var. *tritici*	None	Maurhofer, unpublished
		Wheat	*P. ultimum*	None	Maurhofer, unpublished

[a] Effect on disease suppression: none, not involved in disease suppression; +, involved in disease suppression; improved, improvement of disease suppression; toxic, deleterious to plant growth.

for cress, tobacco and corn (Haas *et al.*, 1991; Maurhofer *et al.*, 1991, 1992).

It is noteworthy that HCN and phloroglucinol derivatives with antifungal activity have been found in certain plant species and may be involved in a biochemical defence mechanism against fungi (Mansfield, 1983; Tomás-Lorente *et al.*, 1989). Therefore, it seems possible that beneficial pseudomonads mimic resistance mechanisms of the plants.

Some *Pseudomonas* strains produce extracellular enzymes including chitinase and laminarase, which may inhibit the pathogen (Lim *et al.*, 1991).

Several antibiotic compounds (e.g. HCN, PCA, Phl) are not only toxic to the target pathogen but also to other micro-organisms and to plants (see review Défago and Haas, 1990). The healthy plants whose roots are colonized by a beneficial *Pseudomonas* strain show in most cases no reduction of yield. This indicates that only a small amount of antibiotic compounds is necessary to protect the plant and that these compounds do not accumulate in active form in the soil (Thomashow *et al.*, 1990).

Siderophore-mediated competition for iron

When grown under iron-limited conditions fluorescent pseudomonads produce fluorescent, yellow-green siderophores (pyoverdine or pseudobactin), which function as high-affinity Fe^{3+} chelators. According to the siderophore hypothesis, pseudomonads suppress disease by sequestering the limited supply of iron in the rhizosphere and thereby limiting the availability of the iron necessary for the growth of the pathogens (Schroth and Hancock, 1982; Leong, 1986; Kloepper *et al.*, 1988; Weller, 1988). The evidence for pyoverdine being associated with biocontrol of soil-borne pathogens has recently been reviewed by Loper and Buyer (1991). In our studies on strain CHA0, we did not find a role for iron competition mediated by pyoverdine in the suppression of black root rot and take-all (Keel *et al.*, 1989; Défago *et al.*, 1990; Keel and Défago, 1991). Several recent reports indicate that bacterial metabolites other than siderophores, i.e. antibiotics, have a key role in disease suppression, (Keel *et al.*, 1989; Thomashow and Weller, 1990; Haas *et al.*, 1991; Hamdan *et al.*, 1991).

Competition for iron and antibiotics may play a role not only in disease suppression but also in rhizosphere competence. In contrast, the mechanisms listed below probably influence only the ability of the pseudomonads to suppress disease and have no role in ecological competence.

Degradation of pathogen toxins and germination factors

Toyada *et al.*, (1988) selected a non-pathogenic mutant of *P. solanacearum* that was able to detoxify fusaric acid, the wilt toxin of *Fusarium oxysporum* f. sp. *lycopersici*. This mutant protected tomato plants from *Fusarium* wilt disease when introduced by wounding the xylem internally. Extracellular proteases of fluorescent *Pseudomonas* strains may inactivate hydrolases and phytotoxins of *Fusarium* spp. (Borowicz *et al.*, 1992; Pietr, 1991).

Pythium survives in soil in the form of oospores, which germinate and infect the plant. The germination is induced by ethanol and other volatile factors of roots and seeds. Some *Pseudomonas* strains, which protect plants against *Pythium* diseases, metabolize the ethanol and inhibit the germination of the oospores (Paulitz, 1991).

Plant growth-promoting compounds

Strain CHA0 and some other beneficial pseudomonads produce indole-3-acetic acid from tryptophan *in vitro*. However, an involvement of this compound in disease suppression has not been demonstrated (Kloepper *et al.*, 1989; Oberhänsli *et al.*, 1991).

Induction of plant defence mechanisms

While colonizing only the root system, some *Pseudomonas* strains reduce the incidence of disease caused by pathogens on stems and leaves (Alström, 1991; van Peer *et al.*, 1991; Wei *et al.*, 1991). Protected leaves and stems show physiological changes (van Peer *et al.*, 1991); some of them are typical for systemic induced resistance (Zdor and Anderson, 1992). It is not clear yet if systemic induced resistance plays a role in the protection of roots against disease.

The bacterial metabolites that are involved in protection of roots against disease are not necessarily identical with the metabolites involved in systemically induced resistance. For example, a *gacA* mutant of strain CHA0, which has lost the ability to protect tobacco roots against black root rot disease, still reduces the incidence of the virus TNV on leaves to the same extent as the wild-type strain. The protected leaves are not colonized by the bacteria. They show typical physiological changes associated with systemic induced resistance: a higher activity of chitinase and β-glucanase and the presence of the PR1 proteins (Maurhofer, unpublished results).

Regulation of *Pseudomonas* metabolites involved in disease suppression

In vitro, the production of antibiotics and other secondary metabolites is influenced by the quality and the quantity of the available nutrients and the presence of oxygen. For example, glucose was found to induce the production of oomycin but to suppress the synthesis of PCA and Phl (James and Gutterson, 1986; Gutterson, 1990; Thomashow and Pierson III, 1991; Shanahan *et al.*, 1992). Iron deficiency induces the production of pyoverdine, but iron sufficiency is necessary for HCN synthesis. Thus, environmental conditions that prevail in the rhizosphere may be crucial for the effectiveness of *Pseudomonas* strains. The influence of these conditions on the production of bacterial metabolites explains, in part, the site-specific activity of the pseudomonads. For example, regulation of cyanogenesis by iron accounts, in part, for the fact that strain CHA0 effectively suppresses black root rot of tobacco in some soils but not in others (see text above). On the other hand, the regulation of cyanogenesis and anaerobic growth by the *anr* gene in *P. fluorescens* strain CHA0, explains why structural *hcn* genes are more important for disease suppression in waterlogged soil than in well-aerated soil (Haas and Keel, unpublished results).

Several isolates of pseudomonads produce pyoverdine, HCN and Phl but they do not suppress diseases (Schippers, 1988; our unpublished results). We hypothesize that the regulation of these metabolites is a key factor in disease suppression and we suggest that compounds from the root exudates may function as signals for these regulatory genes. A similar type of communication

between plants and bacteria is known for *Rhizobium* spp. and *Agrobacterium tumefaciens*.

We have discovered that mutations in a *P. fluorescens* gene named *gacA* (for global antibiotic and cyanide control) pleiotropically block the production of the secondary metabolites Phl HCN, and Plt; in addition, tryptophan side chain oxidase (TSO) activity is suppressed. The *gacA* mutants of strain CHA0 have a drastically reduced ability to suppress black root rot under gnotobiotic conditions. Nucleotide sequence data indicate that the *gacA* gene belongs to a two-component regulatory system in bacteria and that *gacA*-like genes appear to be present in other beneficial and non-beneficial pseudomonads (Laville *et al.*, 1992 and unpublished results).

Environmental impact

When *Pseudomonas* strains are applied to soil, seeds or planting material, sufficient cell numbers must be used to obtain a beneficial effect. These numbers are relatively high, for example about 10^{11} to 10^{15} cfu of strain CHA0 are needed per hectare at the time of planting (Défago *et al.*, 1987 and 1990; Wüthrich and Défago, 1991). The bacteria introduced multiply later in the root system. What happens next is not clear. Are the bacteria disseminated to other econiches and ecosystems (e.g. upper parts of the plants, neighbouring fields, ground water), do these microbes establish themselves and modify the target and the newly occupied niches in a desirable or undesirable way?

The following facts, which are in part described in the text above, may help to answer these questions.

1. The activity of resident and added populations of pseudomonads is contained in distinct areas.
2. *Pseudomonas* strains with obvious deleterious effects on the target crop, or on crops used in a standard rotation, will be eliminated during the screening procedure for field application. However, this procedure does not eliminate strains that have a deleterious effect due to continuous or repeated use of inoculants over several years.
3. The compounds involved in disease suppression (e.g. HCN, Phl) are toxic not only to the target pathogen but also to other micro-organisms. However, it seems possible to select strains which

increase the beneficial effect of other biocontrol agents.

4. *Rhizobium* inoculants have been used for many years on a large scale without negative effects on the environment.
5. *Rhizobium* strains can be exchanged freely between countries and continents.
6. In contrast to *Rhizobium*, fluorescent pseudomonads can be found nearly everywhere, in water, air, soil, food and plants. Some strains are suppressive, most are not. Structural genes important for disease suppression, e.g. genes for HCN and Phl can be found in both types of strains.

Experimental data are needed to develop protocols that will allow evaluation of the environmental impact of added pseudomonads.

First, we wanted to examine if HCN and Phl, which are part of the suppressive ability of strain CHA0, give it an advantage outside the rhizosphere. We used a *gacA* mutant of strain CHA0, which lacks the capacity to produce HCN, Phl and several other extracellular metabolites, and which has a drastically reduced ability to suppress black root rot. We marked the wild-type CHA0 and the *gacA* mutant with resistance to rifampicin, to facilitate the isolation of these strains from the environment. The *gacA* mutation did not influence the survival of *P. fluorescens* in autoclaved soil; in contrast, in natural soil the number of cells of the *gacA* mutant dropped from $10^{6.7}$ to $10^{5.2}$ cfu/g of soil whereas the wildtype cell numbers decreased from $10^{6.7}$ to $10^{6.1}$ in 1 month. Both strains were transported by percolation water in natural, undisturbed soil cores. The cell numbers of the *gacA* mutant recovered from percolation water were about one log unit lower than those of the wild-type. This indicates that the ecological competence is slightly reduced by the *gacA* mutation (Natsch and Pfirter, unpublished results).

Suppressive fluorescent pseudomonads that produce Phl have been isolated at different places in Europe and in the US. Hybridization with probes that contain *phl* genes indicates that some bacteria isolated from both sides of the Atlantic may have a highly similar restriction pattern in *phl* gene regions (Keel, Thomashow, Weller, and Cook unpublished results). This similarity suggests that these organisms are not fundamentally different. We can see no reason *a priori* why such organisms should not be tested in a country other than the one in which they were isolated originally.

Perspectives

Beneficial pseudomonads have the real potential to maintain and restore soil fertility by controlling soil-borne pathogens; they have no adverse effect on humans and animals. The development and implementation of *Pseudomonas* inoculants in agricultural practice needs: (i) improvement of strains by classical or genetic engineering methods to increase the consistency of their effect in field; (ii) a better knowledge of the ecological impact of the added pseudomonads in order to increase their effect and to assess the risks of their deliberate release; (iii) a set of protocols and regulations for release approval based on real risks; (iv) protocols for exchange of wild-type and genetically engineered strains between countries; and (v) a production and commercialization system adapted to living organisms (Cook, 1992). If these points are realized there is no doubt that pseudomonads and other biocontrol micro-organisms can reduce the damage caused by soil-borne pathogens and so contribute to a significant step forward in the agriculture of the twenty-first century.

Acknowledgments

We thank D. Haas, C. T. Bull and M. S. Wolfe for critical reading of part of the manuscript and U. Rosenberger for secretarial assistance. The research of our group was supported mainly by the Swiss National Foundation for Scientific Research (project 31–32473.91) and the Swiss Federal Institute of Technology. The collaboration between C. Keel, L. S. Thomashow, D. M. Weller, and R. J. Cook was possible through a grant of the OECD to C. Keel.

References

Alström, S. (1991). Induction of disease resistance in common bean susceptible to halo blight bacterial pathogen after seed bacterization with rhizosphere pseudomonads. *Journal of General Microbiology* **37**, 495–501.

Ahl, P., Voisard, C. and Défago, G. (1986). Iron bound-siderophores, cyanic acid, and antibiotics involved in suppression of *Thielaviopsis basicola* by a

Pseudomonas fluorescens strain. *Journal of Phytopathology* **116**, 121–134.

Baker, R. and Défago, G. (1988). Aspects of growth promotion resulting from biological control. *Proceedings of the 1st International Workshop on Plant Growth- Promoting Rhizobacteria* (ed. J. Kloepper), pp. 17–22. Allelix Press, Toronto.

Bakker, P. A. H. M., Weisbeek, P. J. and Schippers, B. (1986). The role of siderophores in plant growth stimulation by fluorescent *Pseudomonas* spp. *Mededelingen van de Faculteit Landbouwwetenschappen Rijksuniversiteit Gent* **51**, 1357–1362.

Berling, C.H. (1991). Anwendung der lmmunofluoreszenz-Mikroskopie zur Lokalisation eines krankheitsunterdrückenden *Pseudomonas fluorescens* Stammes auf Tabakwurzeln. Swiss Federal Institute of Technology, Zürich. Ph.D. thesis. No. 9474.

Birch, A., Häusler, A. and Hütter, R. (1990). Genome rearrangement and genetic instability in *Streptomyces* spp. *Journal of Bacteriology*. **172**, 4138–4142.

Borowicz, J. J., Stanislaw, J. P., Stankiewicz, M., Lewicka, T. and Zukowska, Z. (1992). Inhibition of fungal cellulase pectinase and xylanase activity by plant growth-promoting fluorescent *Pseudomonas* spp. In *New Approaches in Biological Control of Soil- Borne Diseases* (ed. D. F. Jensen, J. Hockenhull and N. J. Fokkèma). *International Organisation for Biological Control of Noxious and Plants/West Palaearctic Regional Section Bulletin* **XV/1**, 103–107.

Bull, C. T., Weller, D. M. and Thomashow, L. S. (1991). Relationship between root colonization and suppression of *Gaeumannomyces graminis* var. *tritici* by *Pseudomonas fluorescens* strain 2–79. *Phytopathology* **81**, 954–959.

Burr, T. J. and Caesar, A. (1984). Beneficial plant bacteria. *CRC Critical Review in Plant Science* **2**, 1–20.

Cook, R. J. (1992). Reflections of a regulated biological control researcher. In *Regulations and Guidelines: Critical Issues in Biological Control. Proceedings of a USDA/CSRS National Workshop, 10–12 June 1991, Vienna, Virginia, USA* (ed. R. Charudattan and H. W. Browning), pp. 9–24. Institute of Food and Agricultural Sciences, University of Florida, Gainesville, FL.

Cook, R. J. and Weller, D. M. (1987). Management of take-all in consecutive crops of wheat or barley. In *Innovative Approaches to Plant Disease Control* (ed. L. Chet), pp. 41–76. John Wiley & Sons, Inc., New York.

Davison, J. (1988). Plant beneficial bacteria. *Bio/ Technology* **6**, 282–286.

Défago, G. and Haas, D. (1990). Pseudomonads as antagonists of soilborne plant pathogens: modes of action and genetic analysis. In *Soil Biochemistry* **6**

(ed. J. M. Bollag and G. Stotzky), ch. 5, pp. 249– 291. Marcel Dekker Inc., New York.

Défago, G., Berling, C. H., Henggeler, S., Hungerbühler, W., Kern, H., Schleppi, P., Stutz, E. W. and Zürrer, M. (1987). Survie d'un *Pseudomonas fluorescens* dans le sol et protection du blé contre des maladies d'origine fongique. *Schweizerische Landwirtschaftliche Forschung* **26**, 155–160.

Défago, G., Berling, C. H., Burger, U., Haas, D., Kahr, G., Keel, C., Voisard, C., Wirthner, P. and Wüthrich, B. (1990). Suppression of black root rot of tobacco and other root diseases by strains of *Pseudomonas fluorescens:* potential applications and mechanisms. In *Biological Control of Soil-Borne Plant Pathogens* (ed. D. Hornby, R. J. Cook, Y. Henis, W. H. Ko, A. D. Rovira, B. Schippers and P. R. Scott), ch. 7. pp. 93–108. CAB International, Ascot, Berks.

Duponnois, R. (1992). Les bactéries auxiliaires de la mycorhization du Douglas (*Pseudotsuga menziesii* (Mirb.) Franco) par *Laccaria laccata* souche S238. Ph.D. thesis, Université de Nancy, France.

Duponnois, R. and Garbaye, J. (1991). Mycorrhization helper bacteria associated with the Douglas fir-*Laccaria laccata* symbiosis: effects in aseptic and in glasshouse conditions. *Annales des Sciences Forestières* **48**, 239–251.

Dutrecq, A., Debras, P., Stevaux, J. and Marlier, M. (1991). Activity of 2,4-diacetylphloroglucinol isolated from a strain of *Pseudomonas fluorescens* to *Gaeumannomyces graminis* var. *tritici*. In *Biotic Interactions and Soil-Borne Diseases* (ed. A. B. R. Beemster, G. J. Bollen, M. Gerlagh, M. A. Ruissen, B. Schippers and A. Tempel), pp. 252– 257. Elsevier Science Publishers, Amsterdam.

Fenton, A. M., Stephens, P. M., Crowley, J., O'Callaghan, M. and O'Gara, F. (1992). Exploitation of gene(s) involved in 2,4-diacetylphloroglucinol biosynthesis to confer a new biocontrol capability to a *Pseudomonas* strain. *Applied and Environmental Microbiology* **58**, 3873–3878.

Fravel, D. R. (1988). Role of antibiosis in the biocontrol of plant diseases. *Annual Review of Phytopathology* **26**, 75–91.

Fuchs, J. and Défago, G. (1991). Protection of tomatoes against *Fusarium oxysporum* f. sp. *lycopersici* by combining an apathogenic *Fusarium* with different bacteria in non-sterilized soil. In *Plant Growth-Promoting Rhizobacteria – Progress and Prospects* (ed. C. Keel, B. Koller and G. Défago). *International Organisation for Biological Control of Noxious and Plants/West Palaearctic Regional Section Bulletin* **XIV/8**, 51–56.

Gamliel, A. and Katan, J. (1991). Involvement of

fluorescent pseudomonads and other microorganisms in increased growth response of plants in solarized soils. *Phytopathology* **81**, 494–502.

Gasser, R. and Défago, G. (1981). Mise en évidence de la résistance de certaines terres à la pourriture noire des racines du tabac causée par le *Thielaviopsis basicola*. *Botanica Helvetica* **91**, 75–80.

Gutterson, N. (1990). Microbial fungicides: recent approaches to elucidating mechanisms. *Critical Review in Biotechnology* **10**, 69–91.

Haas, D., Keel, C., Laville, J., Maurhofer, M., Oberhänsli, T., Schnider, U., Voisard, C., Wüthrich, B. and Défago, G. (1991). Secondary metabolites of *Pseudomonas fluorescens* strain CHA0 involved in the suppression of root diseases. In *Advances in Molecular Genetics of Plant–Microbe Interactions, vol. I* (ed. H. Hennecke and D. P. S. Verma), pp. 450–456. Kluwer Academic Publishers, Dordrecht, The Netherlands.

Hamdan, H., Weller, D. M. and Thomashow, L. S. (1991). Relative importance of fluorescent siderophores and other factors in biological control of *Gaeumannomyces graminis* var. *tritici* by *Pseudomonas fluorescens* strains 2–79 and M4–80R. *Applied and Environmental Microbiology* **57**, 3270–3277.

Homma, Y. and Suzui, T. (1989). Role of antibiotic production in suppression of radish damping-off by seed bacterization with *Pseudomonas cepacia*. *Annals of the Phytopathological Society of Japan* **55**, 643–652.

Howie, W. J. and Suslow, T. V. (1991). Role of antibiotic biosynthesis in the inhibition of *Pythium ultimum* in the cotton spermosphere and rhizosphere by *Pseudomonas fluorescens*. *Molecular Plant–Microbe Interactions* **4**, 393–399.

James Jr., D. W. and Gutterson, N. I. (1986). Multiple antibiotics produced by *Pseudomonas fluorescens* HV37a and their differential regulation by glucose. *Applied and Environmental Microbiology* **52**, 1183–1189.

Keel, C. (1992). Bacterial antagonists of plant pathogens in the rhizosphere: mechanisms and prospects. In *New Approaches in Biological Control of Soil-Borne Diseases* (ed. D. F. Jensen, J. Hockenhull and N. J. Fokkema). *International Organisation for Biological Control of Noxious and Plants/West Palaearctic Regional Section Bulletin*. **XV/1**, 93–99.

Keel, C. and Défago, G. (1991). The fluorescent siderophore of *Pseudomonas fluorescens* strain CHA0 has no effect on the suppression of root diseases of wheat. In *Plant Growth-Promoting Rhizobacteria – Progress and Prospects* (ed. C. Keel, B. Koller and G. Défago). *International Organisation for Biological Control of Noxious and Plants/West Palaearctic Regional Section Bulletin* **XIV/8**, 136–142.

Keel, C., Voisard, C., Berling, C. H., Kahr, G. and Défago, G. (1989). Iron sufficiency, a prerequisite for suppression of tobacco black root rot by *Pseudomonas fluorescens* strain CHA0 under gnotobiotic conditions. *Phytopathology* **79**, 584–589.

Keel, C., Wirthner, Ph., Oberhänsli, Th., Voisard, C., Burger, U., Haas, D. and Défago, G. (1990). Pseudomonads as antagonists of plant pathogens in the rhizosphere: role of the antibiotic 2,4-diacetylphloroglucinol in the suppression of black root rot of tobacco. *Symbiosis* **9**, 327–341.

Keel, C., Maurhofer, M., Oberhänsli, Th., Voisard, C., Haas, D. and Défago, G. (1991). Role of 2,4-diacetylphloroglucinol in the suppression of take-all of wheat by a strain of *Pseudomonas fluorescens*. In *Biotic Interactions and Soil-Borne Diseases* (ed. A. B. R. Beemster, G. J. Bollen, M. Gerlagh, M. A. Ruissen, B. Schippers and A. Tempel), pp. 335–338. Elsevier Science Publishers, Amsterdam.

Keel, C., Schnider, U., Maurhofer, M., Voisard, C., Burger, U., Wirthner, Ph., Haas, D. and Défago, G. (1992). Suppression of root diseases by *Pseudomonas fluorescens* CHA0: importance of the bacterial secondary metabolite 2,4-diacetylphloroglucinol. *Molecular Plant–Microbe Interactions* **5**, 4–13.

Kiprianova, E. A. and Smirnov, V. V. (1981). *Pseudomonas fluorescens*, a producer of antibiotic compounds. *Antibiotiki* **26**, 135–143.

Kloepper, J. W. and Schroth, M. N. (1981). Plant growth-promoting rhizobacteria and plant growth under gnotobiotic conditions. *Phytopathology* **71**, 642–644.

Kloepper, J. W., Lifshitz, R. and Schroth, M. N. (1988). Pseudomonas inoculants to benefit plant production. *ISI Atlas of Science – Animal and Plant Sciences*, pp. 60–64.

Kloepper, J. W., Lifshitz, R. and Zablotowicz, R. M. (1989). Free living bacterial inocula for enhancing crop productivity. *Trends in Biotechnology* **7**, 39–44.

Laville J., Voisard, C., Keel, C., Maurhofer, M., Défago, G. and Haas, D. (1992). Global, stationary-phase control in *Pseudomonas fluorescens* mediating antibiotic synthesis and suppression of black root rot of tobacco. *Proceedings of the National Academy of Sciences, USA* **89**, 1562–1566.

Leisinger, T. and Margraff, R. (1979). Secondary metabolites of the fluorescent pseudomonads. *Microbiological Reviews* **43**, 422–442.

Lemanceau, P. and Alabouvette, C. (1991). Biological control of fusarium disease by fluorescent *Pseudomonas* and non-pathogenic *Fusarium*. *Crop Protection* **10**, 279–286.

Lemanceau, Ph., Bakker, P. A. H. M., De Kogel, W. J., Alabouvette, C. and Schippers, B. (1992). Effect of Pseudobactin 358 production by

Pseudomonas putida WCS358 on suppression of *Fusarium* wilt of carnations by nonpathogenic *Fusarium oxysporum* Fo47. *Applied and Environmental Microbiology* 58, 2978–2982.

Leong, J. (1986). Siderophores: their biochemistry and possible role in the biocontrol of plant pathogens. *Annual Review of Phytopathology* 24, 187–209.

Lifshitz, R., Kloepper, J. W. and Kozlowski, W. (1987). Growth promotion of canola (rapeseed) seedlings by a strain of *Pseudomonas putida* under gnotobiotic conditions. *Canadian Journal of Microbiology* 33, 390–395.

Lim, H.-S., Kim, Y.-S. and Kim, S.-D. (1991). *Pseudomonas stutzeri* YPL-1 genetic transformation an antifungal mechanisms against *Fusarium solani*, and agent of plant root rot. *Applied and Environmental Microbiology* 57, 510–516.

Loper, J. E. and Buyer, J. S. (1991). Siderophores in microbial interactions on plant surfaces. *Molecular Plant–Microbe Interactions* 4, 5–13.

Lugtenberg, B. J. J., de Weger, L. A. and Bennett, J. W. (1991). Microbial stimulation of plant growth and protection from disease. *Current Opinion in Biotechnology* 2, 457–464.

Mansfield, J. W. (1983). Antimicrobial compounds. In *Biochemical Plant Pathology* (ed. J. A. Callow), pp. 237–265. John Wiley & Sons, Chichester.

Maurhofer, M., Keel, C., Schnider, U., Voisard, C., Haas, D. and Défago, G. (1991). Does enhanced antibiotic production in *Pseudomonas fluorescens* strain CHA0 improve its disease suppressive capacity? In *Plant Growth-Promoting Rhizobacteria – Progress and Prospects* (ed. C. Keel, B. Koller and G. Défago), pp. 201–202. *International Organisation for Biological Control of Noxious and Plants/West Palaearctic Regional Section Bulletin*, **XIV/8**.

Maurhofer, M., Keel, C., Schnider, U., Voisard, C., Haas, D. and Défago, G. (1992). Influence of enhanced antibiotic production in *Pseudomonas fluorescens* strain CHA0 on its disease suppressive capacity. *Phytopathology* 82, 190–195.

Mazzola, M. and Cook, R. J. (1991). Effects of fungal root pathogens on the population dynamics of biocontrol strains of fluorescent pseudomonads in the wheat rhizosphere. *Applied and Environmental Microbiology* 57, 2171–2178.

Mazzola, M., Cook, R. J., Thomashow, L. S., Weller, D. M. and Pierson III, L. S. (1992). Contribution of phenazine antibiotic biosynthesis to the ecological competence of fluorescent pseudomonads in soil habitats. *Applied and Environmental Microbiology* 58, 2616–2624.

Meyer, J.-M., Azelvandre, P. and Georges, C. (1992). Iron metabolism in *Pseudomonas*: salicylic acid, a siderophore of *Pseudomonas fluorescens* CHA0. *BioFactors* 4, 23–27.

Oberhänsli, Th. and Défago, G. (1991). Spontaneous loss of tryptophan side chain oxidase of *Pseudomonas fluorescens* strain CHA0 – a marker for genetic instability. In *Plant Growth-Promoting Rhizobacteria – Progress and Prospects* (ed. C. Keel, B. Koller and G. Défago). *International Organisation for Biological Control of Noxious and Plants/West Palaearctic Regional Section Bulletin* **XIV/8**, 392–398.

Oberhänsli, Th., Défago, G. and Haas, D. (1991). Indole-3-acetic acid (IAA) synthesis in the biocontrol strain CHA0 of *Pseudomonas fluorescens*: role of tryptophan side chain oxidase. *Journal of General Microbiology* 137, 2273–2276.

Park, C. S., Paulitz, T. C. and Baker, R. (1988). Biocontrol of fusarium wilt of cucumber resulting from interactions between *Pseudomonas putida* and non-pathogenic isolates of *Fusarium oxysporum*. *Phytopathology* 78, 194–199.

Parke, J. L. (1990). Root colonization by indigenous and introduced microorganisms. In *The Rhizosphere and Plant Growth* (ed. D. L. Kleister and P. B. Gregan), pp. 33–42. Kluwer Academic Publishers, Dordrecht, The Netherlands.

Paulitz, T. C. (1991). Effect of *Pseudomonas putida* on the stimulation of *Pythium ultimum* by seed volatiles of pea and soybean. *Phytopathology* 81, 1282–1287.

Pietr, S. J. (1991). Inhibition of culture filtrate toxicity and hydrolase activity of phytopathogenic *Fusarium* spp. by plant growth-promoting *Pseudomonas fluorescens*. In *Plant Growth-Promoting Rhizobacteria – Progress and Prospects* (ed. C. Keel, B. Koller and G. Défago). *International Organisation for Biological Control of Noxious and Plants/West Palaearctic Regional Section Bulletin* **XIV/8**, 147–151.

Pietr, S. J., Karon, B. and Stankiewicz, M. (1991). Influence of rock phosphate-dissolving rhizobacteria on the growth and P-uptake by cereals: preliminary results. In *Plant Growth-Promoting Rhizobacteria – Progress and Prospects* (ed. C. Keel, B. Koller and G. Défago). *International Organisation for Biological Control of Noxious and Plants/West Palaearctic Regional Section Bulletin* **XIV/8**, 81–84.

Rovira, A. D. and Wildermuth, G.B. (1981). The nature and mechanisms of suppression. In *Biology and Control of Take-all* (ed. M. J. Asher and P. J. Shipton), pp. 385–415. Academic Press, New York.

Rovira, A. D., Ryder, M. and Harris, A. (1992). Biological control of root diseases with pseudomonads. In *Biological Control of Plant Diseases – Progress and Challenges for the Future* (ed. E. C. Tjamos, G. C. Papavizas and R. J. Cook). *NATO ASI Series, Series A: Life Sciences* 230, 175–184.

Scher, F. M. and Baker, R. (1980). Mechanism of biological control in a *Fusarium*-suppressive soil. *Phytopathology* 70, 412–417.

Schippers, B. (1988). Biological control of pathogens

with rhizobacteria. *Philosophical Transactions of the Royal Society of London, Series B* **318**, 283–293.

Schippers, B., Bakker, A. W. and Bakker, P. A. H. M. (1987). Interactions of deleterious and beneficial rhizosphere microorganisms and the effect of cropping practices. *Annual Review of Phytopathology* **25**, 339–358.

Schroth, M. N. and Hancock, J. G. (1982). Disease-suppressive soil and root-colonizing bacteria. *Science* **216**, 1376–1381.

Shanahan, P., O'Sullivan, D. J., Simpson, P., Glennon, J. D. and O'Gara, F. (1992). Isolation and characterization of an antibiotic-like compound from a fluorescent pseudomonad and investigation of physiological parameters influencing its production. *Applied and Environmental Microbiology* **58**, 353–358.

Smiley, R. W. (1979). Wheat-rhizoplane pseudomonads as antagonists of *Gaeumannomyces graminis*. *Soil Biology and Biochemistry* **11**, 371–376.

Stutz, E. W., Défago, G., Hantke, R. and Kern, H. (1985). Effect of parent materials derived from different geological strata on suppressiveness of soils to black root rot of tobacco. In *Ecology and Management of Soilborne Plant Pathogens* (ed. C. A. Parker, A. D. Rovira, K. J. Moore, P. T. W. Wong and J. F. Kollmorgen), pp. 215–217. Academic Press, New York.

Stutz, E.W., Défago, G. and Kern, H. (1986). Naturally occurring fluorescent pseudomonads involved in suppression of black root rot of tobacco. *Phytopathology* **76**, 181–185.

Stutz, E. W., Kahr, G. and Défago, G. (1989). Clays involved in suppression of tabacco black root rot by a strain of *Pseudomonas fluorescens*. *Soil Biology and Biochemistry* **21**, 361–362.

Thomashow, L. S. and Pierson III, L. S. (1991). Genetic aspects of phenazine antibiotic production by fluorescent pseudomonads that suppress take-all disease of wheat. In *Advances in Molecular Genetics of Plant-Microbe Interactions*, vol. 1 (ed. H. Hennecke, D. P. S. Verma), pp. 443–449. Kluwer Academic Publishers, Dordrecht. The Netherlands.

Thomashow, L. S. and Weller, D. M. (1988). Role of a phenazine antibiotic from *Pseudomonas fluorescens* in biological control of *Gaeumannomyces graminis* var. *tritici*. *Journal of Bacteriology* **170**, 3499–3508.

Thomashow, L. S. and Weller D. M. (1990). Application of fluorescent pseudomonads to control root diseases of wheat and some mechanisms of disease suppression. In *Biological Control of Soil-Borne Plant Pathogens* (ed. D. Hornby, R. J. Cook, Y. Henis, W. H. Ko, A. D. Rovira, B. Schippers and P. R. Scott), ch. 8, pp. 109–122. *CAB International*, Ascot, Berks.

Thomashow, L. S., Weller, D. M. Bonsall, R. F. and Pierson III, L. S. (1990). Production of the antibiotic phenazine-1-carboxylic acid by fluorescent *Pseudomonas* species in the rhizosphere of wheat. *Applied and Environmental Microbiology* **56**, 908–912.

Tomás-Lorente, F., Iniesta-Sanmartin, E., Tomás-Barberan, F. A., Trowitzsch-Kienast, W. and Wray, V. (1989). Antifungal phloroglucinol derivatives and lipophilic flavonoids from *Helichrysum decumbens*. *Phytochemistry* **28**, 1613–1615.

Toyoda, H., Hashimoto, H., Utsumi, R., Kobayashi, H. and Ouchi, S. (1988). Detoxification of fusaric acid by a fusaric acid-resistant mutant of *Pseudomonas solanacearum* and its application to biological control of Fusarium wilt of tomato. *Phytopathology* **78**, 1307–1311.

van Peer, R., Niemann, G. J. and Schippers, B. (1991). Induced resistance and phytoalexin accumulation in biological control of Fusarium wilt of carnation by *Pseudomonas* sp. strain WCS417r. *Phytopathology* **81**, 728–734.

Vincent, M. N., Harrison, L. A., Brackin, J. M., Kovacevich, P. A., Mukerji, P., Weller, D. M. and Pierson, E. A. (1991). Genetic analysis of the antifungal activity of a soilborne *Pseudomonas aureofaciens* strain. *Applied and Environmental Microbiology* **57**, 2928–2934.

Voisard, C., Keel, C., Haas, D. and Défago, G. (1989). Cyanide production by *Pseudomonas fluorescens* helps suppress black root rot of tabacco under gnotobiotic conditions. *European Molecular Biology Organisation Journal* **8**, 351–358.

Wei, G., Kloepper, J. W. and Tuzun, S. (1991). Induction of systemic resistance of cucumber to *Colletotrichum orbiculare* by selected strains of plant growth-promoting rhizobacteria. *Phytopathology* **81**, 1508–1512.

Weller, D. M. (1988). Biological control of soilborne plant pathogens in the rhizosphere with bacteria. *Annual Review of Phytopathology* **26**, 379–407.

Wüthrich, B. and Défago, G. (1991). Suppression of wheat take-all and black root rot of tobacco by *Pseudomonas fluorescens* strain CHA0, isolated from a *Thielaviopsis-suppressive soil: results of field and pot experiments*. In *Plant Growth-Promoting Rhizobacteria – Progress and Prospects* (ed. C. Keel, B. Koller, and G. Défago). *International Organisation for Biological Control of Noxious and Plants/West Palaearctic Regional Section Bulletin* **XIV/8**, 17–22.

Zdor, R. E. and Anderson, A. J. (1992). Influence of root colonizing bacteria on the defense responses of bean. *Plant Soil* **140**, 99–107.

13

Biological control of soil-borne pathogens of wheat: benefits, risks and current challenges

David M. Weller, Linda S. Thomashow and R. James Cook

Introduction

Why biological control?

The past 15 years have witnessed a dramatic increase in research related to biological control. The current perception that biocontrol will have an important role in commercial agriculture in the future contrasts markedly with previously-held views that biocontrol agents perform too inconsistently, or are too narrow in their spectrum of activity, as compared with chemical pesticides, to be commercially feasible on a large scale. Renewed interest in biological control is in part a response to widespread concern about the potential negative impact of chemical pesticides on public health and the environment. Furthermore, the techniques of molecular biology have revolutionized the field by facilitating the identification of the molecular basis of pathogen suppression and by providing the means for construction of 'superior' biocontrol agents. New biocontrol agents resulting from recent intensive research are slowly becoming available to agriculture, and the trend should accelerate throughout this decade. One example is *Gliocladium virens*, which is being marketed in potting-mix to control *Pythium* and *Rhizoctonia* (see Lumsden and Walter, Chapter 25).

This chapter deals with the potential benefits and risks from the introduction of biocontrol agents for the control of root diseases of wheat, as well as the impediments to the application of this technology in commercial agriculture. The focus of the chapter is on biological control of take-all of wheat by fluorescent *Pseudomonas* spp. because it is a model system for the study of the molecular basis of pathogen suppression, root colonization by introduced bacteria and field application of biocontrol agents.

Root diseases of wheat

In the Pacific Northwest (states of Washington, Oregon and Idaho) of the United States take-all, caused by *Gaeumannomyces graminis* var. *tritici* (Cook and Weller, 1986), Pythium root rot, caused by a complex of *Pythium* spp. (Chamswarng and Cook, 1985; Cook and Veseth, 1991), and Rhizoctonia root rot, caused by *Rhizoctonia solani* AG–8 and *R. oryzae* (Ogoshi *et al.*, 1990) are responsible for serious losses in wheat production. Take-all is probably the most destructive root disease of wheat world-wide and characteristic symptoms include black lesions that expand and coalesce on the roots. Infection of the tiller bases results in premature blighting near heading, due to the restriction of water movement to the tops (Cook and Weller, 1986). Symptoms of Rhizoctonia root rot include brown sunken lesions, which may girdle and sever the roots, and which leave 'pinched-off' pointed brown tips. In the field infected plants are often severely stunted and occur in distinct patches (Cook and Veseth, 1991). *Pythium* spp. infect the embryo of the seed within hours after planting, causing the shoots of seedlings to be shortened and deformed. *Pythium* spp. also destroy root hairs and tips and fine lateral roots, lowering the overall vigour of the plant. Although any one of these pathogens can be the main yield-limiting factor, the other two pathogens usually can also be isolated from the roots. All three diseases are aggravated by reduced tillage and intensive wheat production, and damage is lessened by crop rotation and conventional tillage (Cook and Veseth, 1991). However, it is often economically necessary for farmers to grow two to four crops of wheat before a break crop and to use some form of conservation tillage in order to control soil erosion. As there are no commercial cultivars with resistance to these diseases and the registered chemicals (i.e. metalaxyl and triadimenol) are not widely used by growers because of erratic performance, biological control represents the best alternative method of disease control.

Benefits from the control of root diseases

Increased production is the most obvious benefit from the control of root diseases because losses can commonly reach 25–50%. However, another less considered benefit of good root health is more complete uptake of nutrients, thereby reducing the possibility of nitrates moving below the rooting depth and into ground water. A healthy root system also allows the wheat crop to compete better with weeds, thereby reducing the need for herbicides.

Selection and application of biocontrol agents

Fluorescent *Pseudomonas* spp. have been studied intensively as biocontrol agents of plant pathogens during the last 15 years (Weller, 1988; Handelsman and Parke, 1989). World-wide interest in pseudomonads was initially sparked off in the late 1970s and early 1980s by reports that certain *Pseudomonas* strains, designated as plant growth-promoting rhizobacteria (PGPR), improved the growth of potato and sugar beet when applied as seed or seed-piece treatments (Burr *et al.*, 1978; Kloepper *et al.*, 1980; Schroth and Hancock, 1982; Suslow and Schroth, 1982). Moreover, it was found that antagonistic fluorescent pseudo-

monads were associated with the suppression of take-all in some fields in the state of take-all decline (Smiley, 1979; Weller *et al.*, 1988), a natural form of biological control whereby the severity of take-all declines and wheat yields increase following one or more outbreaks of the disease in monocultured wheat (Cook and Weller, 1986). For example, the fraction of the population of pseudomonads inhibitory to *G. g. tritici in vitro* was greater from wheat roots grown in a Wimmera grey clay (suppressive soil cropped for decades to wheat) than from roots grown in a Rosedale sandy clay loam (conducive soil with a variable cropping history) (Smiley, 1979). In another study, fluorescent *Pseudomonas* spp. isolated from roots of wheat grown in Ritzville and Puget silt loams (conducive soils) were compared with those from a take-all suppressive Shano silt loam (field cropped to wheat for over 22 years). Each soil was first mixed 1:10 with fumigated virgin Ritzville silt loam (to minimize differences in soil properties between soils), augmented with *G. g. tritici* and cropped twice to wheat. After the second crop of wheat the plants in the mix with the suppressive soil, but not in the conducive soils, were protected against take-all. Roots from all three soils had similar populations of total aerobic bacteria; however, roots from the Shano silt loam had significantly more fluorescent *Pseudomonas* spp. inhibitory to *G. g. tritici in vitro* than did roots from the other two conducive soils. Furthermore, significantly more strains isolated from roots in suppressive soils controlled take-all *in vivo* as compared with strains from conducive soils (Weller *et al.*, 1985, 1988). Take-all decline soils represent a reservoir of strains from which to select commercial biocontrol agents for root diseases of wheat.

Pseudomonas fluorescens 2–79, 13–79, Q72a–80, and Q29z–80, *P. aureofaciens* 30–84, Q2–87, Q65c–80, Q1c–80 and Q69c–80, and *P. putida* Q8d–80 were selected from take-all decline soils and have controlled take-all in numerous field trials. Strains Q72a–80 and Q29z–80 also suppressed Pythium root rot of wheat (Weller and Cook, 1986) and Q29z–80 and Q65c–80 have controlled Pythium seed rot of chickpea (Kaiser *et al.*, 1989). Application of these strains as seed treatments (10^7–5×10^8 colony forming units, (cfu) per seed) increased the yield of wheat by an average of 12% (and up to 25%) in commercial fields with a history of take-all (Weller and Cook, 1983; Cook *et al.*, 1988; Weller, 1988; Cook, 1991). Similar yield increases have been achieved with the control of Pythium root rot.

Root colonization

Definition

Root colonization, as used in the context of this chapter, is a process whereby introduced bacteria become distributed along a root that is growing in raw soil and propagate and survive for several weeks in the presence of indigenous rhizosphere microflora. Root colonization includes colonization of the inside and/or surface of the root as well as the rhizosphere soil. Rhizosphere competence describes the relative root-colonizing ability of an introduced strain. The rhizosphere competence of a strain can be quantified by determining: the population it attains on a root; the length or number of roots colonized; and/or the length of time the bacteria survive (Weller, 1988).

Relationship to biocontrol

Root colonization is generally considered essential for suppression of root pathogens by introduced bacteria (Suslow, 1982; Loper *et al.*, 1985; Liddell and Parke, 1989). Bacteria growing in or near infection courts on roots are ideally positioned to limit the establishment and/or secondary spread of major and/or minor pathogens (Schippers *et al.*, 1987). Several studies demonstrated that introduced bacteria reduced populations of root pathogens (Kloepper and Schroth, 1981; Xu and Gross, 1986; Yuen and Schroth, 1986; Caesar and Burr, 1987). However, the threshold population of a biocontrol agent required for pathogen suppression on roots and the duration that the population must be maintained to close the 'window of vulnerability' to infection have not been well defined in most biocontrol systems. These questions were addressed by studying the relationship between the population size of *P. fluorescens* 2–79 (rifampicin and nalidixic acid resistant) and the number of root lesions caused by *G. g. tritici*. Wheat seeds were treated with doses of bacteria ranging from 0 to about 10^8 cfu per/seed and then sown into a natural soil infested with *G. g. tritici*. Approximately 14 days after planting, the popula-

tion size of 2–79 and the number of take-all lesions were determined on sections of root (length 4 cm) taken from the region 2–6 cm below the seed. Linear regression analysis identified an inverse relationship between the population size of 2–79 and the number of lesions on the root segment. Thus, as the population size of 2–79 increased, take-all control improved (Bull *et al.*, 1991).

The colonization process

It was hypothesized that bacteria applied to wheat seeds colonize roots in two phases (Howie *et al.*, 1987). In phase I, bacteria attach to the emerging root tip and are passively transported into the soil. As the root elongates some cells remain associated with the tip while others are left behind as a source of inoculum on older portions of the root. Rapid multiplication allows the bacteria to remain with the tip, but in the absence of multiplication, transport occurs only until the initial inoculum on the tip is diluted as the plant outgrows the bacteria. The bacteria can be dislodged as the root extends through the soil or become adsorbed to soil particles (Bahme and Schroth, 1987). In phase II bacteria deposited along the root spread into nutrient-rich microsites in the rhizoplane and rhizosphere, multiply, compete with indigenous microflora and avoid displacement. Any bacterium applied to seed can be transported into the soil with wheat roots but only those with a high rhizosphere competence can maintain or increase their population once introduced (Bull, 1987). Thus, the ultimate fate of an introduced microorganism is probably determined during phase II. The concept of phase I and II colonization does not define a strict temporal sequence of bacterial distribution since both phases occur simultaneously on different parts of the root. Rather, this concept defines two stages in the life history of a population of introduced bacteria. Percolating water is a major factor in root colonization because it washes bacteria down the root directly from a source of inoculum (i.e. seed) (Liddell and Parke, 1989), it recharges root tips that have outgrown the bacteria with inoculum from older portions of the roots (Bahme and Schroth, 1987) and/or it can redistribute bacteria from established microcolonies into new microsites. The relative importance of root tip and water transport of introduced bacteria is dependent on the bacterial strain, host plant, soil type, temperature and amount of water.

Biotic and abiotic factors

Biotic and abiotic soil factors strongly influence the colonization process. For example, the population size of 2–79 on wheat roots after seed treatment was greatest at a matric potential of −0.3 to −0.7 bars (1 bar $=10^5$ Pa) (depending on the soil) (Howie *et al.*, 1987), a rhizosphere pH of 6.0–6.5 and at temperatures below, rather than above 20°C (Bull, 1987). The wheat genotype also influences the composition of the introduced rhizosphere microflora; populations of 2–79 were 100-fold greater on the most supportive cultivar 'Wampum' than on the least supportive cultivar 'Brevor' (Weller, 1988). Wheat-root pathogens differ in how they affect the multiplication and persistence of fluorescent pseudomonads on wheat roots. The population of 2–79 was larger on roots infected with *G. g. tritici* than on healthy roots (Weller, 1983), probably as a result of bacterial proliferation in lesions that are rich with nutrients. Similarly, root infection by *R. solani* resulted in significantly larger populations of both *P. fluorescens* 2–79 and Q72a–80 than were present on healthy roots (Mazzola and Cook, 1991). This is not surprising since *Rhizoctonia* also causes deep cancerous root lesions that increase the flow of nutrients. In contrast, the population of 2–79 was significantly smaller on roots infected with *Pythium irregulare, P. aristosporum,* or *P. ultimum* var. *sporangiiferum* than on non-infected roots. Interestingly, the effect of *Pythium* was strain specific; the population size of Q72a–80 was reduced only in the presence of *P. irregulare.* Application of metalaxyl (selectively inhibits *Pythium* spp.) to a soil naturally infested with *Pythium* spp. resulted in significantly larger populations of 2–79 or Q72a–80 on roots as compared with roots from soil not treated with the chemical (Mazzola and Cook, 1991).

Population dynamics in the field

Studies of the population dynamics of fluorescent pseudomonads applied to plants have demonstrated that the introduced bacteria can become widely distributed along the root system (Bahme

and Schroth, 1987). Strain 2–79 applied to the seed became distributed along the length of the roots with the population declining along the axis from the seed to the tip and doubling every 15 to 85 h (Weller, 1984). Regardless of the soil type, plot location or kind of wheat (spring or winter), the population dynamics of introduced pseudomonads followed a similar and predictable pattern on wheat; large populations, initially established on the seminal roots, gradually declined in size over the growing season. For example, on wheat grown from seed treated with approximately 10^8 cfu of 2–79 per seed and sown in October at Pullman, Washington, by 18 days after planting the seminal roots were colonized by more than 10^6 cfu/0.1 g seminal root (with adhering soil). However, by day 245 the population declined by three orders of magnitude (Weller, 1988). A similar trend was seen in a 6 month field study of 2–79 on wheat grown at Avon, South Australia.

Field release of a genetically engineered 2–79

The *lacZY* tracking genes (Drahos *et al.*, 1986) were inserted into the chromosome of *P. fluorescens* 2–79 (rifampicin and nalidixic acid resistant), this generated the recombinant strain 2–79RNL3, which utilized lactose as a sole carbon source and produced blue colonies on media amended with 5-bromo-4-chloro-3-indoyl-β-D-galactopyranoside (X-Gal). The blue colour of the colonies facilitated identification of the engineered strain on dilution plates even in the presence of hundreds of other colonies. In the first release of a genetically engineered micro-organism in the Pacific Northwest strains 2–79 and 2–79RNL3 were individually applied to seed of wheat planted 6 October 1988 in a field near Pullman, Washington. Inoculum of the take-all fungus was introduced into the soil at 'high' and 'low' rates to provide two levels of disease severity. The populations of strains 2–79 and 2–79RNL3 were monitored in six samples for a period of 326 days. Roots with adhering soil were washed three times and the first and third washes were plated. Populations of 2–79 and 2–79RNL3 were always greater in the first wash, which removed most of the rhizosphere soil. The population trends of the 2–79RNL3 and 2–79 were identical in both the first and third washes and at the high and low rates of *G. g. tritici* inoculum. The largest populations, approximately 10^8 cfu/g root without adhering soil, were recorded in the first sample 14 days after planting. Thereafter, populations of both strains declined and by the last sample, 13 days after harvest, the population had dropped by five orders of magnitude to about 10^3 cfu/g root. This population trend was very similar to that recorded for 2–79 in an earlier field study (Weller, 1983). No colonies of 2–79 or 2–79RNL3 were detected on roots of plants in border rows, 30 cm away from the rows of inoculated wheat; this indicated that the bacteria remained contained and did not spread from inoculated to uninoculated rows. Interestingly, however, 2–79RNL3 and 2–79 were recovered from the roots of volunteer (previous crop) lentil seedlings growing in the rows of inoculated wheat. The roots of the lentils were entwined with the roots of the wheat (Cook *et al.*, 1990).

Mechanisms of biological control

Phenazines

After roots of wheat are colonized, the presence of the bacteria alone is not sufficient for pathogen suppression to occur. Production of inhibitory compounds such as antibiotics is required. *Pseudomonas fluorescens* 2–79 produces phenazine-1-carboxylic acid (PCA), which is inhibitory to *G. g. tritici* and other soil-borne plant pathogens *in vitro* at concentrations of 1 μg/ml or less (Gurusiddaiah *et al.*, 1986). Both genetic and chemical evidence indicate that production of PCA is the primary mechanism by which 2–79 suppresses take-all; (Thomashow and Weller, 1988; Thomashow *et al.*, 1990). Phenazine-deficient (Phz⁻) mutants of 2–79, generated by Tn5 mutagenesis, failed to inhibit *G. g. tritici in vitro* and were significantly less suppressive of take-all than the parental strain. Furthermore, mutants that were complemented with homologous DNA from a 2–79 genomic library were fully and coordinately restored in ability to produce PCA, inhibit *G. g. tritici in vitro*, and suppress take-all on wheat (Thomashow and Weller, 1988). In a subsequent study (Bull *et al.*, 1991), there was an inverse linear relationship between the population size of 2–79 or a Phz⁺ complemented mutant on roots and the number of take-all lesions that resulted from primary infection. However, no such relationship existed when a phenazine deficient mutant was used (Bull *et al.* 1991).

PCA production accounted for approximately 60 to 80% of the suppressive activity of 2–79 against take-all; this depended on the type of soil used in the experiment. Initially a fluorescent siderophore was reported to contribute to the residual biocontrol activity of 2–79 (Weller *et al.*, 1988). This conclusion was based on studies using chemically-generated, non-fluorescent mutants. More recently (Hamdan *et al.*, 1991; Ownley *et al.*, 1992), abolition of siderophore production in 2–79 by Tn5 mutagenesis had no effect on take-all suppressiveness and a second iron-regulated anti-fungal factor, anthranilic acid, was identified. It is now thought that the siderophore and anthranilic acid make only very minor contributions to the biocontrol activity of 2–79, with the residual suppressiveness resulting mostly from competition for nutrients (Ownley *et al.*, 1992).

P. aureofaciens 30–84 was isolated from roots of wheat grown in a field in Kansas that had been cropped to wheat for 65 years. It produces PCA as well as smaller amounts of 2-hydroxy-phenazine-1-carboxylic acid and 2-hydroxyphenazine. Phz⁻ mutants of 30–84, like those of 2–79, were non-inhibitory to *G. g. tritici in vitro* and less suppressive of take-all than the parental strain; this demonstrates further the importance of the antibiotic in biocontrol of take-all (Pierson and Thomashow, 1992).

PCA was isolated from the roots and rhizosphere of wheat treated with 2–79, 30–84, or their respective Phz⁺ complemented mutants and grown in steamed or natural soil, in the growth chamber, or the field (Thomashow *et al.*, 1990). However, no antibiotic was recovered from the roots of seedlings grown from non-treated seeds or seeds treated with Phz⁻ mutants. In growth-chamber studies in natural soil, comparable amounts of PCA (28–43 ng/g of root with adhering soil) were recovered from roots colonized by 2–79 whether or not *G. g. tritici* was present. Strain 30–84 produced up to 133 ng/g of root with adhering soil. The levels isolated from field-grown plants ranged from 5–27 ng/g root with adhering soil. Roots from which the antibiotic was recovered had significantly less disease than roots from which the antibiotic could not be isolated.

HCN, 2,4-diacetylphloroglucinol and pyoluteorin

P. fluorescens strain CHA0 was isolated from tobacco roots grown in a Swiss soil that was naturally suppressive to black root rot caused by *Thielaviopsis basicola* (see Défago and Keel, Chapter 12). The strain is active against many pathogens (Défago *et al.*, 1990), including *T. basicola* on tobacco, *G. g. tritici* on wheat (Haas *et al.*, 1991; Keel *et al.*, 1992) and *P. ultimum* on cucumber, cress and sweet corn (Maurhofer *et al.*, 1992). It produces at least five potentially bioactive compounds: indoleacetic acid, pyoverdine, pyoluteorin, 2,4-diacetylphloroglucinol (Phl) and HCN. Evidence for their involvement in biocontrol is reviewed in Chapter 12. Vincent *et al.* (1991) reported that production of Phl also is the major mechanism of suppression of take-all by *P. aureofaciens* Q2–87, a strain isolated from roots of wheat growing in a take-all decline soil. Interestingly, at least 20% of the fluorescent pseudomonads from the same soil produced Phl, indicating that the ability to synthesize this antibiotic can be fairly common among pseudomonads from suppressive soils. A Phl-producing strain from Ukraine has been reported to control *Fusarium oxysporum* on wheat (Garagulya, *et al.*, 1974; Pidoplichko and Garagulya, 1974) and another from Ireland (*P. ultimum* on sugar beet (see Fenton *et al.*, 1992)); this suggests that such strains are of broad geographical distribution.

Phenazines and ecological competence

Because many soil micro-organisms produce antibiotics *in vitro* (Gottlieb, 1976) it has been suggested that antibiotics may contribute to the ecological competence of micro-organisms in soil and in the rhizosphere. However, in studies of root colonization that lasted 7 to 10 days, Phz⁻ mutants of 2–79 were as rhizosphere competent as the parental strains (Thomashow and Weller, 1988). Similarly, there were no significant differences between populations of the oomycin-producing strain Hv37aR2 and its antibiotic-deficient mutants in the rhizosphere of cotton (Howie and Suslow, 1991). Mazzola *et al.* (1992) suggested that the advantage of antibiotic production to the ecological competence of a strain may only become apparent in studies that extend over

longer periods of time. He introduced strains 2–79, 30–84, Phz⁻ mutants, or the genetically complemented Phz⁺ mutants individually into Thatuna silt loam with or without *G. g. tritici*. The soils were planted to three or five successive cycles of wheat; each lasted 20 days from planting to harvest. At the end of a cycle the shoots were cut off at the soil line and the soil and roots were removed, mixed, repotted and again sown to wheat. The introduced bacteria were sampled from the soil and roots after each cycle. Populations of Phz⁻ derivatives declined significantly more rapidly than those of their respective parental or genetically restored Phz⁺ strains in both the rhizosphere and bulk soil. The differences between Phz⁺ and Phz⁻ strains appeared more quickly in the absence, than in the presence, of the take-all fungus. In contrast, when studies were conducted in pasteurized soil (to reduce microbial competition), populations of Phz⁻ and Phz⁺ strains remained similar throughout the cycles; this suggested that phenazine production enhances ability to compete with the resident microflora.

Inconsistent performance of biocontrol agents

One characteristic common to most biocontrol systems with introduced bacteria is inconsistency of disease control. Inconsistent performance in the field is the major impediment to the large-scale commercial development and use of biocontrol agents in agriculture. In commercial-scale tests, pseudomonads that suppress take-all and Pythium root rot significantly improved yield for only about 60% of the time (Weller, 1988). Performance of a strain varies from site to site and year to year.

A multitude of factors can account for inconsistent results given the complex interactions among host, pathogen, antagonist and the environment. A positive response to the introduction of a biocontrol agent will not occur when the target pathogen(s) is absent or environmental conditions are not suitable for disease development because improved plant growth is a function of disease control and not direct stimulation. Performance also may be inconsistent when there is unexpected disease pressure from non-target pathogens. If the introduced bacteria suppress the target pathogen but have no effect on a second non-target pathogen, then the treatment will appear ineffective. For example, it is common to isolate *G. g. tritici*, *Pythium* spp. and *R. solani* or *R. oryzae* from the same wheat plant. Unfortunately, most of the biocontrol agents of wheat tend to be highly effective against a single target pathogen and only moderately effective against other pathogens.

Variable root colonization by introduced bacteria is another important reason for inconsistent performance. Populations of introduced bacteria are log-normally distributed among root systems of plants and among individual roots of a single plant (Loper *et al.*, 1984; Bahme and Schroth, 1987), meaning that some roots may be highly colonized while others may have few or no introduced bacteria. For example, populations of introduced *Pseudomonas* strains A1 or SH5 on root systems of individual potato or sugar beet seedlings varied by a factor of 100–1000 (Loper *et al.*, 1984). Further, populations of *P. fluorescens* 2–79 and Q72a–80 on individual roots of wheat seedlings varied up to 1000-fold and bacteria could not be detected on 20–40% of the roots (Bull, 1987). Roots that are not colonized or are only sparsely colonized provide sites for the pathogens to proliferate unchecked, thus allowing secondary spread of the pathogen. The inverse relationship between root colonization and biological control described earlier indicates that any interruption in the colonization process would adversely affect disease suppression. Finally, variable production or inactivation *in situ* of bacterial products active in biocontrol can lead to inconsistent performance. Many of these products are secondary metabolites that are produced under specific conditions only. For example, phenazine production *in vitro* is highly dependent on cultural conditions and is tightly controlled (Turner and Messenger, 1986). In the rhizosphere, the temporal regulation of secondary metabolites such as phenazines is probably even more tightly controlled and very dependent on the environment within a microsite.

Approaches to the improvement of performance

The primary challenge in biocontrol of soil-borne pathogens is the improvement of the consistency of performance of introduced agents. In the biological control of root diseases of wheat this challenge is being addressed with several approaches.

Identify effects of soil factors on biocontrol

Identifying the biotic and abiotic factors that influence biocontrol activity facilitates optimizing the soil environment to enhance antagonism by introduced bacteria. To identify the soil properties associated with enhanced or decreased biocontrol activity of 2–79, ten soils were collected from diverse locations and analysed for 28 physical and chemical characteristics. Strain 2–79 was tested for its ability to suppress take-all in each of the ten soils after steam pasteurization and amendment with *G. g. tritici*. Disease suppression varied considerably among soils. Principal component factor analysis identified 16 of the 28 soil variables as being highly interrelated to root disease (or conversely, to biocontrol activity). Linear correlations between these 16 soil variables and root disease showed that seven (ammonium-nitrogen, sulfate-sulphur, soil pH, extractable and soluble sodium, zinc and the percentage of sand) were directly related to biocontrol activity by 2–79 (reduced root disease), whereas the remaining nine (cation exchange capacity, percentage silt, percentage clay, exchangeable acidity, manganese, iron, percentage organic matter, total carbon, and total nitrogen) were inversely related to biocontrol (increased root disease). A regression model ($R^2 = 0.96$) that included six variables (cation exchange capacity, ammonium-nitrogen, soil pH, iron, zinc, and the percentage of silt) was developed that helps to predict whether 2–79 will be effective in a given soil (Ownley *et al.*, 1990, 1991). The correlation of some soil variables (i.e. zinc) to enhanced biocontrol activity may result from a positive effect on phenazine production in the rhizosphere. As a validation of the model, take-all suppression by 2–79 was significantly greater on wheat grown in a Woodburn silt loam (naturally low in zinc) amended with 50 μg of zinc (as Zn-EDTA)/g soil than on wheat in the non-amended soil (B. H. Ownley and D. M. Weller, unpublished). Interestingly, in a study of the nutritional requirements of both cell growth and PCA production by 2–79 in liquid culture, zinc sulphate enhanced antibiotic production without increasing cell growth (Slininger and Jackson, 1992). Zinc is a structural or catalytic component of nearly 300 enzymes (Valley and Auld, 1990); it regulates gene expression in *Pseudomonas* (Brumlik and Darzins, 1992; Olson and Ohman, 1992); and it is critical in many secondary metabolic processes (Weinberg, 1977).

Strain combinations

Another approach has been the development of strain combinations. In take-all decline soils the suppression of disease no doubt results from the activity of a consortium of micro-organisms. The mixtures, Q2–87 + Q1c–80 + Q8d–80 + Q65c–80 and Q2–87 + Q1c–80 + Q8d–80 + Q69c–80 have provided significantly more control of take-all and greater yield increase in field studies than each strain in the mixture used individually (Pierson and Weller, 1991). For example, in a spring wheat plot at Pullman the combination Q2–87 + Q1c–80 + Q8d–80 + Q65c–80 increased yield 20% over the control but individual strains increased yield by no more than 5%. Random mixing of effective strains did not necessarily yield superior combinations and the use of strains from the same soil resulted in the best performance. The superior performance of certain mixtures may result from the ability of a group of strains to colonize the roots more effectively, to adapt better to the environmental changes that occur throughout the growing season, to present a larger number of pathogen-suppressive mechanisms, and/or to protect against a broader range of pathogens.

Strain improvement by genetic manipulation

Molecular biotechnology already has proven its utility in elucidating the mechanisms responsible for biocontrol activity. Potentially, this understanding can be applied not only to improve the soil environment or to provide a rational basis for the development of strain combinations, but also to help optimize product formulation and to facilitate the selection of new biocontrol strains. Once traits of value have been identified, genetic manipulation also is a means to directed strain improvement, and the first such steps have already been documented (see Dowling *et al.*, Chapter 14). For example, Gutterson (1990) and Gutterson *et al.* (1990), increased production of the antibiotic oomycin A and suppression of *Pythium*-induced damping-off of cotton by placing the oomycin A biosynthetic gene cluster in *P. fluorescens* Hv37a under the control of the constitutive *tac* promoter from *Escherichia coli*. Damping-off of cucumber also was controlled more effectively when additional, plasmid-borne copies of gene(s) involved in synthesis of both Phl and pyoluteorin were present in *P. fluorescens* CHA0. The plasmid-bearing strain was deleterious to cress and sweet corn, however, presumably because the increased production of one or both antibiotics had a phytotoxic effect on these plants (Maurhofer *et al.*, 1992). Phenazine production by strains 2–79 and 30–84 has also been increased *in vitro*, either by the introduction of additional plasmid-borne copies of the entire biosynthetic locus (Thomashow and Pierson, 1991) or of a linked gene that appears to function as an activator of phenazine synthesis (Pierson and Keppenne, 1992). Moreover, the biosynthetic locus from strain 2–79 has been expressed (enabling PCA synthesis) in all pseudomonads (27 strains to date) into which it has been introduced, and the resulting 'transgenic' strains have shown encouraging preliminary evidence of enhanced suppressiveness against take-all. Similarly, a fragment containing genes involved in Phl synthesis from *P. aureofaciens* Q2–87 has been expressed, resulting in Phl production, in ten *Pseudomonas* strains including 2–79 and 5097 (Vincent *et al.*, 1991; M. Bangera and L. S. Thomashow, unpublished). Others (Fenton *et al.*, 1992, Ligon *et al.*, 1992) also have reported that the introduction of biosynthetic or regulatory genes for antibiotic production can convert ineffective strains into agents with biocontrol activity. Apparently, mobilization and heterologous expression of biosynthetic pathways, or the activation of cryptic pathways, can be much more easily achieved than previously might have been anticipated. These initial results support the feasibility of genetic manipulation for strain enhancement and open the way for combining different antagonistic mechanisms within single strains.

Potential risks of introducing fluorescent pseudomonads

There is concern about the effect of the introduction of large populations of a biocontrol agent on non-target, indigenous microflora in an agroecosystem. The concern is even greater when the bacteria are genetically engineered. Potentially, introduced bacteria could displace certain groups of the indigenous microflora or become a significant and permanent part of the microbial community. Studies on wheat clearly argue against this scenario and indicate that although introduced bacteria can dominate portions of roots close to the inoculum source for several weeks, nevertheless, by the end of a growing season they comprise only a small fraction (<2%) of the microbial population (Weller, 1983). Even genetically engineered strains with superior rhizosphere competence or biocontrol ability ultimately will have the same fate as wild-type strains. The knowledge that genes responsible for important biocontrol traits such as PCA or Phl production are common in suppressive soils should lessen concern about the introduction of transgenic strains into an agroecosystem.

References

Bahme, J. B. and Schroth, M. N. (1987). Spatial-temporal colonization patterns of a rhizobacterium on underground organs of potato. *Phytopathology* 77, 1093–1100.

Brumlik, M. J. and Darzins, A. (1992). Zinc and iron regulate translation of the gene encoding *Pseudomonas aeruginosa* elastase. *Molecular Microbiology* 6, 337–334.

Bull, C. T. (1987). Wheat root colonization by disease-suppressive or nonsuppressive bacteria and the effect of population size on severity of take-all caused by *Gaeumannomyces graminis* var. *tritici*. M. Sc. thesis, Washington State University, Pullman.

Bull, C. T., Weller, D. M. and Thomashow, L. S. (1991). Relationship between root colonization and suppression of *Gaeumannomyces graminis* var. *tritici* by *Pseudomonas fluorescens* strain 2–79. *Phytopathology* 81, 954–959.

Burr, T. J., Schroth, M. N. and Suslow, T. (1978). Increased potato yields by treatment of seed pieces with specific strains of *Pseudomonas fluorescens* and *P. putida*. *Phytopathology* 68, 1377–1383.

Caesar, A. J. and Burr, T. J. (1987). Growth promotion of apple seedlings and rootstocks by specific strains of bacteria. *Phytopathology* 77, 1583–1588.

Chamswarng, C. and Cook R. J. (1985). Identification and comparative pathogenicity of *Pythium* species from wheat roots and wheat-field soils in the Pacific Northwest. *Phytopathology* 75, 821–827.

Cook, R. J. (1991). Challenges and rewards of sustainable agriculture research and education. In *Sustainable Agriculture Research and Education in the Field*, pp. 32–76. National Academy Press, Washington, DC.

Cook, R. J. and Veseth, R. J. (1991). *Wheat Health Management*. APS Press, St Paul, MI.

Cook, R. J. and Weller, D. M. (1986). Management of take-all in consecutive crops of wheat or barley. In *Innovative Approches to Plant Disease Control* (ed. I. Chet), pp. 41–76. John Wiley & Sons, Inc. New York.

Cook, R. J., Weller, D. M., and Bassett, E. N. (1988). Take-all and wheat. *Biological and Cultural Tests for Control of Plant Diseases* 3, 53.

Cook, R. J., Weller, D. M., Kovacevich, P., Drahos, D., Hemming, B., Barnes, G. and Pierson, E. L. (1990). Establishment, monitoring, and termination of field tests with genetically altered bacteria applied to wheat for biological control of take-all. In *The Biosafety Results of Field tests of Genetically Modified Plants and Microorganisms* pp. 177–187. Agricultural Research Institute, Bethesda, MD.

Défago, G., Berling, C.-H., Berger, U., Haas, D., Kahr, G., Keel, C., Voisard, C., Wirthner, P. and Wüthrich, B. (1990). Suppression of black root rot of tobacco and other root diseases by strains of *Pseudomonas fluorescens*: potential applications and mechanisms. In *Biological Control of Soil-borne Plant Pathogens* (ed. D. Hornby), pp. 93–108. CAB International, Wallingford.

Drahos, D. J., Hemming, B. C. and McPherson. (1986). Tracking recombinant organisms in the environment: β-galactosidase as a selectable non-antibiotic marker for fluorescent pseudomonads. *Bio/Technology* 4, 439–444.

Fenton, A. M., Stephens, P. M., Crowley, J., O'Callaghan, M. and O'Gara, F. (1992). Exploitation of gene(s) involved in 2,4-diacetylphloroglucinol biosynthesis to confer a new biocontrol capability to a *Pseudomonas* strain. *Applied Environmental Microbiology* 58, 3873–3878.

Garagulya, A. D., Kiprianova, E. A. and Boiko, O. I. (1974). Antibiotic effect of bacteria from the genus *Pseudomonas* on phytopathogenic fungi. *Mikrobiol. Zh. (Kiev)* 36, 197–202.

Gottlieb, D. (1976). The production and role of

antibiotics in soil. *Journal of Antibiotics* **29**, 987–1000.

Gurusiddaiah, S., Weller, D. M., Sarkar, A. and Cook, R. J. (1986). Characterization of an antibiotic produced by a strain of *Pseudomonas fluorescens* inhibitory to *Gaeumannomyces graminis* var. *tritici* and *Pythium* spp. *Antimicrobial Agents and Chemotherapy* **29**, 488–495.

Gutterson, N. (1990). Microbial fungicides: recent approaches to elucidating mechanisms. *Critical Reviews in Biotechnology* **10**, 69–91.

Gutterson, N., Howie, W. and Suslow, T. (1990). Enhancing effects of biocontrol agents by use of biotechnology. In *New Directions in Biocontrol*, (ed. R. R. Baker and P. E. Dunn), pp. 749–765. Alan R. Liss, Inc., New York.

Haas, D., Keel, C., Laville, J., Maurhofer, M., Oberhansli, T., Schnider, U., Voisard, C., Wüthrich, B. and Défago, G. (1991). Secondary metabolites of *Pseudomonas fluorescens* strain CHA0 involved in the suppression of root diseases. In *Advances in Molecular Genetics of Plant-Microbe Interactions*, vol. 1, (ed. H. Hennecke and D. P. S. Verma), pp. 450–456. Kluwer Academic Publishers, Dordrecht, the Netherlands.

Hamdan, H., Weller, D. M. and Thomashow, L. S. (1991). Relative importance of fluorescent siderophores and other factors in biological control of *Gaeumannomyces graminis* var *tritici* by *Pseudomonas fluorescens* 2–79 and M4–80R. *Applied Environmental Microbiology* **57**, 3270–3277.

Handelsman, J. and Parke, J. L. (1989). Mechanisms in biocontrol of soilborne plant pathogens. In *Plant-Microbe Interactions, Molecular and Genetic Perspectives*, vol. 3. (ed. T. Kosuge and E. W. Nester), pp. 27–61. McGraw-Hill, New York.

Howie, W. J. and Suslow, T. V. (1991). Role of antibiotic biosynthesis in the inhibition of *Pythium ultimum* in the cotton spermosphere and rhizosphere by *Pseudomonas fluorescens*. *Molecular Plant-Microbe Interactions* **4**, 393–399.

Howie, W. J., Cook, R. J. and Weller, D. M. (1987). Effects of soil matric potential and cell motility on wheat root colonization by fluorescent pseudomonads suppressive to take-all. *Phytopathology* **77**, 286–292.

Kaiser, W. J., Hannan, R. M. and Weller, D. M. (1989). Biological control of seed rot and preemergence damping-off of chickpea with fluorescent pseudomonads. *Soil Biology and Biochemistry* **21**, 269–273.

Keel, C., Schnider, U., Maurhofer, M., Voisard, C., Laville, J., Burger, U., Wirthner, P., Haas, D. and Défago, G. (1992). Suppression of root diseases by *Pseudomonas fluorescens* CHA0: importance of the bacterial secondary metabolite

2,4-diacetylphloroglucinol. *Molecular Plant-Microbe Interactions* **5**, 4–13.

Kloepper, J. W. and Schroth, M. N. (1981). Relationship of *in vitro* antibiosis of plant growth-promoting rhizobacteria to plant growth and the displacement of root microflora. *Phytopathology* **71**, 1020–1024.

Kloepper, J. W., Leong, J., Teintze, M. and Schroth, M. N. (1980). Enhanced plant growth by siderophores produced by plant growth-promoting rhizobacteria. *Nature* **286**, 885–886.

Liddell, C. M. and Parke, J. L. (1989). Enhanced colonization of pea tap roots by a fluorescent pseudomonad biocontrol agent by water infiltration into soil. *Phytopathology* **79**, 1327–1332.

Ligon, J., Hill, S., Morse, A., Gaffney, T., Gates, K., Lam, S., Frazelle, A., Torkewitz, N. and Becker, J. (1992). Mapping of a genetic region from *Pseudomonas fluorescens* involved in the biosynthesis of several antifungal factors. Abstract 232, *6th International Symposium on Molecular Plant-Microbe Interactions*, Seattle, WA.

Loper, J. E., Suslow, T. V. and Schroth, M. N. (1984). Lognormal distribution of bacterial populations in the rhizosphere. *Phytopathology* **74**, 1454–1460.

Loper, J. E., Haack, C. and Schroth, M. N. (1985). Population dynamics of soil pseudomonads in the rhizosphere of potato (*Solanum tuberosum* L.). *Applied Environmental Microbiology* **49**, 416–422.

Maurhofer, M., Keel, C., Schnider, U., Voisard, C., Haas, D. and Défago, G. (1992). Influence of enhanced antibiotic production in *Pseudomonas fluorescens* strain CHA0 on its disease suppressive capacity. *Phytopathology* **82**, 190–195.

Mazzola, M. and Cook, R. J. (1991). Effects of fungal root pathogens on the population dynamics of biocontrol strains of fluorescent pseudomonads in the wheat rhizosphere. *Applied Environmental Microbiology* **57**, 2171–2178.

Mazzola, M., Cook, R. J., Thomashow, L. S., Weller, D. M. and Pierson, L. S. III. (1992). Contribution of phenazine antibiotic biosynthesis to the ecological competence of fluorescent pseudomonads in soil habitats. *Applied Environmental Microbiology* **58**, 2616–2624.

Ogoshi, A., Cook, R. J. and Bassett, E. N. (1990). *Rhizoctonia* species and anastomosis groups causing root rot of wheat and barley in the Pacific Northwest. *Phytopathology* **80**, 784–788.

Olson, J. C. and Ohman, D. E. (1992). Efficient production and processing of elastase and LasA by *Pseudomonas aeruginosa* require zinc and calcium ions. *Journal of Bacteriology*. **174**, 4140–4147.

Ownley, B. H., Weller, D. M. and Alldredge, J. R. (1990). Influence of soil edaphic factors on suppression of take-all by *Pseudomonas fluorescens* 2–79. *Phytopathology* **80**, 995.

Ownley, B. H., Weller, D. M. and Alldredge, J. R. (1991). Relation of soil chemical and physical factors with suppression of take-all by *Pseudomonas fluorescens* 2–79. In *Plant Growth-Promoting Rhizobacteria-Progress and Prospects* (ed. C. Keel, B. Koller and G. Défago), pp. 299–301. WPRS Bulletin 1991/XIV/8.

Ownley, B. H., Weller, D. M. and Thomashow, L. S. (1992). Influence of *in situ* and *in vitro* pH on suppression of *Gaeumannomyces graminis* var. *tritici* by *Pseudomonas fluorescens* 2–79. *Phytopathology* **82**, 178–184.

Pidoplichko, V. N. and Garagulya, A. D. (1974). Effects of antagonistic bacteria on development of wheat root rot. *Mikrobiol. Zh.* (Kiev) **36**, 599–602.

Pierson, E. A. and Weller, D. M. (1991). Recent work on control of take-all all of wheat by fluorescent pseudomonads. In *Plant Growth-Promoting Rhizobacteria – Progress and Prospects*, (ed. C. Keel, B. Koller and G. Défago), *IOBC/WPRS Bulletin* **XIV/8**, 96–97.

Pierson, L. S. III and Keppenne, V. D. (1992). Identification of a locus that acts in trans to stimulate phenazine gene expression in *Pseudomonas aureofaciens* 30–84. Abstract 197, *6th International Symposium on Molecular Plant–Microbe Interactions*, Seattle, WA.

Pierson, L. S. III and Thomashow, L. S. (1992). Cloning and heterologous expression of the phenazine biosynthetic locus from *Pseudomonas aureofaciens* 30–84. *Molecular Plant–Microbe Interaction* **5**, 330–339.

Schippers, B., Bakker, A. W. and Bakker, P. A. H. M. (1987). Interactions of deleterious and beneficial rhizosphere microorganisms and the effect of cropping practice. *Annual Review of Phytopathology* **25**, 339–358.

Schroth, M. N. and Hancock, J. G. (1982). Disease-suppressive soil and root-colonizing bacteria. *Science* **216**, 1376–1381.

Slininger, P. J. and Jackson, M. A. (1992). Nutritional factors regulating growth and accumulation of phenazine 1-carboxylic acid by *Pseudomonas fluorescens* 2–79. *Applied Microbiology and Biotechnology* **37**, 388–392.

Smiley, R. W. (1979). Wheat–rhizoplane pseudomonads as antagonists of *Gaeumannomyces graminis*. *Soil Biological Biochemistry* **11**, 371–376.

Suslow, T. V. (1982). Role of root-colonizing bacteria in plant growth. In *Phytopathogenic Prokaryotes*, vol. 1. (ed. G. Lacy and M. S. Mount), pp. 187–223. Academic Press, New York.

Suslow, T. V. and Schroth, M. N. (1982). Rhizobacteria of sugarbeets: effects of seed

application and root colonization on yield. *Phytopathology* **72**, 199–206.

Thomashow, L. S. and Pierson L. S. III. (1991). Genetic aspects of phenazine antibiotic production by fluorescent pseudomonads that suppress take-all disease of wheat. In *Advances in Molecular Genetics of Plant–Microbe Interactions* (ed. H. Hennecke and D. P. S. Verma), pp 443–449. Kluwer Academic Publishers, Dordrecht, the Netherlands.

Thomashow, L. S. and Weller, D. M. (1988). Role of a phenazine antibiotic from *Pseudomonas fluorescens* in biological control of *Gaeumannomyces graminis* var. *tritici*. *Journal of Bacteriology* **170**, 3499–3508.

Thomashow, L. S., Weller, D. M., Bonsall, R. F. and Pierson L. S., III. (1990). Production of the antibiotic phenazine-1-carboxylic acid by fluorescent *Pseudomonas* species in the rhizosphere of wheat. *Applied Environmental Microbiology* **56**, 908–912.

Turner, J. M. and Messenger, A. J. (1986). Occurrence, biochemistry and physiology of phenazine pigment production. *Advances in Microbial Physiology* **27**, 211–275.

Valley, B. L. and Auld, D. S. (1990). Zinc coordination, function, and structure of zinc enzymes and other proteins. *Biochemistry* **29**, 5647–5659.

Vincent, M. N., Harrison, L. A., Brackin, J. M., Kovacevich, P. A., Mukerji, P., Weller, D. M. and Pierson, E. A. (1991). Genetic analysis of the antifungal activity of a soilborne *Pseudomonas aureofaciens* strain. *Applied Environmental Microbiology* **57**, 2928–2934.

Weinberg, E. D. (1977). Mineral element control of microbial secondary metabolism. In *Microorganisms and Metals* (ed. E. D. Weinberg), pp. 289–326. Marcel Dekker, Inc., New York.

Weller, D. M. (1983). Colonization of wheat roots by a fluorescent pseudomonad suppressive to take-all. *Phytopathology* **73**, 1548–1553.

Weller, D. M. (1984). Distribution of a take-all suppressive strain of *Pseudomonas fluorescens* on seminal roots of winter wheat. *Applied Environmental Microbiology* **48**, 397–899.

Weller, D. M. (1988). Biological control of soilborne pathogens in the rhizosphere with bacteria. *Annual Review of Phytopathology* **26**, 379–407.

Weller, D. M. and Cook, R. J. (1983). Suppression of take-all of wheat by seed treatments with fluorescent pseudomonads. *Phytopathology* **73**, 463–469.

Weller, D. M. and Cook, R. J. (1986). Increased growth of wheat by seed treatments with fluorescent pseudomonads, and implications of *Pythium* control. *Canadian Journal of Plant Pathology* **8**, 328–334.

Weller, D. M., Howie, W. J. and Cook, R. J. (1988). Relationship between *in vitro* inhibition of *Gaeumannomyces graminis* var. *tritici* and suppression of take-all of wheat by fluorescent pseudomonads. *Phytopathology* **78**, 1094–1100.

Weller, D. M., Zhang, B.-X. and Cook, R. J. (1985). Application of a rapid screening test for selection of bacteria suppressive to take-all of wheat. *Plant Diseases* **69**, 710–713.

Xu, G.-W. and Gross, D. C. (1986). Field evaluations of the interactions among fluorescent pseudomonads, *Erwinia carotovora*, and potato yields. *Phytopathology* **76**, 423–430.

Yuen, G. Y. and Schroth, M. N. (1986). Interactions of *Pseudomonas fluorescens* strain E6 with ornamental plants and its effect on the composition of root-colonizing microflora. *Phytopathology* **76**, 176–180.

14

Genetically engineered fluorescent pseudomonads for improved biocontrol of plant pathogens

David N. Dowling, Bert Boesten, Daniel J. O'Sullivan, Peter Stephens, John Morris, and Fergal O'Gara

Introduction

A major attraction of the application of genetic engineering to microbial inoculant strains is the ability to create new strains with capabilities that may not be easily obtained by a process of natural selection. Such additional features may enhance the performance of an inoculant strain.

The fluorescent *Pseudomonas* species are an important group of bacteria that can play a beneficial role in the protection of crop plants against deleterious pathogenic micro-organisms. These bacteria produce a wide range of factors antagonistic to pathogens (particularly fungi) they include the yellow-green, fluorescent siderophores that characterize this group.

Our work has concentrated on the role of siderophores and antibiotics of fluorescent *Pseudomonas* in the biocontrol of 'damping-off' disease of sugar-beet seedlings (see Table 14.1). The biosynthesis and excretion of iron-binding siderophores are thought to decrease the pool of iron available to pathogenic fungi and other harmful micro-organisms and so contribute to the plant growth promotion effect observed when plants are inoculated with the appropriate *Pseudomonas* strain (Kloepper *et al.*, 1980). The producing micro-organism can use the resulting Fe^{3+} (ferric) siderophore complex via specific receptors located in its outer membrane (Magazin *et al.*, 1986).

Outer membrane receptor proteins can be highly specific and only transport the appropriate ferric siderophore into the producing cell. The structures of these fluorescent (yellow-green) siderophores was reviewed by Leong (1986).

Fluorescent siderophores are only produced under iron-limiting conditions, as are receptor proteins. The expression of these genes is coordinately regulated at the transcriptional level (Marugg *et al.*, 1988; O'Sullivan and O'Gara, 1991a) by ferric iron (Fe^{3+}) and this implies that all the promoters of the biosynthetic and receptor genes contain features recognized by regulatory protein(s). The identification and isolation of these regulatory elements provide a means to understand the expression of siderophores and to identify a potential source of genes with which to manipulate siderophore production genetically.

Manipulating the regulation of siderophore biosynthesis

The regulation by iron of the expression of siderophore genes has been studied in many groups by the use of iron-regulated promoter-*lacZ* fusions. A number of iron-regulated promoters have been cloned and characterized (O'Sullivan and O'Gara, 1991a).

The availability of iron-regulated promoter fusions was exploited by O'Sullivan and O'Gara (1990) to isolate regulatory mutants of *Pseudomonas* strain M114. We have isolated both positive and negative regulatory mutants. An iron-

Table 14.1. Traits for biocontrol in *Pseudomonas*

Trait	Modification	Potential 'Improvement'
Siderophore biosynthesis	Constitutive mutant (repressor gene)	Constitutive siderophore biosynthesis
Siderophore biosynthesis	Activator gene TAF	Expression of siderophore in new hosts
Ferric-siderophore uptake	Additional ferric-siderophore receptor genes	Siderophores host range. Improved competitiveness/colonization
Antibiotic (DAP) biosynthesis	Gene dosage/modified expression	Increased level of expression
Antibiotic (DAP) biosynthesis	Transfer to new hosts	Strains with improved biocontrol properties

regulated promoter-*lacZ* fusion plasmid was introduced into a Tn*5* mutant library of *Pseudomonas* strain M114 and a constitutive mutant was selected that expressed β-galactosidase both in the presence and absence of iron. This Tn*5* mutation (M114FR1) was further characterized and found constitutively to produce siderophore and ferric-siderophore receptor proteins.

Strain M114FR1 could inhibit the growth of fungal pathogens such as *Rhizoctonia solani in vitro* on high iron medium, whereas the parent strain was unable to inhibit this pathogen (O'Sullivan and O'Gara, 1991*b*). These mutants may have considerable potential in the biocontrol of root diseases. We are currently testing this mutant strain in greenhouse experiments. Preliminary results show that the mutant phenotype is stable in the rhizosphere/soil environment over a period of 4 weeks. The competitive and biocontrol ability compared with the parent M114 strain will be evaluated in rhizosphere microcosm experiments in the greenhouse.

Siderophore biosynthesis was also found to be positively regulated in strain M114. A siderophore negative/protease negative Tn*5*-lac mutant was isolated and a recombinant cosmid clone identified that complements this mutation. This plasmid is currently being characterized at the molecular genetic level.

Improved competitiveness: additional ferric siderophore receptors

The specificity of the outer membrane receptor determines whether one *Pseudomonas* strain can use the siderophore synthesized by another strain. The strain M114 ferric-siderophore receptor gene was isolated from a cosmid gene bank by transfer to strain B24, an isolate unable to utilize the M114 siderophore. Strain B24 transconjugants were screened for their ability to utilize the M114 ferric siderophore as a source of iron by a cross-feeding test (O'Sullivan *et al.*, 1990). The gene encoding the M114 ferric-siderophore receptor was located about 10 kbp from the constitutive regulator locus on cosmid pMS639 (Fig. 14.1). This cosmid contains a 27.2 kbp insert and also encodes a ferric-siderophore disassociation function and gene(s) encoding siderophore biosynthesis. The outer membrane receptor protein was deduced to be 89 kd in size from the outer membrane protein profiles (of B24 transconjugants harbouring the siderophore-receptor gene) (Morris *et al.*, 1992). Saturation Tn*5* mutagenesis located the receptor gene to a 1.8 kbp *Kpn*I fragment and allowed the construction of an M114-siderophore receptor negative mutant. The ability of this strain to utilize the M114 siderophore and siderophores from certain other strains was suppressed. However, the mutant could still utilize ferric-siderophores from other *Pseudomonas* strains, this indicates the presence of additional receptor(s). A second M114 receptor gene has been cloned and partially characterized (Morris *et al.*, 1992).

The ferric-siderophore receptor gene of strain B24 can use few siderophores derived from *Pseudomonas* isolates obtained from *Pythium* ('damping-off') infected soil. Transfer of the M114 siderophore receptor gene into strain B24

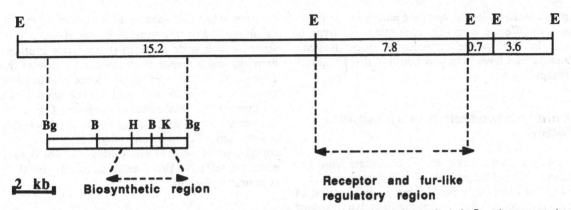

Fig. 14.1 Localization of a *fur*-like regulatory locus and the M114 ferric-siderophore receptor gene in *Pseudomonas* strain M114. E, *Eco*RI; Bg, *Bgl*II; B, *Bam*HI; H, *Hind*III; K, *Kpn*I.

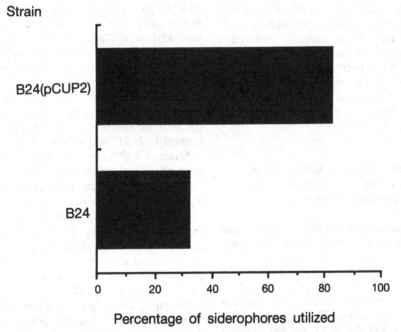

Fig. 14.2 Engineering the utilization of additional ferric-siderophores. The percentage of siderophores from 250 fluorescent pseudomonad isolates from a particular soil type, which were utilized for iron uptake. pCUP2 is a recombinant plasmid encoding the M114 ferric-siderophore uptake receptor.

increased the ability of this strain to utilize siderophores from 32% of 250 fluorescent pseudomonad isolates to 82% of strains tested in a cross-feeding bioassay (Fig. 14.2). This result indicates that either the M114 siderophore receptor has a broad range (or low level of specificity) or

M114-type siderophores are prevalent in these soil isolates.

In nature *Pseudomonas* strains appear to carry additional siderophore receptors for which they do not produce cognate siderophores (Morris *et al.*, 1992). This suggests that siderophore receptors may contribute to competitive ability in the rhizosphere. We are currently evaluating this hypothesis in rhizosphere microcosms using isogenic strains that have acquired additional siderophore receptor genes.

Antibiotic production as a biocontrol factor

Pseudomonas sp. strain F113, an isolate from the rhizosphere of sugar beet, was shown to inhibit a range of plant pathogenic fungi by production of an antibiotic. An antibiotic-negative mutant (F113G22) was generated with Tn5 and had lost the ability to inhibit both bacteria and fungi *in vitro* on high-iron agar plates. The antibiotic was subsequently purified by thin layer chromatography and identified as 2, 4-diacetylphloroglucinol (Phl) by mass spectral analysis (Shanahan *et al.*, 1992). Our laboratory has recently isolated a clone that complements mutant F113G22 and this will facilitate the genetic manipulation of Phl-producing strains for biocontrol purposes. These genes have been introduced into different rhizosphere bacteria where they express Phl in certain strains (Fenton *et al.*, 1992).

Towards a vector system suitable for environmental release of genetically engineered micro-organisms

For proper evaluation of the effectiveness of the genetic modifications described above (i.e. additional siderophore receptor genes or antibiotic biosynthetic genes or positive activating factors), a stable vector system is a prerequisite. Stable maintenance of an introduced genetic trait requires positive selective pressure. Furthermore, this selective pressure will have to be maintained throughout the field life cycle of the micro-organisms. For laboratory purposes, antibiotic resistance genes in conjunction with antibiotics are usually exploited to provide this selective pressure. In the field, however, no external selective pressure can be applied and antibiotic resistance genes are of limited use under these conditions. A solution has been described by Ross *et al.* (1990a,b), where the thymidylate synthase gene (*thyA*) of *Lactococcus lactis* was used as a positive selectable marker in a variety of micro-organisms. The *thy* system is based on a host strain, which is deficient in thymidylate synthase activity, and a vector that contains a copy of the *Lactococcus lactis thyA* gene. Since thymidylate synthase activity is essential for *de novo* DNA synthesis and thus survival of the micro-organism, the vector containing a copy of the gene is stably maintained. Furthermore, thymidine is a rare compound in natural environments, which may ensure a selective pressure to maintain the *thyA* containing plasmid.

Exploitation of the *thy* system in the symbiotic bacterium, *Rhizobium meliloti*

In order to exploit the *thy* system in *R. meliloti*, the first requirement was to isolate a strain deficient in thymidylate synthase activity. Spontaneous *thy* mutants were obtained from *R. meliloti* RM41 by a positive selection procedure, described by Okada *et al.* (1960).

Thymidylate synthase-deficient *R. meliloti* strains were screened for their reversion frequency (growth in absence of thymidine) and a stable mutant (CM21) was selected for further work.

Strain CM21 requires thymidine for growth, even on complex laboratory media such as mannitol salts yeast extract (MSY) and Luria broth agar (LB). The strain does not persist in the rooting medium and is unable to nodulate alfalfa seedlings, unless the rooting medium is supplemented with thymidine (data not shown). All these defects are overcome by the introduction of the heterologous *L. lactis thy* gene on a self-replicating plasmid (pGDT10).

During symbiosis, in the absence of antibiotics, pRK290-derived plasmids are readily lost (Fig. 14.3). In greenhouse laboratory-scale experiments, up to 70% of the bacteria isolated from the nodules of 5-week-old alfalfa seedlings failed to grow in the presence of tetracycline, indicating the loss of pRK290 (Boesten and O'Gara, unpublished). However, plasmids that contained the *thy* gene (pGDT10) were stably maintained in the CM21 background.

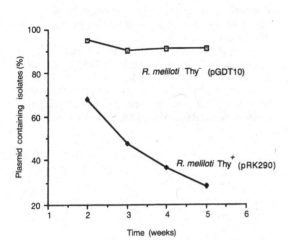

Fig. 14.3 Stability of the Thy⁺ plasmid pGDT10 and its parent pRK290 (Thy⁻) in *R. meliloti* Thy⁻ and Thy⁺ during symbiosis. Nodules were removed from alfalfa plants over a 5-week period. Bacteria were isolated and screened for the presence of the vector plasmid (Tet).

Adaption of the *thy* system for *Pseudomonas* spp.

Attempts to obtain a spontaneous *thyA*-deficient mutant strain of the fluorescent *Pseudomonas* strain M114 by the positive selection method of Okada *et al.* (1960) have been unsuccessful.

Alternative strategies to produce a *thy* M114 derivative are being used. Various *thy* clones from an M114 gene bank have been identified in our laboratory and Tn*5* mutants isolated. Following *in vitro* mutagenesis, the gene will be 'marker exchanged' into the M114 genome.

Acknowledgments

We thank Pat Higgins for technical assistance and Mary Cotter for typing the manuscript. This work was supported, in part, by grants from the Irish Science and Technology Agency (EOLAS) [St/008/89] and the European Community (ECLAIR-AGRE-0019-C, BAP-0413-C-(EDB), ASEAN CI1-0375-IRL (EDB) and BRIDGE BIOT-CT91-0283 and BIOT-CT91-0293).

References

Fenton, A. M., Stephens P. M., Crowley, J., O'Callaghan, M. and O'Gara, F. (1992). Exploiting gene(s) involved in 2, 4-diacetylphloroglucinol biosynthesis in order to improve the biocontrol ability of pseudomonad strains. *Applied and Environmental Microbiology* **58**, 3873–3878.

Kloepper, J. W., Leong, J., Téintze, M. and Schroth, M. N. (1980). *Pseudomonas* siderophores: a mechanism explaining disease suppressive soils. *Current Microbiology* **4**, 317–320.

Leong, J. (1986). Siderophores: their biochemistry and possible role in the biocontrol of plant pathogens. *Annual Reviews of Phytopathology*. **24**, 187–209.

Magazin, M., Moores, J. C. and Leong, J. (1986). Cloning of the gene coding for the outer membrane receptor protein for ferric pseudobactin, a siderophore from a plant growth-promoting *Pseudomonas* strain. *Journal of Biological Chemistry* **261**, 795–799.

Marugg, J. D., Nielander, H. B., Horrevoets, A. J. G., van Megen, I. van Genderen, I. and Weisbeek, P. J. (1988). Genetic organization and transcriptional analysis of a major gene cluster involved in siderophore biosynthesis in *Pseudomonas putida* WCS353. *Journal of Bacteriology* **170**, 1812–1819.

Morris, J., O'Sullivan, D. J., Koster, M., Leong, J., Weisbeek, P. J. and O'Gara, F. (1992). Characterization of fluorescent siderophore-mediated iron uptake in *Pseudomonas* sp. strain M114: evidence for the existence of an additional ferric-siderophore receptor. *Applied and Environmental Microbiology* **58**, 630–635.

Okada, T., Yanagisawa, F. and Ryan, F. J. (1960). Elective production of thymineless mutants. *Nature* **188**, 340–341.

O'Sullivan, D. J. and O'Gara, F. (1990). Iron regulation of ferric iron uptake in fluorescent pseudomonads: cloning of a regulatory gene. *Molecular Plant–Microbe Interactions*. **3**, 86–93.

O'Sullivan, D. J. and O'Gara, F. (1991*a*). Regulation of iron assimilation: nucleotide sequence analysis of an iron-regulated promoter from a fluorescent pseudomonad. *Molecular and General Genetics* **228**, 1–8.

O'Sullivan, D. J. and O'Gara, F. (1991*b*). Genetic improvement of siderophore production aimed at enhancing biocontrol in *Pseudomonas* strains. In *The Rhizosphere and Plant Growth* (ed. D. L. Keister and P. B. Cregan), Kluwer Academic Publishers, Dordrecht, the Netherlands.

O'Sullivan, D. J. Morris, J. and O'Gara, F. (1990). Identification of an additional ferric-siderophore uptake gene clustered with receptor, biosynthesis and *fur*-like regulatory genes in fluorescent *Pseudomonas* sp. strain M114. *Applied and Environmental Microbiology.* **56**, 2056–2064.

Ross, P., O'Gara, F. and Condon, S. (1990*a*). Cloning and characterization of the thymidylate synthase gene from *Lactococcus lactis.* subsp. *lactis. Applied Environmental Microbiology* **56**, 2156–2163.

Ross, P., O'Gara, F. and Condon, S. (1990*b*). Thymidylate synthase gene from *Lactococcus lactis* as a genetic marker: an alternative to antibiotic resistance genes. *Applied and Environmental Microbiology* **56**, 2164–2169.

Shanahan, P., O'Sullivan, D. J., Simpson, P., Glennon, J. D. and O'Gara, F. (1992). Isolation of 2,4-diacetylphloroglucinol from a fluorescent pseudomonad and investigation of physiological parameters influencing its production. *Applied and Environmental Microbiology* **58**, 353–358.

15

Biological control of foliar fungal diseases

Nyckle J. Fokkema

Introduction

Currently the only one commercialized biological control product against foliar fungal diseases exists, namely, 'Trichodex' (Makhteshim Chemical Works) for the control of grey mould of cucumber, tomato and grapevine (Y. Elad, personal communication). A few other products based on fungi are in the final stage of development or registration. Therefore, this review will primarily deal with the state of the art, which ranges from interesting hypotheses to promising research achievements under controlled and natural conditions. Compared with research efforts on biological control of soil-borne diseases, biological control of foliar diseases has up till now received little attention. The main reason for this appears to be the relatively efficient control of many above-ground pathogens by aerial applications of fungicides, while soil-borne pathogens are more difficult targets.

Nowadays, however, the world-wide awareness of the potential danger to the environment from many currently available chemical control agents and the increasing occurrence of fungicide resistance in a number of foliar pathogens has boosted research aimed at alternative means of disease control. For instance, in the Netherlands, a 'Multi-Year Crop Protection Plan' has been launched in 1990. This aims at a reduction in total amount of pesticide used of at least 35% by 1995 and 50% by the year 2000 (Anonymous, 1991). More importantly, the average emissions (the average pesticide emissions (to air, soil/ground water, surface water) to the environment are to be reduced to 50% in 1995 and to 30% in the year 2000.

The following three strategies are defined to meet these objectives:

- reduction of the *dependence* on chemical pesticides;
- reduction of the *use* of chemical pesticides;
- reduction of the *emissions* of chemical pesticides to the environment.

Breeding for disease resistance and biological control are the major tools in the first strategy. Apart from these strategies aimed at quantitative reductions, many fungicides currently used, will loose their registration, because of their adverse effect on the quality of the environment.

During the last 3 years a dramatic change in the political attitude towards pesticide use has occurred. This implies that biological control is no longer an option for disease control in instances where chemical control fails. Instead, it may, in general, be a desirable alternative for chemical control, including control of various leaf pathogens previously satisfactorily controlled by fungicides.

What strategies do plant pathologists have to offer? In this review the potential benefits and risks of three following strategies of microbial suppression in biological control of foliar diseases will be discussed: Microbial suppression of

1. infection
2. sporulation of the pathogen
3. survival of the pathogen

These strategies are chosen because they correspond to the three major stages in the life cycle of the pathogen. Success of biological control agents (BCAs) is not always based on interaction with the most vulnerable stage of the pathogen, but also on interference with the stage that allows a long interaction period between the antagonist and the pathogen.

Microbial suppression of infection

The time between arrival of a spore on the leaf surface, its germination and subsequent penetration of the pathogen into the leaf is short; under favourable climatic conditions less than 12 h. After penetration, the pathogen has generally escaped from the antagonistic interaction in the phyllosphere. Therefore, a successful biocontrol agent should be well established in the phyllosphere before the pathogen arrives. If the antagonist population rapidly declines after introduction, biocontrol is only likely to succeed when: (a) products responsible for inhibition remain present in a concentration high enough for inhibition of the pathogen; or when, (b) in the case of nutrient competition, the phyllosphere remains deprived of nutrients able to stimulate infection.

When discussing the benefits and risks of antagonists that are able to reduce leaf infections, it is useful to differentiate between naturally occurring biological control and introduced antagonists.

Naturally occurring antagonists

In moderate climates, the phyllosphere of field-grown plants is colonized by yeasts, predominantly *Sporobolomyces* and *Cryptococcus* spp., bacteria and a few hyphal fungi, mainly comprising *Cladosporium* spp. (Dickinson, 1976; Blakeman, 1985). In mixed inoculation experiments, individual isolates of the natural phyllosphere microflora interfere with spore germination, hyphal growth and penetration, frequently resulting in 50% or more reduction of infection (Blakeman and Fokkema, 1982). Necrotrophic pathogens, which often cause leaf necroses, utilize exogenous nutrients present in the phyllosphere and are, therefore, antagonized via nutrient competition by almost all microbial isolates able to colonize the phyllosphere. Biotrophic pathogens, such as mildews and rusts, which obtain their nutrients solely from the living host, are not affected by nutrient competition and consequently infection by these pathogens is only occasionally antagonized by specific bacterial isolates.

The importance of the naturally occurring mycoflora in disease suppression of necrotrophs is difficult to establish directly in the field, because fungicides that also affect pathogens, are the only efficient means of creating substantial differences in population densities of phyllosphere yeasts. By the use of broad-spectrum fungicides in a wheat crop, Dik *et al.* (1991*a*) could indirectly demonstrate that suppression of yeast populations resulted in an accumulation of aphid honeydew, an infection stimulating nutrient. Yeasts consume nutrients that otherwise would have benefited necrotrophic pathogens like *Septoria nodorum* (Fokkema *et al.*, 1983). The disease buffering capacity of naturally occurring yeasts has previously also been demonstrated in small-scale field experiments, in which a selective fungicide, benomyl, suppressed yeasts on rye leaves and stimulated infections by the benomyl-insenstive pathogen *Cochliobolus sativus* (Fokkema *et al.* 1975). Phyllosphere nutrients also reduce the sensitivity of pathogens to fungicides and, consequently, removal of nutrients by yeasts increased the efficacy of those fungicides that do not affect the yeast population (Dik, *et al.*, 1991*b*; Dik and Van Pelt, 1992).

The best way of taking advantage of this beneficial microbial buffer is to avoid removing it with broad spectrum fungicides (Fokkema, 1988).

It has been argued, however, that the saprophytic phyllosphere mycoflora may accelerate leaf senescence (Dickinson and Wallace, 1976; Smedegaard-Petersen and Tolstrup, 1985). This suggestion was largely based on increased yields of cereals observed after fungicide applications in the apparent absence of pathogens. Direct positive effects of fungicides on the plant or control of invisible latent infections, however, can equally well account for these observations (Fokkema, 1981). In addition, the assumed detrimental effects of yeasts and *Cladosporium* spp. could not be demonstrated experimentally (Fokkema *et al.*, 1979*b*; Frossard *et al.*, 1983).

In warmer climates, however, fungi that form the sooty mould complex may, in the presence of abundant honeydew, affect the quality of fruits. In general, there seems to be no agricultural risk in the preservation of the natural phyllosphere mycoflora.

Introduction of antagonists

Yeasts and other phyllosphere micro-organisms
The discovery of the disease-suppressing potential of the natural phyllosphere microflora stimulated research to exploit this antagonism further by deliberate introduction of these micro-organisms. It remains questionable, however, whether spraying yeasts onto mature leaves to prevent infection is economically feasible, because the natural yeast population often develops rapidly to an antagonistic level (Fokkema *et al.*, 1979*a*). Moreover, a substantial increase in the yeast population can only be achieved if the introduced yeasts are supplemented with nutrients. Since these nutrients stimulate the pathogen as well, yeasts can only be applied in an uneconomic preventive way well before the pathogen arrives.

A sudden natural excess of nutrients in a sparsely colonized phyllosphere, for example as a result of man-made wounds or a massive release of pollen grains, however, may create a more suitable situation for microbial introductions. Yeasts as well as bacteria are reported to protect wounds of several fruits against infection by post-harvest pathogens in fruit-packing plants (Janisiewicz, 1988; McLaughlin *et al.*, 1992; Wilson and Wisniewski, 1992). Yeasts generally operate through nutrient competion and the production of toxic metabolites is not necessary for antagonism, which makes them attractive for use on consumable products. On young leaves and flower petals with very few natural antagonists, introduced yeasts (Redmond *et al.*, 1987) and hyphal fungi, such as *Epicoccum purpurescens* and *Trichoderma* spp. (Nelson and Powelson, 1988; Boland and Inglish, 1989; Zhou and Reedeler, 1989) controlled infection by the pathogens *Botrytis cinerea* and *Sclerotinia sclerotiorum*, which may have profited from pollen and flower remains. The non-phyllosphere fungi, *Trichoderma* and *Gliocladium* spp. have been studied particularly in these situations.

Trichoderma spp.
Although *Trichoderma* spp. are soil fungi, there are various reports on successful use in the biological control of the grey mould fungus, *Botrytis cinerea*,

in grapes, strawberries, apple and tomato (Tronsmo 1986; and see, Jensen and Wolffhechel, Chapter 16). Dubos (1987) achieved, during 8 years, an average control of grape rot by *B. cinerea* of 65% with four carefully timed applications from flowering till 3 weeks before harvest. Successful integration with the chemical, vinclozolin, has been reported by Gullino and Garibaldi (1983). *Trichoderma viride* could also reduce onion neck rot by *Botrytis cinerea* during storage, when applied to the fresh wounds made by leaf topping (Köhl *et al.*, 1991). The related antagonist, *Gliocladium roseum*, reduced *Botrytis* fruit rot of strawberry (Peng and Sutton, 1991). The infection occurs early during flowering and therefore honeybees can be used for safe and economic delivery of the antagonist (Peng *et al.*, 1992). The positive prospects for the use of *Trichoderma* in *Botrytis* control resulted in a commercial product 'Trichodex' based on *Trichoderma harzianum* (T-39). Particularly in alternation with fungicides a similar degree of protection can be obtained as with fungicides alone. This allows a reduction in fungicide use of 50% (Elad and Zimand, 1991).

The mechanism of interaction is unknown, although it is likely that *Trichoderma* spp. compete with *Botrytis* for nutrients offered as flower remains and pollen grains. These infection-stimulating nutrients may occur suddenly early in the season in such quantities that they cannot be adequately removed by the native mycoflora. Proper timing of the applications seems essential, because in the absence of exogenous nutrients, the survival of *Trichoderma* spp. on green leaves is generally very poor (J. Köhl, personal communication).

Bacteria
Under conditions of very high humidity and regular periods of free water, bacteria, such as *Pseudomonas* spp. (Levy *et al.*, 1989), *Bacillus subtilis* (Rytter *et al.*, 1989) and *Erwinia herbicola* (Kempf and Wolf, 1989) may control necrotrophic as well as biotrophic pathogens, although the mechanisms may differ. Under field conditions, however, introduced bacterial populations normally decline rapidly to densities that are without much effect (Knudsen and Spurr, 1988). In practice at least weekly applications of bacterial antagonists were necessary for sufficient control of bean rust, which also indicates that antibiotics reponsible for

inhibition seem not to be very stable in the phyllosphere (Baker et al., 1985). This instability of introduced bacterial populations is also observed in natural bacterial populations in the phyllosphere, which show a much larger temporal (daily) and spatial variation (from leaf to leaf) than yeast populations (Fokkema and Schippers, 1986; Andrews, 1992; Hirano and Upper, 1991).

An interesting perspective which may overcome the above-mentioned disadvantage, however, is the phenomenon of induced resistance by bacteria or bacterial metabolites. Microbial culture filtrates have been shown to induce resistance against powdery and downy mildew and rusts in a number of crops (Schönbeck and Dehne, 1986). In wheat, penetration by the powdery mildew fungus, haustorium formation and sporulation, but not the germination and appressorium formation, were affected by spraying with bacterial culture filtrate (Schönbeck and Dehne, 1986). The advantages of induced resistance are that it might not be necessary to maintain high bacterial densities in the phyllosphere and that inhibition of the pathogen continues inside the leaf.

Induced resistance may also be one of the mechanisms involved in disease control by spraying with watery compost extract. Field applications of horse manure compost extract controlled downy and powdery mildew in grapevine, Botrytis in strawberries, and (provided micro-organisms were supplemented during the extraction period) Phytophthora infestans in potato (Weltzien, 1991). The mechanisms involved are still largely unknown, but the effect of control is lost after sterilization by heat or filtration, this suggests that viable micro-organisms need to be applied to the leaves. Compost extract has the additional advantage that it can be prepared by (organic) farmers themselves, and thus may also have economic advantages. The observed control of potato late blight by compost extract looks very promising as an alternative to the high input of chemicals in conventional farming. Its control is, however, in contrast to that of the other diseases, not yet consistent. In bioassays on detached potato leaves and whole plants, only one type of compost and several bacterial isolates suppressed Phytophthora infection by more than 75%, but consistent control under field conditions have still to be demonstrated (Jongebloed et al., 1994).

Microbial suppression of the sporulation of the pathogen

Microbial suppression of sporulation, with the subsequent retarded dissemination of the pathogen, is in principle effective against diseases that have many infection cycles per season. It has the advantage of a long interaction period. In this respect the possibilities of the introduction of naturally occurring hyperparasites of biotrophic mildew and rust fungi have already been studied for a long time (Sundheim and Tronsmo, 1988). The same principle can be adopted for the suppression of sporulation of necrotrophic pathogens that may occur on the leaf necroses as well as on dead plant parts and plant debris. With respect to necrotrophs not only hyperparasitisms but also substrate competition by saprophytes should be considered as a suitable approach.

Control of biotrophic pathogens

By definition, biotrophic pathogens sporulate on living host tissue only. This implies that some disease should be tolerated when hyperparasites are used for control. This may be no problem in diseases where the leaves, but not the marketable products, are attacked as is for instance the case with mildew in cucumber, tomato and fruit trees.

Research on hyperparasites to control rusts has concentrated on the hyperparasites Eudarluca caricis, Verticillium lecanii, Aphanocladium album, and Cladosporium spp. (Kranz, 1981). Some of these are also reported as inhibitors of uredospore germination and thus reduce infection, but all seem to have their major effect through suppression of sporulation. The required high humidity for the parasitism seems to be the major constraint against their use. Considering the achievements with Verticillium lecanii as a microbial pesticide in insect control (Hall, 1982), further exploration of this hyperparasite in rust control might be rewarding (Mendgen, 1981; Spencer and Atkey, 1981).

Powdery mildews, of which the biomass is almost entirely superficial, are perhaps an easier target for hyperparasites than rusts. Of the numerous hyperparasitic fungi known (Hijwegen, 1988), Ampelomyces quisqualis (Philipp et al., 1984; Sundheim, 1986), Tilletiopsis spp., Verticillium lecanii (Hijwegen, 1988) and Stephanoascus spp. (Hajlaoui and Bélanger, 1991) are being investigated intensively for practical application against mildews. Formulations, e.g. based on paraffin oil (Philipp *et al.*, 1990), as well as selection for drought resistant mutants, may reduce the dependency on humid conditions. The common belief that glasshouse conditions are more suitable for hyperparasitism than the outdoor environment is not substantiated. Successful field experiments with *A. quisqualis* demonstrated control of powdery mildew in Israel on carrots, cucumber and mango (Sztejnberg *et al.*, 1989) and in New York State on grapevine (Gadoury, 1991).

Control of necrotrophic pathogens

Many necrotrophs sporulate abundantly on leaf necroses as well as on dead leaves and other plant remains. Nevertheless, microbial interference with this stage in the development of the pathogen has barely been explored for necrotrophs. An exception is research on the control of apple scab by the application of urea, which stimulates natural antagonism and decomposition of leaf litter and thus reduces the ascospore production of *Venturia inaequalis* in spring by more than 90% (Burchill, 1972; Margraf *et al.*, 1972; Latorre and Marin, 1982). Applications of the hyperparasites *Chaetomium globosum* and *Athelia bombacina* to leaf litter have a similar effect (Heye and Andrews, 1983; Miedtke and Kennel, 1990). However, the apple scab pathogen has a rapid secondary conidial dissemination that may eventually neutralize the initial reduction of primary inoculum. Even so, attempts to combine microbial suppression of ascospore production with fungicides or resistant varieties may be rewarding. These, and other studies on the inhibition of pathogen populations on straw by natural saprophytes clearly indicate that in general interactions on dead plant material may strongly interfere with the inoculum production (Cook, 1970; Pfender, 1988).

Research along this line has recently been initiated in Canada and in the Netherlands with respect to *Botrytis* control in strawberries (Peng and Sutton, 1990) and onions (Köhl *et al.*, 1992*a*) respectively. Conidia of *Botrytis cinerea* as well as of more specialized *Botrytis* spp. that originate from dead plant remains and necrotic leaf areas in the crop have a major impact on disease development (Sutton, 1990). This continuously produced inoculum may be restricted by the introduction of saprophytes that normally colonize dead plant material. *Gliocladium* spp., *Trichoderma* spp. and other fungi have been tested in bioassays and under field conditions. In strawberry, there was a good correlation between suppression of sporulation on dead leaf discs in the laboratory and with the reduction of fruit rot in the field. *Gliocladium roseum* reduces *B. cinerea* in overwintering strawberry leaves and was equally as effective as the fungicides captan or chlorothalonil (Peng and Sutton, 1990). The same antagonist also reduced latent infection of young strawberries during flowering (see above). In the Netherlands, biological control of *Botrytis* leaf blight of onions is attempted by suppression of sporulation on dead leaves. Bioassays at different water potentials enable selection of antagonists resistant to dry conditions. *Gliocladium roseum* performed better at low water potential than *Trichoderma viride*, with little difference between individual isolates (Köhl *et al.* 1992*b*). However, in the first field experiment *G. roseum* applications did not result in sufficient colonization to suppress sporulation. Contrary to some other saprophytes *G. roseum* is sensitive to regularly occurring changes in wetness, which seems to be a more important parameter for ecocompetence than drought resistance (J. Köhl, personal communication). Nevertheless, a treatment in which 30–50% of the necrotic tissue was artificially removed from the field reduced the concentration of *B. cinerea* conidia in the air and the number of lesions on onion leaves by 40% (Köhl *et al.*, 1992*a*). This demonstrated the validity of the hypothesis that substrate removal, either by hand or by micro-organisms, is a promising strategy in biocontrol of *Botrytis* spp. Since biocontrol on dead leaf material is aimed at giving naturally occurring saprophytes an advantage in competition for the available substrate, this strategy seems to have limited risks for the environment and the consumers. *Botrytis cinerea* causes a disease of world-wide importance which is unlikely to be controlled by breeding for host-plant resistance and control therefore demands an enormous input of chemicals of which several are of limited use because of development of fungicide-resistance in the pathogen. Consequently, there is an economic as well as an environmental need for alternative methods to control this pathogen. The general principle of microbial suppression of sporulation might be applicable to a variety of necrotrophic pathogens.

Microbial suppression of the survival of the pathogen

Foliar pathogens of the genera *Botrytis* and *Sclerotinia* often produce sclerotia in infected tissue. These are melanized thick-walled mycelial structures by which the pathogen may survive in the soil for several years outside the host. The primary inoculum for newly planted crops may originate from these sclerotia. Biological control aimed at these survival structures mainly focuses on enhanced degradation by hyperparasites in the soil (Adams, 1990; Whipps and Gerlagh, 1992). However, when formed on and in aerial plant parts, interference with the production and viability of freshly formed sclerotia by spray application of hyperparasites is also an attractive option for control of sclerotial diseases (Trutmann *et al.*, 1982).

Sclerotinia sclerotiorum infects all crops except graminaceous species by ascospores produced in small toadstools (apothecia) that arise from soil-borne sclerotia. In field-grown snap beans, the pathogen colonizes the whole plant, initially from lesions caused by ascospore infections, which results in large rotten areas in which sclerotia are formed. In contrast to the earlier mentioned apple scab fungus, this disease has no secondary dissemination by conidia. Therefore, microbial interference with the production of the sclerotia may be expected to reduce disease in the following seasons. *Coniothyrium minitans*, a well-known hyperparasite of sclerotia and associated with natural disease decline, has been applied to the soil and to the aerial parts of plants with considerable success (Whipps and Gerlagh, 1992). Only a few sprays of field-grown snap beans with suspensions of *C. minitans* conidia reduced the viability of the sclerotia by 85% (Fokkema *et al.*, 1992), which had a significant impact on the number of apothecia in the following seasons (M. Gerlagh, personal communication). *C. minitans* is able to colonize and to fructify on plant tissue infected by *S. sclerotiorum*. A five-year field experiment currently being carried out in the Netherlands will reveal whether a continous reduction in the numbers of sclerotia formed will reduce the ascospore infections to a tolerable level.

Sclerotial diseases in arable crops often gradually become a problem over time, so why not be realistic and aim at an even gradual disappearance?

Conclusion

There are three major strategies available for biological control of foliar fungal pathogens. This does not mean that it is always clear along which strategy a successful biological control agent is effective. Inhibition of infection as well as suppression of sporulation for instance can both be involved in disease control by *Trichoderma* spp.

In biocontrol we should not automatically try to imitate the effect of fungicides by spraying healthy leaves with BCAs to prevent infection. In fact, prevention of infection seems to be the most difficult approach for biocontrol because of the very short period of interaction time available. Use of induced resistance, however, may reduce this disadvantage.

Suppression of sporulation or survival has the great advantage of a long interaction period. Contrary to protection of leaf infection where preventive sprayings are necessary, these strategies allow BCA-application when the disease really appears. Treatments may continue after harvest on crop remains to reduce future inoculum pressure.

In spite of valuable texts on the microbial ecology of the phyllosphere (Andrews and Hirano, 1991; Blakeman, 1981; Blakeman and Fokkema, 1982; Fokkema and Van den Heuvel, 1986), the limited knowledge of the ecology of the BCA's seems to be a major constraint for successful biocontrol of foliar diseases. As long as insight into the characteristics necessary for colonization is lacking, the progress will remain poor. Genetic modification of BCAs may serve as an experimental tool for discovering such traits and for genetically marking introduced BCAs in population studies (Lindow, 1991). As far as practical application is concerned, there is currently no reason to believe that genetically modified BCAs will have an advantage over natural isolates. Development of dynamic simulation models of substrate colonization by antagonists as well as of pathogen development will further improve the understanding of the potential for control under a variety of natural conditions (Dik, 1991; Knudsen and Hudler, 1987).

In contrast to specific reports on the potential benefits of introducing microbial antagonists for disease control, information on the potential risks of such introductions is much more general in nature, and is a basis for scientific speculation. Risks may be directed towards the farmer, the consumer and the environment. Microbial ecologists should be able to estimate the risks for the environment. In all the above-mentioned examples, biocontrol is attempted with non-exotic naturally occurring micro-organisms. Introduction has only a temporary effect on the natural balance of micro-organisms including the pathogen. Although, perhaps desired in view of disease control, long-term effects have never been noticed. Whether genetically altered micro-organisms create additional risks, totally depends on the nature of the genetic modification. The added risks inherent in the use of deletion mutants or genetically marked micro-organisms are much less than those involved in combinations of traits that are unlikely ever to occur in nature. For instance, adding genes for antibiotic production to phyllosphere yeasts may create enormous problems with respect to natural microbial succession and leaf decomposition.

With respect to the farmer's safety, the potential hazard of exposure to an excess of BCAs should be evaluated. Literature exists on agriculture-related diseases caused by common fungal saprophytes and details the risks of allergies and eye irritations during the handling of mouldy hay and harvesting of cereal crops (Burge *et al.* 1991). Whether similar risks may occur during the application of BCAs depends on the nature of the BCA and the concentration in aerosols.

Human and animal health may be a subject of concern if the products treated with the BCAs are subsequently consumed. These products should be checked for harmful levels of antibiotics and other secondary metabolites (e.g. mycotoxins). If the BCAs are able to grow at body temperature, they may have the potential to infect humans and animals. Apart from these sensible precautions, there is no reason for any particular concern from the use of micro-organisms in biocontrol, because micro-organisms, in contrast to pesticides, are naturally associated with plant products. For instance, the numbers of phyllosphere yeasts on ripe strawberries and apples may amount to 200 000 cells/cm^2 (N. J. Fokkema, unpublished).

Currently, microbial biocontrol agents are treated by most regulation authorities as 'chemicals'. This implies extensive testing procedures before registration, which is often too expensive for commercialization of BCAs with a small potential market. There is a growing awareness, however, that depending on the nature and means of application of the BCA, a more simple registration procedure should be possible (Cook, 1992).

Acknowledgment

I would like to thank my colleagues Thijs Gerlagh and Jürgen Köhl for their helpful suggestions during the preparation of this review.

References

Adams, P. B. (1990). The potential of mycoparasites for biological control of plant diseases. *Annual Review of Phytopathology* **28**, 59–72.

Andrews, J. H. (1992). Biological control in the phyllosphere. *Annual Review of Phytopathology* **30**, 603–635.

Andrews, J. H. and Hirano, S. S. (1991). *Microbial Ecology of Leaves*. Springer-Verlag, New York.

Anonymous (1991). *Meerjarenplan Gewasbescherming. Regeringsbeslissing*. SDU Uitgeverij, Den Haag.

Baker, C. J., Stavely, J. R. and Mock, N. (1985). Biocontrol of bean rust by *Bacillus subtilis* under field conditions. *Plant Disease* **69**, 770–772.

Blakeman, J. P. (1981). *Microbial Ecology of the Phylloplane*. Academic Press, London.

Blakeman, J. P. (1985). Ecological succession of leaf surface micoorganisms in relation to biological control. In *Biological Control on the Phylloplane* (ed. C. E. Windels and S. E. Lindow), pp. 6–30. American Phytopathological Society, St Paul, MI.

Blakeman, J. P. and Fokkema, N. J. (1982). Potential for biological control of plant diseases on the phylloplane. *Annual Review of Phytopathology* **20**, 167–192.

Boland, G. J. and Inglish, G. D. (1989). Antagonism of white mold (*Sclerotinia sclerotiorum*) of bean by fungi from bean and rapeseed flowers. *Canadian Journal of Botany* **67**, 1775–1781.

Burchill, R. T. (1972). Comparison of fungicides for surpressing ascospore production by *Venturia inaequalis* (Cke.) Wint. *Plant Pathology* **21**, 19–22.

Burge, H. A., Muilenberg, M. L. and Chapman, J. A. (1991). Crop plants as a source of fungus spores of medical importance. In *Microbial Ecology of Leaves* (ed. J. H. Andrews and S. S. Hirano), pp. 222–236. Springer-Verlag, New York.

Cook, R. J. (1970). Factors affecting saprophytic colonization of wheat straw by *Fusarium roseum* f. sp. *cerealis* 'culmorum'. *Phytopathology* **60**, 1672–1676.

Cook, R. J. (1992). Reflections of a regulated biological control researcher. In *Regulations and Guidelines: Critical Issues in Biological Control* (Proceedings of a USDA/CSRS National Workshop) (ed. R. Charudattan and H. W. Browning), pp. 9–24. Institute of Food and Agricultural Sciences, University of Florida, Gainesville.

Dickinson, C. H. (1976). Fungi on the aerial surfaces of higher plants. In *Microbiology of Aerial Plant Surfaces* (ed. C. H. Dickinson and T. F. Preece), pp. 293–324. Academic Press, London.

Dickinson, C. H. and Wallace, B. (1976). Effects of late applications of foliar fungicides on activity of micro-organisms on winter wheat flag leaves. *Transactions of the British Mycological Society* **76**, 103–112.

Dik, A. J. (1991). Interactions among fungicides, pathogens, yeasts, and nutrients in the phyllosphere. In *Microbial Ecology of Leaves* (ed. J. H. Andrews and S. S. Hirano), pp. 412–429. Springer-Verlag, New York.

Dik, A. J. and Van Pelt, J. A. (1992). Interaction between phyllosphere yeasts, aphid honeydew and fungicide effectiveness in wheat under field conditions. *Plant Pathology* **41**, 661–675.

Dik, A. J., Fokkema, N. J. and Van Pelt, J. A. (1991a). Consumption of aphid honeydew, a wheat yield reducing factor, by phyllosphere yeasts under field conditions. *Netherlands Journal of Plant Pathology* **97**, 209–232.

Dik, A. J., Fokkema, N. J. and Van Pelt, J. A. (1991b). Interference of nutrients with fungicide activity against *Septoria nodorum* on wheat leaves. *Plant Pathology* **40**, 25–37.

Dubos, B. (1987). Fungal antagonism in aerial agrobiocenoses. In *Innovative Approaches to Plant Disease Control* (ed. I. Chet), pp. 107–135. Wiley & Sons, New York.

Elad, Y. and Zimand, G. (1991). Experience in integrated chemical-biological control of grey mould (*Botrytis cinerea*). *IOBC/WPRS Bulletin* **XIV/5**, 195–199.

Fokkema, N. J. (1981). Fungal leaf saprophytes, beneficial or detrimental? In *Microbial Ecology of the Phylloplane* (ed. J. P. Blakeman), pp. 433–454. Academic Press, London.

Fokkema, N. J. (1988). Agrochemicals and the beneficial role of phyllosphere yeasts in disease control. *Ecological Bulletins* **39**, 91–93.

Fokkema, N. J. and Van den Heuvel, J. (1986). *Microbiology of the Phyllosphere*. Cambridge University Press, Cambridge.

Fokkema, N. J. and Schippers, B. (1986). Phyllosphere versus rhizosphere as environments for saprophytic colonization. In *Microbiology of the Phyllosphere* (ed. N. J. Fokkema and J. van den Heuvel), pp. 137–159. Cambridge University Press, Cambridge.

Fokkema, N. J., Van Laar, J. A. J., Nelis-Blomberg, A. L. and Schippers, B. (1975). The buffering capacity of the natural mycoflora of rye leaves to infection by *Cochliobolus sativus*, and its susceptibility to benomyl. *Netherlands Journal of Plant Pathology* **81**, 176–186.

Fokkema, N. J., Den Houter, J. G., Kosterman, Y. J. C. and Nelis, A. L. (1979a). Manipulation of yeasts on field-grown wheat leaves and their antagonistic effect on *Cochliobolus sativus* and *Septoria nodorum*. *Transactions of the British Mycological Society* **72**, 19–29.

Fokkema, N. J., Kastelein, P. and Post, B. J. (1979b). No evidence for acceleration of leaf senescence by phyllosphere saprophytes of wheat. *Transactions of the British Mycological Society* **72**, 312–315.

Fokkema, N. J., Riphagen, I., Poot, R. J. and De Jong, C. (1983). Aphid honeydew, a potential stimulant of *Cochliobolus sativus* and *Septoria nodorum* and the competitive role of saprophytic mycoflora. *Transactions of the British Mycological Society* **81**, 355–363.

Fokkema, N. J., Gerlagh, M. and Köhl, J. (1992). Biological control of *Sclerotinia sclerotiorum* and *Botrytis* spp. In *Biological Control of Plant Diseases* (ed. E. C. Tjamos, G. C. Papavizas and R. J. Cook), pp. 267–271. Plenum Press, New York.

Frossard, R., Fokkema, N. J. and Tietema, T. (1983). Influence of *Sporobolomyces roseus* and *Cladosporium cladosporioides* on leaching of ^{14}C-labelled assimilates from wheat leaves. *Transactions of the British Mycological Society* **80**, 289–296.

Gadoury, D. M. Pearson, R. C. and Seem, R. C. (1991). Reduction of the incidence and severity of grape powdery mildew by *Ampelomyces quisqualis*. *Phytopathology* **81**, 122.

Gullino, M. L. and Garibaldi, A. (1983). Situation actuelle et perspectives d'avenir de la lutte biologique et integrée contre la pourriture grise de la vigne en Italie. *Les Colloques de l'INRA* **18**, 91–97.

Hajlaoui, M. R. and Bélanger, R. R. (1991). Comparative effects of temperature and humidity on the activity of three potential antagonists of rose powdery mildew. *Netherlands Journal of Plant Pathology* **97**, 203–208.

Hall, R. A. (1982). Control of whitefly *Trialeurodes vaporariorum* and cotton aphid, *Aphis gossypii*, in glasshouses by two isolates of the fungus, *Verticillium lecanii. Annals of Applied Biology* **101**, 1–11.

Heye, C. C. and Andrews, J. H. (1983). Antagonism of *Athelia bombacina* and *Chaetomium globosum* to the apple scab pathogen, *Venturia inaequalis. Phytopathology* **73**, 650–654.

Hijwegen, T. (1988). Effect of seventeen fungicolous fungi on sporulation of cucumber powdery mildew. *Netherlands Journal of Plant Pathology* **94**, 185–190.

Hirano, S. S. and Upper, C. D. (1991). Bacterial community dynamics. In *Microbial Ecology of Leaves* (ed. J. H. Andrews and S. S. Hirano), pp. 271–294. Springer-Verlag, New York.

Janisiewicz, W. (1988). Biological control of diseases of fruit. In *Biocontrol of Plant Diseases*, vol. 2 (ed. K. G. Mukerji and K. L. Garg), pp. 153–165. CRC Press, Boca Raton.

Jongebloed, P. H. J., Kessel, G. J. T., Molhoek, W. M. L., Van der Plas, C. H. and Fokkema, N. J. (1993). Biological control of *Phytophthora infestans* with compost extracts and selected bacterial antagonists. *IOBC/WPRS Bulletin* **XVI/11**, 16–20.

Kempf, H. J. and Wolf, G. (1989). *Erwinia herbicola* as a biocontrol agent of *Fusarium culmorum* and *Puccinea recondita* f. sp. *tritici* on wheat. *Phytopathology* **79**, 990–994.

Knudsen, G. R. and Hudler, G. W. (1987). Use of a computer simulation model to evaluate a plant disease biocontrol agent. *Ecological Modelling* **35**, 45–62.

Knudsen, G. R. and Spurr, H. W. Jr. (1988). Management of bacterial populations for foliar disease biocontrol. In *Biocontrol of Plant Diseases*, vol. I (ed. K. G. Mukerji and K. L. Garg), pp. 83–92. CRC Press, Boca Raton.

Köhl, J., Molhoek, W. M. L. and Fokkema, N. J. (1991). Biological control of onion neck rot (*Botrytis aclada*): protection of wounds made by leaf topping. *Biocontrol Science and Technology* **1**, 261–269.

Köhl, J., Molhoek, W. M. L., Van der Plas, C. H., Kessel, G. J. T. and Fokkema, N. J. (1992*a*). Biological control of *Botrytis* leaf blight on onions: significance of sporulation suppression. In *Recent advances in Botrytis research*, (ed. K. Verhoeff, N. E. Malathrakis and B. Williamson), pp. 192–196. PUDOC, Wageningen.

Köhl, J., Krijger, M. C. and Kessel, G. J. T. (1992*b*). Drought tolerance of *Botrytis squamosa, B. aclada* and potential antagonists. In *Recent Advances in Botrytis Research* (ed. K. Verhoeff, N. E. Malathrakis and B. Williamson), pp. 206–210. PUDOC, Wageningen.

Kranz, J. (1981). Hyperparasitism of biotrophic fungi. In *Microbial Ecology of the Phylloplane* (ed. J. P.

Blakeman), pp. 327–352. Academic Press, London.

Latorre, B. A. and Marin, G. (1982). Effect of bitertanol, fenarimol, and urea as fall treatments on *Venturia pirina* ascospore production. *Plant Disease* **66**, 585–586.

Levy, E., Eyal, Z., Carmely, S., Kashman, Y. and Chet, I. (1989). Suppression of *Septoria tritici* and *Puccinia recondita* of wheat by an antibiotic-producing fluorescent pseudomonad. *Plant Pathology* **38**, 564–570.

Lindow, S. E. (1991). Determinants of epiphytic fitness in bacteria. In *Microbial Ecology of Leaves* (ed. J. H. Andrews and S. S. Hirano), pp. 295–314. Springer-Verlag, New York.

McLaughlin, R. J., Wilson, C. L., Droby, S., Ben-Arie, R. and Chalutz, E. (1992). Biological control of postharvest diseases of grape, peach, and apple with the yeasts *Kloeckera apiculata* and *Candida guilliermondii. Plant Disease* **76**, 470–473.

Margraff, K., Hoffmann, I. and Koberstein, S. (1972). Harnstoff-Blattfallbespritzungen gegen Obstscharf Erreger. *Nachrichtenblatt für den Pflanzenschutz in der DDR* **26**, 255–256.

Mendgen, K. (1981). Growth of *Verticillium lecanii* in pustules of stripe rust (*Puccinia striiformis*). *Phytopathologische Zeitschrift* **102**, 301–309.

Miedtke, K. and Kennel, W. (1990). *Athelia bombacina* and *Chaetomium globosum* as antagonists of the perfect stage of the apple scab pathogen (*Venturia inaequalis*) under field conditions. *Zeitschrift für Pflanzenkrankheiten und Pflanzenschutz* **97**, 24–32.

Nelson, M. E. and Powelson, M. L. (1988). Biological control of grey mold of snap beans by *Trichoderma hamatum. Plant Disease* **72**, 727–729.

Peng, S. and Sutton, J. C. (1990). Biological methods to control grey mould of strawberry. *Proceedings of the British Crop Protection Conference – Pests and Diseases*, 233–240.

Peng, G. and Sutton, J. C. (1991). Evaluation of micro-organisms for biocontrol of *Botrytis cinerea* in strawberry. *Canadian Journal of Plant Pathology* **13**, 247–257.

Peng, G., Sutton, J. C. and Kevan, P. G. (1992). Effectiveness of honey bees for applying the biocontrol agent *Gliocladium roseum* to strawberry flowers to suppress *Botrytis cinerea. Canadian Journal of Plant Pathology* **14**, 117–129.

Pfender, P. F. (1988). Suppression of ascocarp formation in *Pyrenophora triticirepentis* by *Limonomyces roseipellis*, a basidiomycete from reduced-tillage wheat straw. *Phytopathology* **78**, 1254–1258.

Philipp, W.-D., Grauer, U. and Grossmann, F. (1984). Ergänzende Untersuchungen zur biologischen und integrierten Bekämpfung von

Gurkenmehltau unter Glas durch *Ampelomyces quisqualis*. *Zeitschrift für Pflanzenkrankheiten und Pflanzenschutz* **91**, 438–443.

Philipp, W.-D., Beuther, D. H., Hermann, D., Klinkert, F., Oberwalder, C., Schmidtke, M. and Staub, B. (1990). Zur Formulierung des Mehltauhyperparasiten *Ampelomyces quisqualis* Ces. *Zeitschrift für Pflanzenkrankheiten und Pflanzenschutz* **97**, 120–132.

Redmond, J. C., Marois, J. J. and MacDonald, J. D. (1987). Biological control of *Botrytis cinerea* with epiphytic microorganisms. *Plant Disease*. **71**, 799–802.

Rytter, J. L., Lukezic, F. L., Craig, R. and Moorman, G. W. (1989). Biological control of geranium rust by *Bacillus subtilis*. *Phytopathology* **79**, 367–370.

Schönbeck, F. and Dehne, H. W. (1986). Use of microbial metabolites inducing resistance against plant pathogens. In *Microbiology of the Phyllosphere* (ed. N. J. Fokkema and J. van den Heuvel), pp. 363–375. Cambridge University Press, Cambridge.

Smedegaard-Petersen, V. and Tolstrup, K. (1985). The limiting effect of disease resistance on yield. *Annual Review of Phytopathology* **23**, 475–490.

Spencer, D. M. and Atkey, P. T. (1981). Parasitic effects of *Verticillium lecanii* on two rust fungi. *Transactions of the British Mycological Society* **77**, 535–542.

Sundheim, L. (1986). Use of hyperparasites in biological control of biotrophic plant pathogens. In *Microbiology of the Phyllosphere* (ed. N. J. Fokkema and J. van den Heuvel), pp. 333–347. Cambridge University Press, Cambridge.

Sundheim, L. and Tronsmo, A. (1988).

Hyperparasites in biological control. In *Biocontrol of Plant Diseases, vol. I* (ed. K. G. Mukerji and K. L. Garg), pp. 53–69. CRC Press, Boca Raton, FL.

Sutton, J. C. (1990). Epidemiology and management of botrytis leaf blight of onion and grey mold of strawberry: a comparative analysis. *Canadian Journal of Plant Pathology* **12**, 100–110.

Sztejnberg, A., Galper, S., Mazar, S. and Lisker, N. (1989). *Ampelomyces quisqualis* for biological and integrated control of powdery mildews in Israel. *Journal of Phytopathology* **124**, 285–295.

Tronsmo, A. (1986). Use of *Trichoderma* spp. in biological control of necrotrophic pathogens. In *Microbiology of the Phyllosphere* (ed. N. J. Fokkema and J. van den Heuvel), pp. 348–362. Cambridge University Press, Cambridge.

Trutmann, P., Keane, P. J. and Merriman, P. R. (1982). Biological control of *Sclerotinia sclerotiorum* on aerial parts of plants by the hyperparasite *Coniothyrium minitans*. *Transactions of the British Mycological Society* **78**, 521–529.

Weltzien, H. C. (1991). Biocontrol of foliar fungal diseases with compost extracts. In *Microbial Ecology of Leaves* (ed. J. H. Andrews and S. S. Hirano), pp. 430–450. Springer-Verlag, New York.

Whipps, J. M. and Gerlagh, M. (1992). Biology of *Coniothyrium minitans* and its potential for use in disease control. *Mycological Research* **96**, 897–907.

Wilson, C. L. and Wisniewski, M. E. (1992). Biological control of postharvest diseases of fruits and vegetables: an emerging technology. *Annual Review of Phytopathology* **27**, 425–441.

Zhou, T. and Reedeler, R. D. (1989). Application of *Epicoccum purpurescens* spores to control white mold of snap bean. *Plant Disease* **73**, 639–642.

16

The use of fungi, particularly *Trichoderma* spp. and *Gliocladium* spp., to control root rot and damping-off diseases

Dan Funck Jensen and Hanne Wolffhechel

Introduction

Root rot and damping-off pathogens often cause severe problems both in extensively grown field crops and in protected crops such as vegetables and ornamental plants. New control measures are being sought both for environmental reasons and since the possibilities for chemical control are limited. Among these, biological control (Sewell, 1965) seems promising: it is generally believed that the use of biological control will be beneficial and that risks and unwanted side effects are limited. In this chapter, the use of fungal antagonists as biocontrol agents (BCAs) will be discussed.

Most fungi used for biological control of damping-off and root rot are Hyphomycetes, and among these the genera *Penicillium*, *Trichoderma* and *Gliocladium* have received most attention (Kommedahl and Windels, 1981). Mycoparasitic *Pythium* spp. (Paulitz and Baker, 1987; Lewis *et al.*, 1989; Hockenhull *et al.*, 1992) and non-pathogenic *Fusarium* spp. (Alabouvette, 1990; Komada, 1990) and other species are also potential BCAs. Although the potential for using fungal antagonists for biological control is well documented (e.g. Papavizas, 1985), there are as yet only a few examples of their use on a commercial scale. However, BCAs such as non-pathogenic

Fusarium oxysporum (Rajnchapel-Messai, 1990), *Pythium oligandrum* (Vesely, 1989), *Gliocladium virens* (Lumsden *et al.*, 1991; see Lumsden and Walter, Chapter 25) and *Trichoderma harzianum* (Harman and Tronsmo, 1992) are examples of agents that are expected to be on the market in the near future.

This review will be focused mainly on the use of *Trichoderma* spp. and *Gliocadium* spp. for the control of root rot and damping-off caused by soil-borne pathogens. Possibilities for using biological control for seed-borne diseases will also be mentioned briefly. The perspectives concerning genetically modified antagonists will be addressed as well as some other aspects that need to be considered before use on a commercial scale can be successful.

Damping-off diseases

A great deal of effort has been put into the biological control of damping-off diseases. The advantage of this disease – seen from a biocontrol point of view – is that the plants are only vulnerable to attack for a short period so the antagonist has to be effective for only a relatively short period after sowing. Damping-off is caused mainly by soil-borne, unspecialized, fungi like *Rhizoctonia solani*, *Pythium* spp. or *Fusarium* spp., which attack the germinating seed and the emerging seedling (Garrett, 1970). By applying antagonistic micro-organisms to the seed, the objective is that pathogens are out-competed in the spermosphere, and in this way the germinating seed is protected from attack. This is similar to the strategy for the control of seed-borne diseases where the antagonist must be found in close association with the seed surface (see below).

Harman *et al.* (1980) treated seeds of radish and pea with *Trichoderma hamatum* which, in pot experiments, protected the seedlings against damping-off caused by *Rhizoctonia solani* or *Pythium* spp. almost as effectively as fungicide treatment. Likewise, Lutchmeah and Cooke (1985) achieved control of *Pythium ultimum* in sugar beet and cress, and of *Mycocentrospora acerina* in carrots by coating the seeds with *Pythium oligandrum*. Lifshitz *et al.* (1986) controlled pre-emergence damping-off in peas caused by *P. ultimum* by coating the seeds with different isolates of *Trichoderma* spp.

Antagonists applied as seed treatments were, in some experiments, able to control pre- but not post-emergence damping-off (Windels and Kommedahl, 1978; Martin and Hancock, 1987), or they could control damping-off, but not root rot (Kraft and Papavizas, 1983). This reflects the difficulty of the antagonist in extending its growth from the spermosphere to the rhizosphere and/or remaining active for a longer period than it takes for seed germination.

Root rot diseases

To control root rot diseases, it is important that the antagonist is established in the infection sites along the roots where pathogen attack takes place. For example, *Pythium* spp. mainly attack directly behind the root cap and at the base of lateral roots (e.g. Wester *et al.*, 1991). Antagonists applied to the seeds must, therefore, have the ability to grow from the seed and colonize the developing root. The importance of root colonization by antagonistic bacteria in biological control is discussed by Weller *et al.* (see Chapter 13).

In experiments carried out by Chao *et al.* (1986) none of the fungal strains used was capable of colonizing pea roots when the treated seeds were planted in raw soil. The same was observed by Papavizas (1982) when he applied several isolates of *Trichoderma harzianum* to pea and bean seeds.

In experiments with antagonistic bacteria, Mendez-Castro and Alexander (1983) induced resistance to the fungicide mancozeb in a strain of *Pseudomonas*. When the bacterium was coated on corn seeds in combination with the fungicide, it colonized the roots much more readily than if it was applied alone. Mancozeb has the ability to suppress the rhizosphere bacterial population, and so the competition from other organisms was reduced significantly.

The objective of the research reported by Ahmad and Baker (1987*a*) was to induce *T. harzianum* to colonize the rhizosphere from the seed by using the same strategy as Mendez-Castro and Alexander (1983). Mutants of *T. harzianum* resistant to benomyl were obtained after mutagenic treatments, and conidia of the mutant strains were coated on seeds of different plant species. The mutants were shown to be able to colonize the rhizosphere when the seeds were planted in soil treated with benomyl. Although this might be expected as a result of lower competition from other fungi, this ability was retained even in raw soil without benomyl. Wild types of *T. harzianum* were not able to colonize the roots in the tests. Ahmad and Baker (1987*b*, 1988) presented results that supported the hypothesis that the rhizosphere competence of the mutants was a consequence of an increased cellulase production and hence an enhanced ability to utilize the mucilage layer on the root surface as a substrate. We did not succeed in demonstrating rhizosphere competence on cucumbers grown in sphagnum peat using the *Trichoderma* mutants from Ahmad and Baker (Wolffhechel and Jensen, unpublished). However, Harman and coworkers obtained new strains by fusing protoplasts derived from a mutant (strain T95) from Ahmad and Baker with one of their own isolates (strain T12). Two of their new fusants had even better rhizosphere competence than the parental strain, T95, when tested in a sandy soil (Harman *et al.*, 1989).

It can be concluded that it is generally difficult to establish a fungal antagonist in the rhizosphere by planting seeds coated with the organism. However, it is possible to give the antagonists a competitive advantage over other micro-organisms in the rhizosphere. Such methods must be developed before seed treatment with antagonists can be used for the control of root rot diseases. Alternatively, the antagonist could be incorporated directly in the soil, as discussed below in the section concerning delivery systems.

Seed-borne diseases

Seed-borne pathogens can spread from the seed and infect the new plants in several ways (Neergaard, 1977). In some of the seed-borne diseases, as for example loose smut (*Ustilago nuda*) of barley, the seeds are infected within the embryo (Malik and Batts, 1960*a*). With this type of infection, the mycelium will normally be activated and cause a systemic infection as soon as the seed germinates (e.g. Malik and Batts, 1960*b*). A pathogen present within the embryo might be expected to be a difficult target and therefore difficult to control by biological agents. There is, however, at least one report of the successful control of an embryo-borne pathogen (*Drechslera avenae*) (Gerhardson, personal communication). The mechanism of control of such deep-seated pathogens is not known. One possibility is that the antagonist triggers a resistance response in the host. Alternatively, perhaps the antagonist produces compounds that suppress the pathogen in the embryo.

Many seed-borne pathogens are found outside the embryo in the seed coat, pericarp or endosperm. Others are transmitted as contaminants on the seed surface. Some pathogens can be found both inside and outside the embryo. A few of these pathogens grow directly into the young plant and cause a systemic infection during seed germination. A high proportion, however, will have a period of saprophytic growth and/or sporulation on the seed surface and in the soil surrounding the seed before they infect the plant and cause disease. This seems to be the case for *Tilletia caries* (Neergaard, 1977), *Alternaria zinniae* (Tarp, 1978) and *Septoria* leaf spot of celery (Sheridan, 1966). It should be possible to control this type of seed-borne pathogen by treating the seeds with an antagonist.

Experiments in this field are scarce and have often been conducted with artificially infected seeds (e.g. Tyner and McKinnon, 1964), but there are some reports of the successful control of seed-borne diseases originating from naturally infected seeds. Thus, *Drechslera sorokiniana* (syn. Bipolaris sorokiniana) on barley has been controlled by seed treatment with *Trichoderma* spp. (Vannacci and Pecchia, 1986), and *Helminthosporium victoriae* (syn. Bipolaris victoriae) on oats by treatment with a *Chaetomium* sp. (Tveit and Moore, 1954). *Fusarium culmorum* has been successfully controlled in field experiments by coating artificially and naturally infected wheat seeds with *Gliocladium* spp. both in Denmark (Inge Knudsen, personal communication) and in Finland (Tahvonen, personal communication).

Occurrence of *Trichoderma* and *Gliocladium* and influence of environmental conditions

Trichoderma spp. and *Gliocladium* spp. are found all over the world in many different soils and habitats (reviewed by Domsch *et al.*, 1980, and Papavizas, 1985). The species of *Trichoderma* and *Gliocladium* which dominate differ between habitats and is influenced by environmental conditions (Roiger *et al.*, 1991). The density of *Trichoderma* and *Gliocladium* is usually reported as being below 10^2 propagules/g soil (Chet, 1987). However, concentrations as high as 8×10^5 propagules/g soil have been reported for *Trichoderma* spp. in an organic soil from Columbia (Chet and Baker, 1981). The influence of soil type on the 'carrying capacity' for *Trichoderma* spp. is further discussed by Alabouvette and Steinberg (see Chapter 1).

The ability to control damping-off and root rot is also highly influenced by environmental conditions. *Trichoderma* and *Gliocladium* are favoured by low soil pH (Chet and Baker, 1980) – a fact that has been exploited in the formulation of biocontrol agents as discussed below. We tested the effect of different water contents in sphagnum peat moss on the antagonistic activity of isolates of *Trichoderma* and *Gliocladium*. Differences were found among isolates of *Trichoderma* and *Gliocladium* in their ability to control *P. ultimum* at different matric potentials (Wolffhechel and Jensen, 1991). The biocontrol abilities of different isolates are also influenced by temperature, Lifshitz *et al.* (1986) found that strain T95 of *T.*

harzianum only showed good biocontrol of *Pythium* damping-off when the temperature was above 19 °C whereas Tronsmo (1989) found that one isolate of *T. harzianum* (strain P1) could control a post-harvest disease on carrots in storage at 2 °C. Isolates of *Gliocladium roseum* were selected, which in greenhouse experiments controlled *F. culmorum* on barley at 15 °C (Knudsen *et al.*, 1992). Effective control was obtained with the same isolates on wheat under field conditions where the maximum temperature in the daytime ranged between 10 °C and 15 °C (Knudsen, personal communication).

Mechanisms of action

Several mechanisms are thought to be important in antagonistic interactions. Mycoparasitism and in that the role of different types of chitinase (Harman, personal communication) and β-1,3-glucanase (Chet, 1987; Jacobs *et al.*, 1991) from *Trichoderma* and *Gliocladium* are being investigated. Competition for substrates (Ahmad and Baker, 1987b; Nelson *et al.*, 1988) and for infection sites (i.e. rhizosphere competence) is also being studied.

The role of antibiosis in biocontrol and the production of secondary metabolites by antagonistic bacteria is discussed elsewhere (see Défago *et al.*, Chapter 12; and Weller *et al.*, Chapter 13). Strains of *Trichoderma* and *Gliocladium* produce many different secondary metabolites. Some of these seem to play a key role in many interactions causing antibiosis and lysis of the pathogen (Howell, 1991). A review concerning the antifungal antibiotics produced by *Trichoderma* spp. has recently been published (Ghisalberti and Sivasithamparam, 1991).

Regulation of microbial pesticides and risk assessment

The requirements for registration of microbial pesticides in the European Community (EC; now European Union, EU) are found in directive 91/414/EEC (Annexes II and III parts B) (Anonymous, 1991) (see Klingauf, Chapter 28). The directive concerns both natural organisms and genetically modified organisms (GMOs). Additional documentation is, however, required for the deliberate release of GMOs (cf. directive 90/220/EEC (Anonymous, 1990)). The directives are at present being implemented by the different EU members.

Several areas need to be documented concerning, for example, toxicology, ecotoxicological aspects, and the fate and behaviour of the micro-organism in the environment before a BCA can be approved according to directive 91/414/EEC. If an organism produces a toxin, additional data are required concerning the fate and behaviour of the toxin in the environment as well as ecotoxicological data. These data are, however, only required if they are considered to be relevant.

There is still a need for good methods for obtaining many of the required data – especially those for testing ecotoxicological effects, and the fate and behaviour of the micro-organism in the environment.

There are a few reports indicating that some strains of *Trichoderma* may be able to produce toxic metabolites (e.g. Irmscher *et al.*, 1978), and it is well established that some strains of *Gliocladium* can produce secondary metabolites which are toxic to mammals (Taylor, 1986). The taxonomy of the genera *Trichoderma* and *Gliocladium* has been very confused especially in the earlier literature. Therefore, information concerning toxins produced by species from these genera should be treated carefully when taken from old references. Some *Gliocladium* spp. may also have herbicidal effects (Howell and Stipanovic, 1984). It has been discussed as to whether *Trichoderma* and *Gliocladium* or their metabolites have a deleterious effect on other beneficial micro-organisms in the soil. Linderman *et al.* (1991) evaluated the effect of several BCAs including the effect of *Trichoderma* and *Gliocladium* on mycorrhizal fungi. Their results indicated that mycorrhizal fungi are not very sensitive to metabolites from antagonists, and that mycorrhiza will be compatible with *Trichoderma* and *Gliocladium* if they are applied in combination (Linderman *et al.*, 1991; Paulitz and Linderman, 1991). However, there seems to be a need for further research concerning the effect of antagonists on the other beneficial micro-organisms in the soil.

Trichoderma and *Gliocladium* are considered to be true saprophytes although there are examples that some strains can cause lesions in corn seedlings (McFadden and Sutton, 1975). Strains of *Trichoderma* spp. – especially *T. harzianum* strain Th2 – can be pathogenic to mushroom and cause problems in mushroom production (Fletcher *et al.*, 1989). Other *Trichoderma* spp. are harmless or can be used for biological control of important fungal diseases in mushroom production (Ricard cited in Connick *et al.*, 1990). The risk that a *Trichoderma* sp. applied in another agrosystem should spread and give additional problems in mushroom production seems to be minimal as *Trichoderma* spp. are ubiquitous in soil.

The risks involved in the deliberate release of an antagonist must be evaluated in each case. If toxin producers are considered, however, it seems likely that secondary metabolites are only produced in very small amounts in the soil and rhizosphere. Besides, many of the metabolites are unstable and most will probably be broken down very quickly by the soil microflora (e.g. Howell and Stipanovic, 1983). It might be reasonable, therefore, to allow the release into the environment of all strains except those that are capable of producing metabolites highly toxic to mammals. Also, there might be a need to restrict the concentration of toxic metabolites present in formulated biocontrol products. Another aspect concerns the safety precautions necessary for the protection of the people handling the biocontrol agent during its production, formulation and delivery.

We are not aware of any fungal pesticide having been approved in the EU on the basis of the new directive 91/414/EEC, and it is not clear how the directive will be handled in practice by the authorities. Lumsden and Walter (see Chapter 25) have discussed how the commercial preparation of their *G. virens* strain was risk assessed and approved for horticultural use by the American authorities (EPA and APHIS). This example from the USA illustrates that many tests might be considered to be irrelevant when a BCA is evaluated. The regulation of microbial pesticides is further discussed by Klingauf (see Chapter 28).

Monitoring introduced fungal antagonists

For risk evaluation, methods will be required that facilitate the monitoring of antagonistic fungi deliberately released into the environment. Such methods are important in relation to the documentation of their fate and behaviour in the environment (directive 91/414/EEC). In the directive concerning release of GMOs for commercial purposes (90/220/EEC), one of the requirements is that monitoring, control techniques and emergency response plans exist in the case of an unexpected spread of the organism. There is also a need for monitoring introduced micro-organisms for experimental purposes (both GMOs and natural organisms).

Various staining techniques and immunological methods are being tried for fungi (Alabouvette *et al.*, 1992). Immunological methods have already proved useful in monitoring bacteria (see Tsuchiya, Chapter 20). For many of these methods, however, there are problems with specificity or cross-reactions and they are difficult to use for monitoring organisms in the soil and the rhizosphere. In the vital fluorescent staining techniques, only short time-course experiments can be carried out. Measurements of ATP, chitin and esterase activity have been shown to be useful for following the biomass of *Trichoderma* and its biocontrol activity against *P. ultimum* in pot experiments (Lumsden *et al.*, 1990).

With respect to *Trichoderma* and *Gliocladium*, dilution plating on selective media has mainly been used (Davet, 1979; Elad *et al.*, 1981; Papavizas and Lumsden, 1982). However, in these methods, it is common to run into problems with a background microflora that is also capable of growing on the selective substrate. Furthermore, it is difficult to distinguish between introduced and indigenous *Trichoderma* spp. and *Gliocladium* spp. on these selective media. The introduction of resistance to fungicides by traditional mutation (Papavizas *et al.*, 1982; Ahmad and Baker, 1987*a*) or by conditioning the antagonist to fungicide tolerance (Moity *et al.*, 1982), and then quantifying the introduced antagonist on a selective medium that contains the fungicide has improved the usability of dilution plating.

New molecular approaches, similar to those now being developed with antagonistic bacteria (e.g. Steffan and Atlas, 1988; Weger *et al.*, 1991) may be expected in the near future. With fungi it may be possible to monitor the introduced strain by insertion of reporter genes. Alabouvette *et al.*

(1992) have studied root colonization by a pathogenic mutant of *Fusarium oxysporum* f. sp. *lini* in which a reporter gene coding for β-glucuronidase was inserted (i.e. the GUS system). The transformed fungus turns blue when it is exposed to the appropriate substrate. Root colonization by the transformant was quantified by fluorimetric methods, and the competitive interaction between the transformant and a non-pathogenic *Fusarium* sp. on the root was determined.

Good transformation systems for *Trichoderma* have been developed (Pentillä *et al.*, 1987; Goldman *et al.*, 1990; Herrera-Estrella *et al.*, 1990), and strains of *Gliocladium* have also been transformed (Ossana and Mischke, 1990; Mette Lübeck, personal communication). Genes that carry resistance to the antibiotic hygromycin B (Goldman *et al.*, 1990), a selectable marker gene *amdS* (Pe'er *et al.*, 1991) and the GUS reporter genes (Mette Lübeck, personal communication) have been inserted into strains of *Trichoderma* by transformation. *AmdS* has already been used in ecological studies (Pe'er *et al.*, 1991). The other genetic markers mentioned might also be useful in studying the ecology and population dynamics of introduced *Trichoderma* spp. and *Gliocladium* spp. in relation to biological control.

Stasz *et al.* (1989) used starch gel electrophoresis for analysing isoenzymes from different strains of *Trichoderma* and *Gliocladium*. They showed that the method could be used for grouping strains of morphologically defined species of *Trichoderma* and *Gliocladium*, and that it can be used for monitoring an introduced antagonist (Harman *et al.*, 1989). The method is, however, time consuming and therefore difficult to use in quantitative studies except for checking samples of colonies isolated in the dilution plating technique. The DNA based methods such as PCR/RFLP or rDNA and RAPD may be used in a similar way for monitoring an introduced strain of *Trichoderma* and *Gliocladium* (Bruns *et al.*, 1991) or it may be possible to use specific probes.

Many of the methods mentioned above can give problems or be inadequate when used alone. However, as discussed by Alabouvette *et al.* (1992), when several methods are used in combination, useful information concerning the population dynamics and activities of the released organisms can be obtained.

The improvement of antagonistic strains of *Trichoderma* and *Gliocladium* by genetic modification

Harman and Tronsmo (1992) have summarized methods for genetic manipulation for the production of superior strains of *Trichoderma* and *Gliocladium*. UV-treatments (Papavizas *et al.*, 1982), chemical mutation (Ahmad and Baker, 1987a) and protoplast fusion (Harman *et al.*, 1989) have been used successfully. Protoplast fusion was thought to have the advantage that characters from different parents could be combined in the fusants. Genetic analysis, however, indicated that parasexuality is absent in the progeny of protoplast fusions and no true asexual hybrids are produced (Stasz and Harman, 1990). Therefore, Stasz and Harman (1990) concluded that the method is not useful for combining parental characteristics. However, a large variation was found among the progeny, and, based on RFLP and isoenzyme analysis, Harman and Tronsmo (1992) suggested that small parts of the genome may be recombined in the fusants. Thus, due to the large variability, fusants can be used as a source for the selection of improved strains if the selection is based on an effective screening procedure.

Transformation may be used for improving antagonistic strains. At present, it is not fully understood as to which genes will be valuable to insert or regulate in the genome of the antagonists. However, transformation techniques can be used to study single genes that might be important for the antagonistic potential of an organism. For example, work is in progress for cloning chitinase genes into *T. harzianum* (Gary Harman and Arne Tronsmo, personal communication). Heidenreich and Kubicek (1991) are studying other genes involved in mycoparasitism, and van Montagu and coworkers are also using cloning and transformation techniques for studying genes in *Trichoderma* spp. (Geremia *et al.*, 1991; Goldman *et al.*, 1991). These investigations will probably lead to a better understanding of the role of different genes in the antagonistic activities of *Trichoderma* strains and be the basis for creating improved strains by transformation.

One concern is that new or modified genes may spread and be transferred to other organisms in the environment (see Ryder and Correll, Chapter 19). Similarities in gene sequences might be indirect evidence for natural gene transfer between distantly related organisms, and an important factor in horizontal gene transfer between microbes in nature may be promiscuous plasmids (Mazodier and Davies, 1991). Little is known, however, about gene transfer between micro-organisms in nature – especially with regard to distantly related fungal populations or between fungi and other micro-organisms.

Production of antagonists

Some of the main points concerning large-scale production and delivery systems for *Trichoderma* and *Gliocladium* will be mentioned. The subjects are further discussed by others (e.g. Harman, 1992; Papavizas, 1992).

The methods used for the production of the antagonists, the strain and the kind of propagules used (conidia, chlamydospores, mycelium etc.) are important for their efficacy in biological control (Papavizas *et al.*, 1984). Organisms such as *Trichoderma* or *Gliocladium* can be produced in either liquid culture or on solid media. The amount of biomass and the ratio between conidia and chlamydospores, as well as the viability of the product after drying will depend on the method of production and the media used (Lewis and Papavizas, 1983). According to Papavizas *et al.* (1984) and Harman *et al.* (1991) it will be preferable in most cases to produce the antagonists by liquid fermentation rather than solid fermentation, and procedures for the production of *Trichoderma* and *Gliocladium* with a high yield of active biomass have been developed (Papavizas *et al.*, 1984; Harman, 1992).

Delivery systems

Soil incorporation

Various methods have been used for applying antagonistic fungi to soil to control root rot and damping-off. Antagonists have been mixed into or sprayed on the surface of the soil in many cases. This is especially valuable if the aim is to destroy resting structures of the pathogen, such as sclerotia (Ayers and Adams, 1981; Davet *et al.*, 1981; Trutmann and Keane, 1990). Mixing antagonists into the soil has also been used in combination with soil fumigation to establish a favourable soil microflora and in this way prevent the build-up of pathogen populations (Elad *et al.*, 1981). The formulations that have been used for direct application to soil of fungal antagonists include suspensions of spores (Marois *et al.*, 1982) and powdery preparations of mycelium and sclerotia (Lewis and Papavizas, 1980). A peat wheat bran formulation (Sivan *et al.*, 1984) and a vermiculite bran formulation (Lewis *et al.*, 1991) of *Trichoderma* spp. and *Gliocladium* spp. have been developed that allow incorporation of antagonists into soil or soil-less substrates like peat.

Nowadays, special carrier materials are commonly used. These include clay granules soaked in molasses (Backman and Rodrigues-Kabana, 1975), alginate-skim milk beads (Bashan, 1986) and alginate pellets with wheat bran as a food base (Lewis and Papavizas, 1987). Although some of these carrier materials are lightweight compared with organic materials (Backman and Rodriguez-Kabana, 1975), large quantities are needed if the antagonists are to be mixed into the bulk soil. Hence, this method will mainly be of interest in protected crops and container-grown plants. This is, for example, the case with the preparation of *Gliocladium virens* developed by Lumsden and Walter (see Chapter 25). The possibilities for the incorporation of antagonistic fungi in peat substrates used in glasshouses has recently been reviewed (Jensen and Wolffhechel, 1992).

The most efficient way to utilize antagonists in the field will probably be by application close to possible infection courts. Thus, if the germinating seed and the seedling are to be protected, the antagonist should be added to the soil close to the seed. Several attempts have been made to apply the antagonist directly into the seed furrow, for example by adding the antagonist to the gel used for fluid-drilling of vegetable seeds (Conway, 1986). However, most efforts have been put into coating the seeds with antagonists.

Seed treatments

Seeds can be treated in various ways to ensure optimal chemical and physical conditions for germination. Seed coating with fungicides or physical treatments for the control of plant diseases are normal procedures. In addition to coating, high value seeds are usually pelleted with a filler to increase the size, to ensure seed separation and thereby facilitate handling by machines. Different aspects of seed-treatment technologies have recently been reviewed (Taylor and Harman, 1990). Most of the technologies have not been aimed at delivery of antagonistic micro-organisms along with the seeds. However, many of the traditional methods may be adapted or modified in a way that will ensure both optimal conditions for seed germination and for the biocontrol agent delivered with the seed. These aspects have also been discussed by Taylor and Harman (1990) and Harman (1992). Thus, in this chapter some of the main points only will be briefly mentioned.

The main objective is to make the antagonist capable of protecting the seed and seedlings against soil- and/or seed-borne diseases. The intention is to give the antagonist a competitive advantage over the indigenous microflora by ensuring favourable conditions for the antagonist around the seed at the time of planting. Inclusion of a food base in the formulation will in many cases favour the antagonist (Papavizas et al., 1984). However, the food base must be fully occupied by the antagonist at the time of planting or it must be separated physically from the indigenous microflora until it becomes fully colonized by the antagonist. This strategy was used in the preparation of alginate beads as discussed by Harman and Lumsden (1990). Another strategy would be to use a food base that can only be utilized by the antagonists as described by Nelson et al. (1988).

Trichoderma spp. are normally favoured by low pH. Harman and Taylor (1988) incorporated lignite with a pH of 4.1 in the seed pelleting material. In this way they were able to improve the activity of the antagonist, thereby gaining a better disease control. Addition of pesticides to which the antagonists are resistant could also be a promising approach as demonstrated by Ahmad and Baker (1987a) and Papavizas et al. (1982). This could add to the effect of the antagonist in the control of the disease, as shown by Howell (1991).

In many cases, as for example with protection against *Pythium*, the antagonist must be active within a few hours after planting. This may be obtained by using liquid coating, seed priming or solid matrix priming in which the seed can be colonized by the antagonist before planting (Har-

man, 1992). Activation before planting is also possible in the fluid drilling technique mentioned above and in the application method described by Lewis et al. (1991).

A high inoculum density on the seed is important for successful disease control (Harman et al., 1981). In terms of small seeds, this can be ensured by using the pelleting technique. However, there is some evidence that competition for oxygen between the seed and the antagonists applied at too high an inoculum density results in reduced seed germination (Lynch and Pryn, 1977).

Summary

Root rot and damping-off diseases caused by seed- and soil-borne pathogens are important in many crops. For environmental reasons and because the possibilities for chemical control are limited, new control measures are being sought. Among these biolgogical control seems promising. In this chapter fungal antagonists have been considered – especially *Trichoderma* spp. and *Gliocladium* spp.

There are only a few examples of the use of BCAs for the control of plant diseases on a commercial scale. One reason is the lack of consistent results in many experiments. In many cases this may be due to insufficient knowledge about the antagonist itself, and to its interactions with the microflora in a natural soil system. However, within the last 10 to 15 years, research in this area has intensified, and new methods have been developed. These include methods for studying antibiosis, mycoparasitism and competition, and methods for monitoring antagonists deliberately released into the environment. Methods for improving antagonistic strains of fungi are also being developed. Production and delivery systems have been devised for the application of fungal antagonists either with the seed or by incorporation into the soil. Thus, the methods and scientific background seem to be available for obtaining improved results.

Directives for the regulation of microbial pesticides have now been set up in many countries including the USA and the EU. BCAs based on *Gliocladium virens* and *Trichoderma harzianum* have recently been approved by the USA authorities (EPA and APHIS). According to these regulations a risk assessment is required before a BCA can be approved. Some fungal isolates may have characteristics that are unacceptable from a toxicological or ecotoxicologial point of view, e.g. the production of secondary metabolites which are highly toxic to mammals. With fungal antagonists, however, it is generally believed that the risks and unwanted side effects are limited, and that the technology is environmentally safe. These perspectives should prompt the industry to increase the number of BCAs approved by the authorities. Thus, new microbial pesticides for biological control of plant diseases might be expected on the market in the near future.

Acknowledgment

We thank our colleagues Dr John Hockenhull, Dr David Collinge, Helge Green and Inge M.B. Knudsen for valuable comments on the manuscript.

References

Ahmad, J. S. and Baker, R. (1987*a*). Rhizosphere competence of *Trichoderma harzianum*. *Phytopathology* 77, 182–189.

Ahmad, J. S. and Baker, R. (1987*b*). Competitive saprophytic ability and cellulytic activity of rhizospherecompetent mutants of *Trichoderma harzianum*. *Phytopathology* 77, 358–362.

Ahmad, J. S. and Baker R. (1988). Growth of rhizosphere-competent mutants of *Trichoderma harzianum* on carbon substrates. *Canadian Journal of Microbiology* 34, 807–814.

Alabouvette, C. (1990). Biological control of Fusarium wilt pathogens in suppressive soils. In *Biological Control of Soil-borne Plant Pathogens* (ed. D. Hornby), pp. 27–43. CAB International, Wallingford, UK.

Alabouvette, C., Eparvier, A., Couteaudier, Y. and Steinberg, C. (1992). Methods to be used to study the competitive interactions between pathogenic and nonpathogenic *Fusarium* spp. in the rhizosphere and at the root surface. In *New Approaches in Biological Control of Soil-Borne Diseases* (ed. D. Funck Jensen, J. Hockenhull and N. J. Fokkema. *IOBC/WPRS Bulletin* **XV/1**, 1–7.

Anonymous (1990). Council Directive of 23 April 1990 on the deliberate release into the environment of genetically modified organisms (90/220/EEC). *Official Journal of the European Communities*, 8 May 1990, No. L117:23–27.

Anonymous (1991). Council Directive of 15th July 1991 concerning the placing of plant protection products on the market (91/414/EEC). Official Journal of the European Communities, 19 August 1991, No. L230:1–30.

Ayers, W. A. and Adams, P. B. (1981). Mycoparasitism and its application to biological control of plant diseases. In *Biological Control in Crop Production* (ed. G. C. Papavizas), pp. 91–103. *Beltsville Symposia in Agricultural Research 5*, Allanheld, Osmun, Totowa.

Backman, P. A. and Rodriguez-Kabana, R. (1975). A system for the growth and delivery of biological control agents to the soil. *Phytopathology* **65**, 819–821.

Bashan, Y. (1986). Alginate beads as synthetic inoculant carriers for slow release of bacteria that affect plant growth. *Applied and Environmental Microbiology* **51**, 1089–1098.

Bruns, T. D., White, T. J. and Taylor, J. W. (1991). Fungal molecular systematics. *Annual Review of Ecological Systematics* **22**, 525–564.

Chao, W. L., Nelson, E. B., Harman, G. E. and Hoch, H. C. (1986). Colonization of the rhizosphere by biological control agents applied to seeds. *Phytopathology* **76**, 60–65.

Chet, I. (1987). *Trichoderma* – application, mode of action, and potential as a biocontrol agent of soilborne plant pathogenic fungi. In *Innovative Approaches to Plant Disease Control* (ed. Ilan Chet), pp. 137–160. Wiley Series in Ecological and Applied Microbiology, New York.

Chet, I. and Baker, R. (1980). Induction of suppressiveness to *Rhizoctonia solani* in soil. *Phytopathology* **70**, 994–998.

Chet, I. and Baker, R. (1981). Isolation and biocontrol potential of *Trichoderma hamatum* from soil naturally suppressive to *Rhizoctonia solani*. *Phytopathology* **71**, 286–290.

Connick, W. J., Lewis, J. A. and Quimby, P. C. (1990). Formulation of biocontrol agents for use in plant pathology. In *New Directions in Biological Control: Alternatives for Suppressing Agricultural Pests and Diseases* (ed. R. Baker and P. E. Dunn), pp. 345–372. Alan R. Liss, Inc., New York.

Conway, K. E. (1986). Use of fluid-drilling gels to deliver biological control agents to soil. *Plant Disease* **70**, 835–839.

Davet, P. (1979). Technique pour l'analyse des populations de *Trichoderma* et de *Gliocladium virens* dans le sol. *Annales de Phytopathologie* **11**, 529–533.

Davet, P., Artigues, M. and Martin, C. (1981). Production en conditions non aseptiques d'inoculum de *Trichoderma harzianum* Rifai pour des essais de lutte biologique. *Agronomie* **1**, 933–936.

Domsch, K. H., Gams, E. and Anderson, Y. (1980).

Compendium of Soil Fungi, pp. 794–809. Academic Press, New York.

Elad, Y., Chet, I. and Henis, Y. (1981). A selective medium for improving quantitative isolation of *Trichoderma* spp. from soil. *Phytoparasitica* **9**, 59–67.

Fletcher, J. T., White, P. F. and Gaze, R. H. (1989). *Mushrooms: Pest and Disease Control. 2nd edn*, p. 174. Intercept, Andover, Hants.

Garrett, S. D. (1970). *Pathogenic Root-infecting Fungi*, p. 294. Cambridge University Press, Cambridge.

Geremia, R. A., Goldman, G. H., Jacobs, D., van Montagu, M. and Herrera-Estrella, A. (1991). Role and specificity of different proteinases in pathogen control by *Trichoderma harzianum*. *Petria* **1**, 131.

Ghisalberti, E. L. and Sivasithamparam, K. (1991). Antifungal antibiotics produced by *Trichoderma* spp. *Soil Biology and Biochemistry* **23**, 1011–1020.

Goldman, G. H., van Montagu, M. and Herrera-Estrella, A. (1990). Transformation of *Trichoderma harzianum* by high voltage electric pulse. *Current Genetics* **17**, 169–174.

Goldman, G. H., Demolder, J., Villarroel, R., Dewaele, S., van Montagu, M., Contreras, R. and Herrera-Estrella, A. (1991). Cloning and characterization of metabolic genes of *Trichoderma* spp. *Petria* **1**, 128.

Harman, G. E. (1992). Production and delivery systems for biocontrol agents. In *New Approaches in Biological Control of Soil-Borne Diseases* (ed. D. Funck Jensen, J. Hockenhull and N. J. Fokkema. *IOBC/WPRS Bulletin* **XV/1**, 201–205.

Harman, G. E. and Lumsden, R. D. (1990). Biological disease control. In *The Rhizosphere* (ed. J. M. Lynch), pp. 259–280. John Wiley & Sons, Chichester.

Harman, G. E. and Taylor, A. G. (1988). Improved seedling performance by integration of biological control agents at favorable pH levels with solid matrix priming. *Phytopathology* **78**, 520–525.

Harman, G. E. and Tronsmo, A. (1992). Methods of genetic manipulation for the production of improved bioprotectant fungi. In *New Approaches in Biological Control of Soil-Borne Diseases* (ed. D. Funck Jensen, J. Hockenhull and N. J. Fokkema). *IOBC/WPRS Bulletin* **XV/1**, 181–187.

Harman, G. E., Chet, I. and Baker, R. (1980). *Trichoderma hamatum* effects on seed and seedling disease induced in radish and pea by *Pythium* spp. or *Rhizoctonia solani. Phytopathology* **70**, 1167–1172.

Harman, G. E., Chet, I. and Baker, R. (1981). Factors affecting *Trichoderma hamatum* applied to seeds as a biocontrol agent. *Phytopathology* **71**, 569–572.

Harman, G. E., Taylor, A. G. and Stasz, T. E. (1989). Combining effective strains of *Trichoderma harzianum* and solid matrix priming to improve biological seed treatments. *Plant Disease* **73**, 631–637.

Harman, G. E., Jin, X., Stasz, T. E., Peruzzotti, G., Leopold, A. C. and Taylor, A. G. (1991). Production of conidial biomass of *Trichoderma harzianum* for biological control. *Biological Control* **1**, 23–28.

Heidenreich, E. and Kubicek, C. P. (1991). Towards cloning of biocontrol genes of *Trichoderma harzianum. Petria* **1**, 125–126.

Herrera-Estrella, A., Goldman, G. H. and Van Montagu, M. (1990). High-efficiency transformation system for the biocontrol agents, *Trichoderma* spp. *Molecular Microbiology* **4**, 839–843.

Hockenhull, J., Jensen, D. F. and Yudiarti, T. (1992). The use of *Pythium periplocum* to control damping-off of cucumber seedlings caused by *Pythium aphanidermatum*. In *Biological Control of Plant Diseases: Progress and Challenges for the future* (ed. E. C. Tjamos, G. C. Papavizas and R. J. Cook), pp. 203–206. Plenum Press, New York.

Howell, C. R. (1991). Biological Control of Pythium damping-off of cotton with seed-coating preparations of *Gliocladium virens. Phytopathology* **81**, 738–741.

Howell, C. R. and Stipanovic, R. D. (1983). Gliovirin, a new antibiotic from *Gliocladium virens*, and its role in the biological control of *Pythium ultimum. Canadian Journal of Microbiology* **29**, 321–324.

Howell, C. R. and Stipanovic, R. D. (1984). Phytotoxicity to crop plants and herbicidal effects on weeds of viridiol produced by *Gliocladium virens. Phytopathology* **74**, 1346–1349.

Irmscher, G., Bovermann, G., Boheim, G. and Jung, G. (1978). Trichotoxin A-40, a new membrane-exciting peptide part A. Isolation, characterization and conformation. *Biochimica et Biophysica Acta* **507**, 470–484.

Jacobs, D., Geremia, R. A., Goldman, G. H., Kamoen, O., van Montagu, M., and Herrera-Estrella, A. (1991). Study of the expression of β-(1,3) glucanase of *Trichoderma harzianum. Petria* **1**, 125.

Jensen, D. F. and Wolffhechel, H. (1992). Biological control of *Pythium ultimum* by incorporation of antagonistic fungi in peat substrates. In *New Approaches in Biological Control of Soil-Borne Diseases* (ed. D. Funck Jensen, J. Hockenhull and N. J. Fokkema). *IOBC/WPRS Bulletin* **XV/1**, 149–155.

Knudsen, I. M. B., Hockenhull, J. and Jensen, D. F. (1992). *In vivo* screening of potential antagonists against *Fusarium culmorum* in barley. In *New Approaches in Biological Control of Soil-Borne Diseases* (ed. D. Funck Jensen, J. Hockenhull and N. J. Fokkema). *IOBC/WPRS Bulletin* **XV/1**, 21–23.

Komada, H. (1990). Biological control of Fusarium wilts in Japan. In *Biological Control of Soil-borne Plant Pathogens* (ed. D. Hornby), pp. 65–75. CAB International, Wallingford.

Kommedahl, T. and Windels, C. (1981). Introduction of microbial antagonists to specific courts of infection: Seeds, seedlings, and wounds. In *Biological Control in Crop Production* (ed. G. C. Papavizas), pp. 227–248. Allanheld, Osmun, Totowa.

Kraft, J. M. and Papavizas, G. C. (1983). Use of host resistance, *Trichoderma*, and fungicides to control soilborne diseases and increase seed yields of peas. *Plant Disease* 67, 1234–1237.

Lewis, J. A. and Papavizas, G. C. (1980). Integrated control of *Rhizoctonia* fruit rot of cucumber. *Phytopathology* 70, 85–89.

Lewis, J. A. and Papavizas, G. C. (1983). Production of chlamydospores and conidia by *Trichoderma* spp. in liquid and solid growth media. *Soil Biology and Biochemistry* 15, 351–357.

Lewis, J. A. and Papavizas, G. C. (1987). Application of *Trichoderma* and *Gliocladium* in alginate pellets for control of *Rhizoctonia* damping-off. *Plant Pathology* 36, 438–446.

Lewis, K., Whipps, J. M. and Cooke, R. C. (1989). Mechanisms of biological disease control with special reference to the case study of *Pythium oligandrum* as an antagonist. In *Biotechnology of Fungi for Improving Plant Growth* (ed. J. M. Whipps and R. D. Lumsden), pp. 191–217. Cambridge University Press, Cambridge.

Lewis, J. A., Papavizas, G. C. and Lumsden, R. D. (1991). A new formulation system for the application of biocontrol fungi to soil. *Biocontrol Science and Technology* 1, 59–69.

Lifshitz, R., Windham, M. T. and Baker, R. (1986). Mechanism of biological control of preemergence damping-off of pea by seed treatment with *Trichoderma* spp. *Phytopathology* 76, 720–725.

Linderman, R. G., Paulitz, T. C., Mosier, N. J., Griffiths, R. P., Loper, J. E., Caldwell, B. A. and Henkels, M. E. (1991). Evaluation of the effects of biocontrol agents on mycorrhizal fungi. In *The Rhizosphere and Plant Growth* (ed. D. L. Keister and P. B. Cregan), p. 379. Kluwer Academic Publishers, Dordrecht, the Netherlands.

Lumsden, R. D., Carter, J. P., Whipps, J. M. and Lynch, J. M. (1990). Comparison of biomass and viable propagule measurements in the antagonism of *Trichoderma harzianum* against *Pythium ultimum*. *Soil Biology and Biochemistry* 22, 187–194.

Lumsden, R. D., Locke, J. C. and Walter, J. F. (1991). Approval of *Gliocladium virens* by the US Environmental Protection Agency for biological control of Pythium and Rhizoctonia damping-off. *Petria* 1, 138.

Lutchmeah, R. S. and Cooke, R. C. (1985). Pelleting of seed with the antagonist *Pythium oligandrum* for biological control of damping-off. *Plant Pathology* 34, 528–531.

Lynch, J. M. and Pryn, S. J. (1977). Interaction between a soil fungus and barley seed. *Journal of General Microbiology* 103, 193–196.

McFadden, A. G. and Sutton, J. C. (1975). Relationship of populations of *Trichoderma* spp. in soil to disease in maize. *Canadian Journal of Plant Science* 55, 579–586.

Malik, M. M. S. and Batts, C. C. V. (1960a). The infection of barley by loose smut (*Ustilago nuda* (Jens.) Rostr.). *Transactions of the British Mycological Society* 43, 117–125.

Malik, M. M. S. and Batts, C. C. V. (1960b). The development of loose smut (*Ustilago nuda*) in barley plant, with observations on spore formation in nature and in culture. *Transactions of the British Mycological Society* 43, 126–131.

Marois, J. J., Johnston, S. A., Dunn, M. T. and Papavizas, G. C. (1982). Biological control of *Verticillium* wilt of eggplant in the field. *Plant Disease* 66, 1166–1168.

Martin, F. N. and Hancock, J. G. (1987). The use of *Pythium oligandrum* for biological control of pre-emergence damping-off caused by *P. ultimum*. *Phytopathology* 77, 1013–1020.

Mazodier, P. and Davies, J. (1991). Gene transfer between distantly related bacteria. *Annual Review of Genetics* 25, 147–171.

Mendez-Castro, F. A. and Alexander, M. (1983). Method for establishing a bacterial inoculum on corn roots. *Applied and Environmental Microbiology* 45, 248–254.

Moity, Abd-El T. H., Papavizas, G. C. and Shatla, M. N. (1982). Induction of new isolates of *Trichoderma harzianum* tolerant to fungicides and their experimental use for control of white rot of onion. *Phytopathology* 72, 396–400.

Neergaard, P. (1977). *Seed Pathology*, vol. 1. Macmillan Press Ltd., London.

Nelson, E. B., Harman, G. E. and Nash, G. T. (1988). Enhancement of *Trichoderma*-induced biological control of Pythium seed rot and pre-emergence damping-off of peas. *Soil Biology and Biochemistry* 20, 145–150.

Ossanna, N. and Mischke, S. (1990). Genetic transformation of the biocontrol fungus *Gliocladium virens* to benomyl resistance. *Applied and Environmental Microbiology* 56, 3052.

Papavizas, G. C. (1982). Survival of *Trichoderma harzianum* in soil and in pea and bean rhizosphere. *Phytopathology* 72, 121–125.

Papavizas, G. C. (1985). *Trichoderma* and *Gliocladium*:

Biology, ecology, and potential for biocontrol. *Annual Review of Phytopathology* **23**, 23–54.

Papavizas, G. C. (1992). Biological control of selected soilborne plant pathogens with *Gliocladium* and *Trichoderma*. In *Biological Control of Plant Diseases: Progress and Challenges for the future* (ed. E. C. Tjamos, G. C. Papavizas and R. J. Cook), pp. 223–230. Plenum Press, New York.

Papavizas, G. C. and Lumsden, R. D. (1982). Improved medium for isolation of *Trichoderma* spp. from soil. *Plant Disease* **66**, 1019–1020.

Papavizas, G. C., Lewis, J. A. and Abd-El Moity, T. H. (1982). Evaluation of new biotypes of *Trichoderma harzianum* for tolerance to Benomyl and enhanced biocontrol capabilities. *Phytopathology* **72**, 126–132.

Papavizas, G. C., Dunn, M. T., Lewis, J. A. and Beagle-Ristaino, J. (1984). Liquid fermentation technology for experimental production of biocontrol fungi. *Phytopathology* **74**, 1171–1175.

Paulitz, T. C. and Baker, R. (1987). Biological control of Pythium damping-off of cucumbers with *Pythium nunn*: influence of soil environment and organic amendments. *Phytopathology* **77**, 341–346.

Paulitz, T. C. and Linderman, R. G. (1991). Lack of antagonism between the biocontrol agent *Gliocladium virens* and vesicular arbuscular mycorrhizal fungi. *New Phytologist* **117**, 303–308.

Pe'er, S., Barak, Z., Yarden, O. and Chet, I. (1991). Stability of *Trichoderma harzianum amd*S transformants in soil and rhizosphere. *Soil Biology and Biochemistry* **23**, 1043–1046.

Pentillä, M., Nevalainen, H., Rättö, M., Salminen, E. and Knowles, J. (1987). A versatile transformation system for the cellulolytic filamentous fungus *Trichoderma reesei*. *Gene* **61**, 155–164.

Rajnchapel-Messai, J. (1990). Les biopesticides. *Biofutur* **92**, 23–34.

Roiger, D. J., Jeffers, S. N. and Caldwell, R. W. (1991). Occurrence of *Trichoderma* species in apple orchard and woodland soils. *Soil Biology and Biochemistry* **23**, 353–359.

Sewell, G. W. F. (1965). The effect of altered physical condition of soil on biological control. In *Ecology of Soil-borne Plant Pathogens* (ed. K. F. Baker and W. C. Snyder), pp. 479–494. University of California Press, Berkeley, Los Angeles.

Sheridan, J. E. (1966). Celery leaf spot: sources of inoculum. *Annals of Applied Biology* **57**, 75–81.

Sivan, A., Elad, Y. and Chet, I. (1984). Biological control effects of a new isolate of *Trichoderma harzianum* on *Pythium aphanidermatum*. *Phytopathology* **74**, 498–501.

Stasz, T. E. and Harman, G. E. (1990). Nonparental progeny resulting from protoplast fusion in *Trichoderma* in the absence of parasexuality. *Experimental Mycology* **14**, 145–159.

Stasz, T. E., Nixon, K., Harman, G. E., Weeden, N. F. and Kuter, G. A. (1989). Evaluation of phenetic species and phylogenetic relationships in the genus *Trichoderma* by cladistic analysis of isozyme polymorphism. *Mycologia* **81**, 391–403.

Steffan, R. J. and Atlas, R. M. (1988). DNA amplification to enhance detection of genetically engineered bacteria in environmental samples. *Applied and Environmental Microbiology* **54**, 2185–2191.

Tarp, G. (1978). *Alternaria zinniae* on *Zinnia elegans*. The Royal Veterinary and Agricultural University, Copenhagen, Denmark, Yearbook 1978, pp. 13–20, Copenhagen.

Taylor, A. (1986). Some aspects of the chemistry and biology of the genus *Hypocrea* and its anamorphs, *Trichoderma* and *Gliocladium*. *Proceedings of the Nova Scotia Institute of Science* **36**, 27–58.

Taylor, A. G. and Harman, G. A. (1990). Concepts and technologies of selected seed treatments. *Annual Review of Phytopathology* **28**, 321–339.

Tronsmo, A. (1989). *Trichoderma harzianum* used for biological control of storage rot on carrots. *Norwegian Journal of Agricultural Science* **3**, 157–161.

Trutmann, P. and Keane, P. J. (1990). *Trichoderma koningii* as a biological control agent for *Sclerotinia sclerotiorum* in Southern Australia. *Soil Biology and Biochemistry* **22**, 43–50.

Tveit, M. and Moore, M. B. (1954). Isolates of *Chaetomium* that protect oats from *Helminthosporium victoriae*. *Phytopathology* **44**, 686–689.

Tyner, L. E. and McKinnon, A. (1964). Fungi of barley seed and their associative effects. *Phytopathology* **54**, 506–508.

Vannacci, G. and Pecchia, S. (1986). Evaluation of biological seed treatment for controlling seed-borne inoculum of *Drechslera sorokiniana* on barley. *Mededelingen van de Faculteit Landbouwwetenschappen. Rijksuniversiteit, Gent* **51/2b**, 741–750.

Vesely, D. (1989). Biological control of damping-off pathogens by treating sugar-beet seed with a powdery preparation of the mycoparasite *Pythium oligandrum* in large-scale field trials. In *Interrelationships between Microorganisms and Plants in Soil* (ed. V. Vancura and F. Kunc), pp. 445–449. Academia, Czechoslovakia.

Weger, de L. A., Dunbar, P., Mahafee, W. F., Lugtenberg, B. J. J. and Sayler, G. S. (1991). Use of bioluminescence markers to detect *Pseudomonas* spp. in the rhizosphere. *Applied and Environmental Microbiology* **57**, 3641–3644.

Wester, E., Jensen, D. F. and Møller, J. D. (1991). Anatomical studies of cucumber roots inoculated with *Pythium* zoospores. In *Biotic Interactions and Soil-borne Diseases* (ed. A.B.R. Beemster, G.J. Bollen, M. Gerlagh, M.A. Ruissen, B. Schippers and A. Tempel), pp. 401–407. Elsevier Science Publishers B.V., Amsterdam.

Windels, C. E. and Kommedahl, T. (1978). Factors affecting *Penicillium oxalicum* as a seed protectant against seedling blight of pea. *Phytopathology* **68**, 1656–1661.

Wolffhechel, H. and Jensen, D. F (1991). Influence of the water potential of peat on the ability of *Trichoderma harzianum* and *Gliocladium virens* to control *Pythium ultimum*. In *Biotic Interactions and Soil-borne Diseases* (ed. A.B.R. Beemster, G.J. Bollen, M. Gerlagh, M.A. Ruissen, B. Schippers and A. Tempel), pp. 392–397. Elsevier Science Publishers B.V., Amsterdam.

17

Bacillus thuringiensis in pest control

Raymond J. C. Cannon

Introduction

The spore-forming, Gram-positive bacterium, *Bacillus thuringiensis* (*Bt*), is a ubiquitous soil micro-organism with a world-wide distribution (Martin and Travers, 1989). Many thousands of isolates have been discovered since Ishiwata first isolated *Bt* from diseased silkworm larvae in 1901 (Nakamura and Dulmage, 1988; Beegle and Yamamoto, 1992; Feitelson *et al.*, 1992), and it is possible to group these isolates into at least 34 serovars (also called subspecies) (de Barjac and Frachon, 1990). However, biomolecular techniques, such as multiplex polymerase chain reaction (PCR), permit a more rapid identification of strains (Bourque *et al.*, 1993), and it is now possible to target specific DNA sequences with the use of this technology.

The delta-endotoxin produced by *Bt* is the most widely used biological pesticide – 2.3 million kg was used annually according to Rowe and Margaritis (1987) – and it is the 'front-runner' in attempts to commercialize biological control agents (BCAs); albeit that *Bt* is only a BCA in a limited sense, since it is the product of a living organism (the delta-endotoxin) which is utilized, rather than the organism itself.

The precise ecological role of this cosmopolitan invertebrate pathogen in the natural environment remains speculative (Smith and Couche, 1991). *Bt* reportedly does not produce spores or crystals in infected host cadavers, it is relatively poor at spreading natural infections in the field and rarely causes epizootics (Fuxa, 1989). Its ubiquitous occurrence in the soil (Martin and Travers, 1989), phylloplane microflora (Smith and Couche, 1991), and in man-made ecosystems (Meadows *et al.*, 1992) suggests that it may have a primary function other than as a pathogen.

Mode of action

The insecticidal delta-endotoxins, which are pro-toxin proteins deposited in the cytoplasm of the bacterium during sporulation, react with the brush-border membrane of the insect mid-gut epithelial cells, after an activation process that involves solubilization in the (typically) alkaline environment of the mid-gut and cleavage by proteases (Andrews *et al.*, 1987). The mode of action involves binding to putative membrane receptors, thus forming cation-selective channels (pores), which cause swelling and lysis of the cells and eventual death of the organism (English and Slatin, 1992).

History of *Bt* research

First isolated at the turn of the century in Japan, *Bt* was not characterized until a decade later, when E. Berliner named it after the region in Germany, Thuringia, where it was discovered on diseased Mediterranean Flour moth larvae, *Anagasta kuehniella* (Zeller) (Beegle and Yamamoto, 1992).

Preparations of this insect pathogen were produced during the late 1920s and early 1930s ('Sporeine') and tested in field trials in Hungary and Yugoslavia against the European corn borer, *Ostrinia nubilalis* (Hubner). However, it was not until the late 1950s ('Thuricide' was available in 1957 in the USA) that wide-scale commercial production began (Beegle and Yamamoto, 1992).

It was only after the discovery of the considerably more active *Bt* var. *kurstaki* (then termed *alesti*) HD-1 isolate (Dulmage, 1970), first commercialized in the USA as 'Dipel' in 1970 by Abbott Laboratories, that widespread agricultural usage commenced. Usage reportedly peaked in the mid-1970s, after which competition from the synthetic pyrethroids led to a decline in demand for *Bt* in the agricultural sector (Beegle and Yamamoto, 1992).

The opening of a new era of *Bt* research, stimulated by progress in biotechnology, started at the beginning of the 1980s, when Schnepf and Whiteley (1981) cloned a *Bt* crystal toxin gene from *Bt kurstaki* into *E. coli*. Since then, advances in recombinant (rDNA) technology have resulted in the development of more efficacious strains, e.g. with improved target spectra and enhanced activity (e.g. Baum *et al.*, 1990). In addition, *Bt* genes have been cloned into a variety of different organisms (these include baculoviruses, non-pathogenic bacteria, algae, protozoa, and higher plants – both monocots and dicots) in an effort to provide improved delivery systems for the toxins and more precise targeting of the pests (Martens *et al.*, 1990; Gelernter and Schwab, 1991). Research is continuing to add to the list of *Bt* toxins, (so-called Cry proteins), which promises to broaden the host range of *Bt*-based biopesticide products (Feitelson *et al.*, 1992).

Commercial development of *Bt*

Niche markets

Although biological control methods have made significant inroads into some areas of agriculture (e.g. glasshouse crops), synthetic chemical pesticides continue to dominate the majority of Crop Protection (CP) markets. Estimates for the volume of *Bt* product sales vary, e.g. from $US60m–$US120 million but even the upper end of this range accounts for only about 0.5% of the global market of CP products (Cannon, 1993). However, the market for *Bt* products is reportedly increasing, at a rate of 20% per year in some countries (Anonymous, 1992*a*), which if sus-

tained would project sales into the $US300–$US500 million range by the turn of the century.

The lack of market penetration by *Bt* products in the past 35 years was largely a consequence of their inability to compete with synthetic chemical insecticides such as the synthetic pyrethroids, in all but limited niche markets. The reason for this was primarily cost-performance: the chemical alternatives have simply been cheaper and more effective (Cannon, 1993). Other reasons given for the relative lack of commercial success for *Bt* include variable field performance, poor persistence, a narrow target spectrum, a lack of contact activity and a slow rate of action (Cannon, 1990; Gelernter, 1990*a*).

Keys to development of second-generation products

Three key developments were identified by Lisansky (1989) as being critical to the process of addressing the limitations of so-called 'first-generation' *Bt* products. First, the scale of investment in the research and development of biopesticides, without which no product would be successfully commercialized. Low development costs and short time-frames are frequently cited as being the commercial advantages of biological products, nevertheless, considerable investment is still required to overcome some of their traditional limitations, and the ultimate success of this ongoing investment has yet to be decided.

The second key development is the emergence of a variety of new markets for biological products, such as *Bt*, as a result of: (i) public concerns regarding the safety and environmental impact of synthetic chemical pesticides; and (ii) the emergence of insecticide-resistance in a number of important pest species (Cheng, 1988).

The third factor highlighted by Lisansky (1989), was the biomolecular revolution, which has provided powerful new tools with which to develop 'second-generation' genetically-modified products.

Companies and products

The *Bt* market is dominated by three major producers: Sandoz Inc. and Novo Nordisk with approximately 25% each, and Abbott Laboratories with about 45% of the market (Anonymous,

Table 17.1. Examples of products incorporating *Bacillus thuringiensis* delta-endotoxins

Product	*Bt* variety	Company (Country)
A. Products for the control of Lepidopterous larvae		
DIPEL	*kurstaki* (HD-1)	Abbott Laboratories
THURICIDE	*kurstaki*	Sandoz AG
JAVELIN/DELFIN	*kurstaki* (NRD-12)	Sandoz AG
BACTEC BERNAN I, III, V	*kurstaki*	Bactec Corporation
CATERPILLAR ATTACK	*kurstaki*	Ringer Corporation
BIOBIT	*kurstaki*	Novo Nordisk
BACTOSPEINE	*kurstaki*	Novo Nordisk
TOAROW CT	*kurstaki*	Taogosei Chem.
TUREX/AGREE[a]	*kurstaki/aizawai*	Ciba Geigy
CUTLASS[a]	*kurstaki*	Ecogen Inc.
MVP[b]	*kurstaki*	Mycogen Corporation
M-PERIL[b]	*kurstaki*	Mycogen Corporation
CERTAN PLUS	*aizawai*	Sandoz AG
FLORBAC	*aizawai*	Novo Nordisk
XENTARI	*aizawai*	Abbott Laboratories
BACTEC BERNAN II	*morrisoni*	Bactec Corporation
B. Products for the control of Dipterous (mosquito and blackfly) larvae		
VECTOBAC	*israelensis*	Abbott Laboratories
TECHNAR	*israelensis*	Sandoz AG
ACROBE	*israelensis*	American Cyanamid
MOSQUITO ATTACK	*israelensis*	Ringer Corporation
SKEETAL	*israelensis*	Novo Nordisk
BACTIMOS	*israelensis*	Novo Nordisk
C. Products for the control of Coleopteran (e.g. Colorado Potato beetle) larvae		
NOVODOR	*tenebrionis*	Novo Nordisk
BK-100	*tenebrionis*	Novo Nordisk
FOIL[a]	*kurstaki/tenebrionis*	Ecogen Inc.
M-TRAK[b]	*san diego*	Mycogen Corporation
D. Products for the control of forest defoliating caterpillars (e.g. Spruce budworm and Gypsy moth)		
THURICIDE (32LV)	*kurstaki*	Sandoz AG
FORAY	*kurstaki*	Novo Nordisk
BIODART	*kurstaki*	ICI (Zeneca)
CONDOR[a]	*kurstaki*	Ecogen Inc.

[a]transconjugant.
[b]Killed microbial (transgenic).

1991). However, a number of other companies (such as Ecogen Incorporated and Mycogen Corporation) have emerged as specialists in the biotechnology field, and are committed to developing CP products based on naturally-occurring and 'environmentally compatible' agents (Anonymous, 1992*b*). These so-called 'Agbiotech' companies are in the process of introducing a number of innovative products (see below and Table 17.1), and to achieve world-wide sales have in many cases formed alliances with established CP companies (e.g. Cannon, 1992).

A biotechnology hierarchy?

The past few years has witnessed the development of a range of 'second- generation' products, based on genetic modifications to the traditional *Bt* strains. Indeed, in theory the customer can choose from a variety of different products, not all of which are fully commercialized, but which nevertheless can be produced by existing (bio)technology and which in effect span a gamut of technical sophistication (Cannon, 1990).

Traditional *Bt* products, such as Dipel (Abbott

Laboratories) and Javelin (Sandoz AG), contain a mixture of spores, crystals and lysed cells and are based on naturally occurring isolates. A variation on this theme is the production of spore-killed (e.g. Toarow) or sporeless ('spore-minus') varieties, which potentially overcome concerns which exist in certain countries regarding *Bt* spores, that they are difficult to distinguish from those of the pathogen, *Bacillus cereus*.

Transconjugants

Moving up the ladder of technical sophistication we come to products, such as those developed by Ecogen Inc. These are based on strains, which although genetically altered, are the product of non-recombinant, transconjugation (or plasmid exchange) techniques (Carlton *et al.*, 1990). In other words, combinations of toxin genes from different parental source strains have been produced without using genetic modification techniques. Ecogen Inc. have registered two such products, 'Cutlass' (for caterpillar pests of vegetables) and 'Condor' (for caterpillar pests of forests), in 1989 in the USA. A third product, 'Foil', combines activity against the Colorado potato beetle, *Leptinotarsa decemlineata* (Say), with activity against European corn borer (*O. nubilalis*), and was granted regulatory approval in the USA in 1990. The production of these novel strains reportedly took 4 years 'from microbe to marketplace' (B. C. Carlton, personal communication), which was possible because of the streamlined, and relatively inexpensive, regulatory review process required by the US Environmental Protection Agency (EPA) (Gelernter, 1990*b*). This is a significantly shorter development time than for most chemical insecticides (Lethbridge, 1989).

Ciba-Geigy has recently entered the *Bt* market with a new product, which is also a transconjugant, for lepidopteran pests of fruit and vegetables ('Turex' in Europe and 'Agree' in the USA).

Transgenic microbes – dead or alive

Mycogen Corporation has adopted a different approach, which involves the transfer of the toxin-producing (*cry*) genes from *Bt* into another bacterial host. This CellCap technology (Fig. 17.1) is based on a non-pathogenic bacterium, *Pseudomonas fluorescens* (*Pf*), which following production of

the biotoxins within the transgenic cells, is killed in the fermentor by a proprietary physical and chemical fixing process (Gelernter, 1990*a*). The dead *Pf* cells (Fig. 17.2) serve as a biological microencapsulation system for the *Bt* toxins, thus providing increased protection against environmental degradation, as compared with the 'naked' delta-endotoxin crystals (Soares and Quick, 1991). There are a number of other advantages provided by this bioencapsulation system which are summarized in Table 17.2.

The first of Mycogen Corp.'s genetically-modified bioinsecticide products (MVP and M-Trak) were granted US EPA approval in June 1991, and since then regulatory approval for MVP has also been obtained in Europe (in Spain in 1992). In addition, a third CellCap product, M-Peril, was registered in the USA in February 1993, for the control of European corn borer, *O. nubilalis*, in maize. These examples of 'killed microbial' products demonstrate the benefits that can be gained from the use of rDNA technology, i.e. to derive highly effective, safe and environmentally friendly products.

The next rung on the ladder could be said to involve the release of live, genetically-modified, transgenic micro-organisms; an approach being developed by Crop Genetics International. They have genetically transformed a xylem-inhabiting endosymbiotic bacterium, *Clavibacter xyli* subsp. *cynodontis*, a species that naturally occurs in the vascular system of Bermuda grass (*Cynodon dactylon* L.), with a gene from a *Bt* subsp. *kurstaki* HD-73 strain. The resultant transformed bacteria (containing the expressed *Bt* (CryIA(c)) delta-endotoxin) are inoculated into maize plants where populations spread throughout the vascular system, thereby providing protection against tunnelling pests such as *O. nubilalis*. To address concerns about environmental safety, CGI are reportedly developing strains in which the delta-endotoxin gene is eventually lost from the introduced *C. xyli* populations (Turner *et al.*, 1991).

Transgenic crops

The final rung on the ladder involves the production of insect-resistant, transgenic crops that express selected *Bt* genes, either constitutively (throughout the plant) or in specific tissues, where pests are most likely to inflict damage. A wide variety of dicot species (such as tomato, potato and tobacco) and also more recalcitrant monocot species such as maize have now been successfully transformed into varieties resistant to some insects via the expression of *Bt* toxins, in most cases using a version of the *cry*IA(b) gene from *Bt kurstaki* HD-1 strain (Vaeck *et al.*, 1987; Koziel *et al.*, 1993).

This system does have the distinct advantage of producing the pesticide at exactly the right time and place (for example, when the neonate larvae commence feeding they are at their most vulnerable stage). Indeed, it may even be possible to control the timing of expression of the toxins within plant tissues, for example by utilizing plants modified to synthesize *Bt* only in response to an inducer substance, which would be sprayed immediately prior to an invasion of the pest (Williams *et al.*, 1992).

Many resources are being expended in this area, the technology is already here and much is promised for the future (Collins, 1992). However, concern centres around the possibility that resistance could develop rapidly within pest populations if they were presented with fields of plants that all contained the same toxin. Strategies to avoid such resistance include providing 'refuges' where pests can avoid coming under continual selection pressure, sowing mixtures of transgenic and untransformed seed, and developing plants that express two different types of toxin.

Safety considerations

The examples described above involve (bio)technologies, which are recent inventions. However, *Bt* preparations have been registered as insecticides for at least 35 years, during which time a

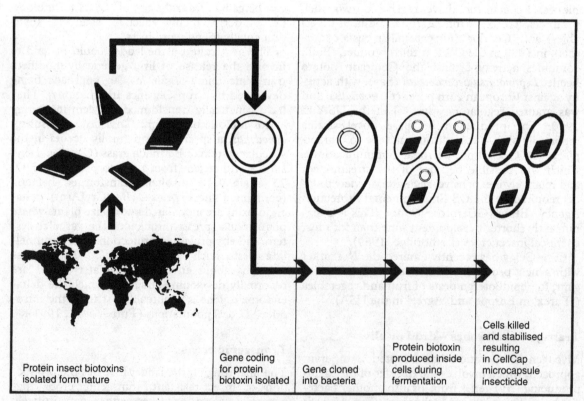

| Protein insect biotoxins isolated form nature | Gene coding for protein biotoxin isolated | Gene cloned into bacteria | Protein biotoxin produced inside cells during fermentation | Cells killed and stabilised resulting in CellCap microcapsule insecticide |

NB CellCap is Mycogen Corporation's proprietory bioencapsulation technology

Fig. 17.1 The CellCap system for the production of microencapsulated *Bt* toxins.

Table 17.2. Advantages provided by Mycogen Corporation's proprietary CellCap technology (after Gelernter, 1990*a*)

1	Increased persistence – bioencapsulation provides greater residual activity and increased efficacy
2	Gene flexibility – single gene products expressed in a transgenic host enable a flexible approach to product development
3	Environmental safety – harmless to non-target organisms (including beneficial insects) and no spores are present
4	Regulatory advantages – two CellCap products received regulatory approval from the US EPA in 1991, under the category 'killed microbial'
5	Shelf life – increased due to the absence of living organisms (i.e. including microbial contaminants)
6	Formulation advantages – due to increased robustness of the microencapsulated crystal toxins

considerable body of toxicological data has been collected, and studies have demonstrated a consistent lack of toxicity to mammals, birds, fish, aquatic invertebrates and other non-target organisms (Burgess, 1981; 1982). Nevertheless, the mode of action of these (ingestion) toxicants is still not fully understood, and while there is no

evidence that *Bt* delta-endotoxins are hazardous, some authors have called for studies into whether activated toxins resist proteases in the human gut (Goldberg and Tjaden, 1990).

New opportunities for *Bt*-based products

Forestry

Public pressure to reduce the use of chemical insecticides in so-called eradication programmes against forest-defoliating caterpillar pests, such as Spruce budworm (*Choristoneura fumiferana* Clemens) and Gypsy moth (*Lymantria dispar* (L.)), has encouraged the development of biological agents as alternatives (Dubois, 1981). Although it took time and a considerable investment to improve reliability and efficacy, *Bt* is now cost-competitive with chemical insecticides in forest-insect control programmes (Irland and Rumpf, 1987), and has displaced chemical insecticides in many regions of Canada and the USA, in this sector. The key to this process was the development of more concentrated, high potency formulations, with volume

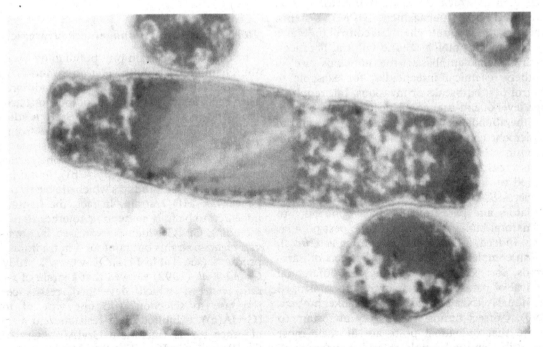

Fig. 17.2 *Bacillus thuringiensis* var *kurstaki* single gene product encapsulated in genetically manipulated, dead, *Pseudomonas fluorescens* cells.

application rates equivalent to those of chemicals (van Frankenhuyzen, 1991). North American forestry is now one of the major outlets for *Bt* products, with the Canadian *Bt* market alone accounting for $Can 10 million in 1990 (Anonymous, 1992c).

Novel isolates

The host range of most *Bt* isolates is limited (e.g. to Lepidoptera), but screening for novel isolates has led to the discovery of strains with activity against pests from a broad range of invertebrate groups, including several orders of insects (Lepidoptera, Diptera, Hymenoptera and Coleoptera), mites, nematodes, flatworms and protozoa (Goldberg and Margalit, 1977; Krieg *et al.*, 1983; Feitelson *et al.*, 1992). The toxins produced by these novel strains can still retain the advantages of specificity (i.e. safe to non-target organisms) associated with traditional *Bt* products.

Integrated pest management

One area in which *Bt* usage is increasing is in integrated pest management (IPM) systems, which aim to integrate chemical control measures with a range of biological and cultural practices. Such systems emphasize the judicious use of synthetic chemical insecticides, for example to control pest outbreaks or invasions, but require a high level of end-user sophistication, e.g. to monitor the abundance and life-stages of pests and to implement control measures in accordance with recommended timings and thresholds.

For reasons of target specificty – probably related to differences in toxin receptors between species – *Bt* does not affect beneficial insects (i.e. predators and parasitoids), which contribute to the natural biological control of the pest populations. Indeed, if the application timing is optimal, *Bt* can complement the beneficial effects of parasitoids, for example by causing the differential survival of parasitized hosts over non-parasitized individuals (Nealis and van Frankenhuyzen, 1990). Correct timing is also very important to ensure that individual pests are at their most susceptible and vulnerable stage (i.e. neonate or early instar larvae) when sprays are applied.

Insecticide resistance management

Insect resistance management (IRM) systems, such as the strategy employed in Australia for *Heliothis armigera* (Hubner) control in cotton (Cox and Forrester, 1992), present additional emerging opportunities for *Bt*. *Bt* products may be used early in the season (e.g. mixed with chemical ovicides) during periods when pyrethroid use is prohibited, or later in the season to 'mop up' any pyrethroid-resistant larvae. In 1987/8, *Bt kurstaki*-based products accounted for a 2% market share of all insecticides used on Australian cotton (Powles and Rogers, 1989). It can be expected that the increase in pyrethroid resistance of *H. armigera* populations in Australia, and other cotton-growing countries, may provide greater opportunity for *Bt* use, for example in rotations, and in mixtures with chemicals (Tabashnik, 1989).

It is to be expected that *Bt* use will increase, as the benefits of utilizing biological and chemical products in a rational and efficient way (i.e. one which enhances their effectiveness and longevity while reducing undesirable environmental effects) are increasingly recognized.

Bt resistance and management strategies

The capacity of certain pest populations (such as the Diamondback moth, *Plutella xylostella* L.) to develop resistance to *Bt* under field conditions has been reported from a number of locations around the world (e.g. Tabashnik *et al.*, 1990). A study of a population of resistant *P. xylostella* from the Philippines (Ferre *et al.*, 1991) demonstrated that resistance appeared to be limited to one particular type of *Bt* protein toxin (CryIA(b)), found in the most widely-used products which are based on the *Bt kurstaki* HD-1 strain. In fact, the population appeared to be fully susceptible to other *Bt* toxins, including CryIC which is expressed by some *Bt* var. *aizawai* strains but not found in traditional *Bt* products (see Table 17.1). However, a study by Gould *et al.* (1992) showed that a strain of *Heliothis virescens*, which developed resistance in response to selection with one type of toxin (CryIA(c)), exhibited cross-resistance to a range of other toxins (including CryIA(a), CryIA(b), CryIB and CryIC). Finally, McGaughey and Johnson (1992), concluded, on the basis of laboratory selection experiments with Indianmeal moths, *Plodia interpunctella* (Hubner), that mixtures of *Bt* strains do not preclude the possibility of resistance, and the authors questioned their value in slowing the development of resistance.

Thus, it appears that *Bt* resistance is not straightforward, and until the mode of action and the mechanisms of resistance to the different toxins are better understood, strategies for avoiding the development of *Bt* resistance need to be implemented; for example, avoiding excessive applications and using *Bt* in rotation with chemicals and other biocontrol methods (Gelernter 1990*b*).

To investigate the potential for *Bt* resistance in key pests, a *Bt* Management Working Group, comprising a consortium of interested companies, was set up in 1988 under the auspices of the Insecticide Resistance Action Committee (IRAC). The aims of this Group are to fund research targeted at elucidating the mechanisms of resistance, and to promote strategies that minimize the risk of it developing. These include: adopting an IPM approach (see above); the use of within-crop 'refuges'; and the utilization of some novel toxins selected from the diverse range of *Bt* strains now available (Commandeur and Komen, 1992; Feitelson *et al.*, 1992).

Discussion

If biocontrol agents are to realize their full commercial potential, then it will be necessary to look more realistically at the demands required by the market-place. For example, some of the inherent properties of biological control agents such as specificity, can be disadvantageous in purely commercial terms, so some compromises may be necessary as regulatory demands (even for biological control agents) often preclude the development of niche market products. However, where a favourable regulatory environment exists, there may be a potential for the development of specialized *Bt* products for use against individual pests, for example in high value cash crops (Smits, 1987).

To some extent, and for a variety of different reasons, the crop protection market is becoming more open to biocontrol. However, it is probable that the rate of increase in this sector will continue to be gradual in overall terms, and unless some significant 'quantum leap' occurs will probably remain below 5% of the total CP market (currently estimated at about $US24 billion for some time (Anon, 1993).

Genetically modified crops are probably a 'joker in the pack' in terms of future CP market trends. The (bio)technology involved is innovative and is evolving rapidly, but the likelihood of commercial success is difficult to predict, since it is dependent both upon the public acceptance of this novel technology as well as on the full realization of the promises offered by the technology itself.

The CP industry aims to supply customers with safe and cost-effective products and remain profitable. If safety testing and environmental assessments are sufficiently rigorous, it may be 'chemophobic' to argue that naturally-occurring products are inherently more desirable than synthetic chemicals produced by human ingenuity. However, Nature has produced a remarkable range of biocontrol agents, which are perhaps best suited to a role within the framework of a more sophisticated approach to protecting our crops, livestock and produce, i.e. IPM.

The ever-increasing environmental, safety and performance demands that are applied to CP products inevitably require technological solutions. Given world population trends, and disregarding anomalous local and temporary reversal, the demands for increased levels of agricultural production are likely to continue as we move into the twenty-first century (Pimentel, 1991). If biocontrol agents are to flourish in this environment, they must be reliable and highly effective, and retain their original beneficial features of safety coupled with a low risk of environmental perturbation.

Conclusions

In the past, *Bt* has failed to make major inroads into CP markets, despite the benefits of specificity and environmental safety, largely for reasons of cost-performance in relation to synthetic chemical alternatives. However, the development of new, more efficacious products (that span a range of biotechnological sophistication), combined with the emergence of new markets and opportunities (e.g. created by insecticide resistance and wider implementation of IPM), promises growth for such products in the future.

In addition, the continuing discovery of novel strains and a growing list of *Bt* toxins active against a wide range of pests (including insects, mites, nematodes and flatworms), offers the prospect of *Bt*-based products in outlets beyond those currently on offer. However, careful usage and a creative approach to pest management will be needed to promote the longevity of products and to avoid the onset of *Bt* resistance.

References

Andrews, R. E. Jr, Faust, R. M., Wabiko, H., Raymond, K. C. and Bulla, L.A. (1987). The biotechnology of *Bacillus thuringiensis*. *CRC Critical Reviews in Biotechnology* 6, 163–232.

Anonymous (1991). *Agrow* 139, 1. PJB Publications Ltd.

Anonymous (1992a). *Agrow* 167, 4. PJB Publications Ltd.

Anonymous (1992b). *Mycogen Corporation Annual Report 1991*.

Anonymous (1992c). *Agrow* 172, 3. PJB Publications Ltd.

Anonymous (1993). *Agrow* 179, 19. PJB Publications Ltd.

Baum, J. A., Coyle, D. M., Gilbert, M. P., Jany, C. S. and Gawron-Burke, C. (1990). Novel cloning vectors for *Bacillus thuringiensis*. *Applied and Environmental Microbiology* 56, 3420–3428.

Beegle, C. C. and Yamamoto, T. (1992). Invitation paper (C. P. Alexander fund): History of *Bacillus thuringiensis* Berliner research and development. *Canadian Entomologist* 124, 587–616.

Bourque, S. N., Valero, J. R., Mercier, J., Lavoie, M. C. and Levesque, R. C. (1993). Multiplex polymerase chain reaction for detection and differentiation of the microbial insecticide *Bacillus thuringiensis*. *Applied and Environmental Microbiology* 59, 523–527.

Burgess, H. D. (1981). Safety, safety testing and quality control of microbial pesticides. In *Microbial Control of Pests and Plant Diseases 1970–1980* (ed. H. D. Burgess), pp. 737–767. Academic Press, London.

Burgess, H. D. (1982). Control of insects by bacteria. *Parasitology* 84, 79–117.

Cannon, R. J. C. (1990). Commercial perspectives of biological control. In *Biological Control of Pests and Diseases* (ed. A. R. McCracken and P. C. Mercer), pp. 48–53. Agriculture and Food Science Centre, Belfast.

Cannon, R. J. C. (1992). Biopesticides: Shell and Mycogen forge ahead. *Shell Agriculture* 13, 15–16.

Cannon, R. J. C. (1993). Prospects and progress for Bt-based pesticides. *Pesticide Science* 37, 331–335.

Carlton, B. C., Gowron-Burke, C., Johnson, T. B. (1990). Exploiting the genetic diversity of *Bacillus thuringiensis* for the creation of new bioinsecticides. In *Proceedings of the Vth International Colloquium on Invertebrate Pathology and Microbial Control, August 1990* (ed. D. E. Pinnock), pp. 18–22. Society for Invertebrate Pathology, Adelaide, Australia.

Cheng, E. Y. (1988). Problems of control of insecticide-resistant *Plutella xylostella*. *Pesticide Science* 23, 177–188.

Collins, H. (1992). Biotechnology promises big future for cotton plant breeding. *The Australian Cottongrower* 13, 33–34. Berekua Pty Ltd, Brisbane.

Commandeur, P. and Komen, J. (1992). Biopesticides: Options for biological pest control increase. *Biotechnology and Development Monitor* 13, 6–8.

Cox, P. and Forrester, N. W. (1992). Economics of insecticide resistance management in *Heliothis armigera* (Lepidoptera: Noctuidae) in Australia. *Journal of Economic Entomology* 85, 1539–1550.

de Barjac, H. and Frachon, E. (1990) Classification of *Bacillus thuringiensis* strains. *Entomophaga* 35, 233–240.

Dubois, N. R. (1981). Microbials: *Bacillus thuringiensis*. In *The Gypsy Moth: Research Toward Integrated Pest Management* (ed. C. C. Doane and M. L. McManus), pp. 445–453. US Department of Agriculture Technical Bulletin 1584. Washington, DC.

Dulmage, H. T. (1970). Insecticidal activity of HD-1, a new isolate of *Bacillus thuringiensis* var. *alesti*. *Journal of Invertebrate Pathology* 15, 232–239.

English, L. and Slatin, S. L. (1992). Mode of action of delta-endotoxins from *Bacillus thuringiensis*: a comparison with other bacterial toxins. *Insect Biochemistry and Molecular Biology* 22, 1–7.

Feitelson, J. S., Payne, J. and Kim, L. (1992). *Bacillus thuringiensis* and beyond. *Bio/technology* 10, 271–275.

Ferre, J., Real, M. D., Van Rie, J. and Peferoen, M. (1991). Resistance to *Bacillus thuringiensis* bioinsecticide in a field population of *Plutella xylostella* is due to a change in a midgut membrane receptor. *Proceedings of the National Academy of Sciences of the USA* 88, 5119–5123.

Fuxa, J. R. (1989). Fate of released entomopathogens with reference to risk assessment of genetically engineered micro-organisms. *Bulletin of the Entomological Society of America* 35, 12–24.

Gelernter, W. D. (1990a). *Bacillus thuringiensis*, bioengineering and the future of bioinsecticides. In

Proceedings of the 1990 Brighton Crop Protection Conference 3, 617–624.

Gelernter, W. D. (1990*b*). Targeting insecticide-resistant markets: New developments in microbial-based products. In *Managing Resistance to Agrochemicals: From Fundamental Research to Practical Strategies* (ed. M. B. Green, W. K. Moberg and H. LeBaron), pp. 105–117. American Chemical Society Symposium Series, No. 421.

Gelernter, W. and Schwab, G. E. (1991). Transgenic bacteria, viruses, algae and other microorganisms as *Bacillus thuringiensis* delivery systems. In Bacillus thuringiensis, *An Environmental Biopesticide: Theory and Practice* (ed. P. F. Entwhistle, J. S. Cory, M. J. Bailey and S. R. Higgs), pp. 89–104. John Wiley, Chichester.

Goldberg, L. J. and Margalit, J. (1977). A bacterial spore demonstrating rapid larvicidal activity against *Anopheles sergentii, Uranotaenia unguiculata, Culex univerittus, Aedes aegyptii* and *Culex pipiens. Mosquito News* 37, 355–358.

Goldberg, R. J. and Tjaden, G. (1990). Are B.t.k. plants really safe to eat? *Bio/technology* 8, 1011–1015.

Gould, F., Martinez-Ramirez, A., Anderson, A., Ferre, J., Silva, F. J. and Moar, W. J. (1992). Broad-spectrum resistance to *Bacillus thuringiensis* toxins in *Heliothis virescens. Proceedings of the National Academy of Sciences, USA* 89, 7986–7990.

Irland, L. C. and Rumpf, T. A. (1987). Cost trends for *Bacillus thuringiensis* in the Maine Spruce budworm control program. *Bulletin of the Entomological Society of America* 33, 86–90.

Koziel, M. G., Beland, G. L., Bowman, C., Carozzi, N. B., Crenshaw, R., Crossland, L., Dawson, J., Desai, N., Hill, M., Kadwell, S., Launis, K., Lewis, K., Maddox, D., McPherson, K., Meghji, M. R., Merlin, E., Rhodes, R., Warren, G. W., Wright, M. and Evola, S. V. (1993). Field performance of elite transgenic maize plants expressing an insecticidal protein derived from *Bacillus thuringiensis. Bio/Technology* 11, 194–200.

Krieg, A., Huger, A. M., Langenbruch, G. A. and Schnetter, W. (1983). *Bacillus thuringiensis* var. tenebrionis: en neuer, gegenuber Larven von Coleopteran wirksamer Pathotyp. *Zeitschrift für angewandante Entomologie* 96, 500– 508.

Lethbridge, G. (1989). An industrial view of microbial inoculants for crop plants. In *Microbial Inoculation of Crop Plants* (ed. R. M. Macdonald), pp. 11–28. *Special Publications of Scientific General Microbiology* 25, IRL Press, Oxford.

Lisansky, S. (1989). Biopesticides: the next revolution? *Chemistry and Industry* 17, 478–482.

McGaughey, W. H. and Johnson, D. E. (1992). Indianmeal moth (Lepidoptera: Pyralidae) resistance to different strains and mixtures of *Bacillus*

thuringiensis. Journal of Economic Entomology 85, 1594–1600.

Martens, J. W. M., Honee, G., Zuidema, D., van Lent, J. W. M., Visser, B. and Vlak, J. M. (1990). Insecticidal activity of a bacterial crystal protein expressed by a recombinant baculovirus in insect cells. *Applied and Environmental Microbiology* 56, 2764–2770.

Martin, P. A. W. and Travers, R. S. (1989). Worldwide abundance and distribution of *Bacillus thuringiensis* isolates. *Applied and Environmental Microbiology* 55, 2437–2442.

Meadows, M. P., Ellis, D. J., Butt, J., Jarrett, P. and Burgess, H. D. (1992). Distribution, frequency, and diversity of *Bacillus thuringiensis* in an animal feed mill. *Applied and Environmental Microbiology* 58, 1344– 1350.

Nakamura, L. K. and Dulmage, H. T. (1988). *Bacillus thuringiensis* cultures available from the US Department of Agriculture. *US Department of Agriculture, Technical Bulletin,* No. 1738, 38 pp.

Nealis, V. and van Frankenhuyzen, K. (1990). Interactions between *Bacillus thuringiensis* Berliner and *Apantales fumiferanae* Vier. (Hymenoptera: Braconidae), a parasitoid of the Spruce budworm, *Choristoneura fumiferana* (Clem.) (Lepidoptera: Tortricidae). *Canadian Entomologist* 122, 585–594.

Pimentel, D. (1991). World resources and food losses to pests. In *Ecology and Management of Food-industry Pests* (FDA Technical Bulletin 4) (ed. J. R. Gorham), pp. 5–11. Association of Official Analytical Chemists, Arlington, VA.

Powles, R. J. and Rogers, P. L. (1989). Bacillus toxins for insect control – a review. *Australian Journal of Biotechnology* 3, 223–228.

Rowe, G. E. and Margaritis, A. (1987). Bioprocess developments in the production of bioinsecticides by *Bacillus thuringiensis. CRC Critical Reviews in Biotechnology* 6, 87–127

Schnepf, H. E. and Whiteley, H. R. (1981). Cloning and expression of the *Bacillus thuringiensis* crystal protein gene in *Escherichia coli. Proceedings of the National Academy of Sciences USA* 78, 2893–2897.

Smith, R. A. and Couche, G. A. (1991). The phylloplane as a source of *Bacillus thuringiensis* variants. *Applied and Environmental Microbiology* 57, 311–315.

Smits, P. H. (1987). Possibilities and constraints in the development of new *Bacillus thuringiensis* strains. *Mededelingen Faculteit Landbouwwetenschappen Rijksuniversiteit Gent* 53, 155–157.

Soares, G. G. and Quick, T. C. (1991). MVP, a novel bioinsecticide for the control of Diamondback moth. In *Diamondback Moth and Other Crucifer Pests* (ed. N. S. Talekar), pp. 129–137. Proceedings of the

Second International Workshop, Tainan, Taiwan, 10–14 December 1990.

Tabashnik, B. E. (1989). Managing resistance with multiple pesticide tactics: theory evidence and recommendations. *Journal of Economic Entomology* 82, 1263–1269.

Tabashnik, B. E. Cushing, N. L. Finson, N. and Johnson, M. W. (1990). Field development of resistance to *Bacillus thuringiensis* in Diamondback moth (Lepidoptera: Plutellidae). *Journal of Economic Entomology* 83, 1671–1676.

Turner, J. T., Lampel, J. S., Stearman, R. S., Sundin, G. W., Gunyuzlu, P. and Anderson, J. J. (1991). Stability of the delta-endotoxin gene from *Bacillus thuringiensis* subsp. *kurstaki* in a recombinant strain of *Clavibacter xyli* subsp. *cynodontis*. *Applied and Environmental Microbiology* 57, 3522–3528.

Vaeck, M., Reynaerts, A., Hofte, H., Jansens, S., De Beuckeleer, M., Dea, C., Zabeau, M. Van Montagau, M. and Leemans, J. (1987). Transgenic plants protected from insect attack. *Nature* 328, 33–37.

van Frankenhuyzen, K. (1991). The challenge of *Bacillus thuringiensis*. In *Bacillus thuringiensis, An Environmental Biopesticide: Theory and Practice* (ed. P. F. Entwhistle, J. S. Cory, M. J. Bailey and S. Higgs), pp. 1–35. John Wiley, Chichester.

Williams, S., Friedrich, L., Dincher, S., Carrozi, N., Kessman, H., Ward, E. and Ryals, J. (1992). Chemical regulation of *Bacillus thuringiensis* delta-endotoxin expression in transgenic plants. *Bio/Technology* 10, 540–542.

18

Opportunities with baculoviruses

Jürg Huber

Introduction

Today, about 1200 viruses are known to infect insects (Martignoni and Iwai, 1986). These do not form a taxonomic unit but belong to a range of different virus families including the Iridoviridae, Parvoviridae, Poxviridae, Reoviridae and Baculoviridae. Most of these families also have representatives that infect not only insects, but also vertebrates and even plants. There is one exception to this rule: viruses that belong to the family Baculoviridae have so far only been isolated from arthropod hosts. More than 60% of all insect viruses known today belong to this family. Some baculoviruses have been found in shrimps and mites, but most of them have been isolated from insects, in particular lepidopterans.

Baculoviruses are characterized by double-stranded circular DNA, which is included in rod-shaped nucleocapsids (Federici, 1986). They are formed mainly in the nucleus of the host cells. In common with many insect viruses from other virus families, the virions of most baculoviruses are contained within proteinacious para-crystalline formations, the so called occlusion bodies, which are often polyhedral (hence the name polyhedrosis virus for one group of the baculoviruses). The thick layers of polyhedral protein provide protection against adverse physical and chemical factors within the environment, allow the viruses to survive outside the host cell, and enable them to kill their host rapidly, without jeopardizing their own existence (Jaques, 1977). It is obvious that protection of the virus particles by

the occlusion bodies is a great advantage for the use of these viruses as biological insecticides.

The natural infection process is by ingestion of food contaminated with virus. In the gut of the susceptible host the protein matrix of the occlusion bodies is dissolved and virus particles are released. They enter the epithelial cells of the gut, where they multiply and subsequently spread to other tissues of the host (Granados and Williams, 1986). Baculoviruses do not produce toxins, but their massive multiplication in vital tissues eventually leads to the death of the insect, usually within 1 or 2 weeks after infection.

Mainly based on the morphology of the occlusion bodies, baculoviruses are divided into three subgroups (Bilimoria, 1986).

1. Nuclear polyhedrosis viruses (NPV), where each occlusion body contains many virions.
2. Granulosis viruses (GV), where every virion has its own occlusion body.
3. The small group of non-occluded baculoviruses, where in general no occlusion bodies are formed.

The molecular biology, in particular the structure of the viral genome, and the regulation of the gene-expression in baculoviruses are very well studied. Occlusion-body-forming baculoviruses have a number of unique features, which make them very attractive vectors for the expression of foreign genes in a eukaryotic environment. As a consequence, baculovirus expression vector systems have evolved to become standard tools in modern biochemical research (Luckow and Summers, 1988). As a spin-off, a considerable amount of information has been collected on the functioning of the baculovirus genome. This knowledge is being used today to improve some traits of baculoviruses for their use in pest control (see Crook and Winstanley, Chapter 21).

Use of baculoviruses in pest control

Several approaches have evolved for the use of baculoviruses in pest management programmes. They can be mainly grouped into three different techniques.

1. Inundative control (insecticidal approach): The virus is used as a chemical insecticide and is not meant to persist in the population of the target pest. It might have already been present in the population, but in an insufficient amount. Since it is applied in large quantities, it usually gives relatively quick control. It is not maintained in the pest population, but gradually disappears and has to be applied several times within the same season, if necessary.
2. Inoculative release of viruses: A new virus is brought into the pest population in relatively low numbers. It multiplies there and is horizontally and sometimes even vertically transmitted. In many cases it initiates an epizootic, which leads to the collapse of the pest population. Since the virus has to multiply and spread in the population, it needs more time for its effect to become noticeable. It can, on the other hand, give long-term control for several years.
3. Manipulation of naturally occurring viruses: Through adequate cultural manipulation it is ensured that the virus can reproduce more than usual, or can persist better into the next generation of the pest insect. Viruses depend in many aspects upon a favourable environment for their survival and their infectivity. Sometimes a small reduction in natural attrition can greatly improve their efficacy (Ignoffo, 1985).

There are more than 700 baculoviruses known today and there are estimates that they can be used against nearly 30% of the major pests of fruit and fibre crops. It has even been claimed that, in Central America, by replacing chemical insecticides with insect viruses, pesticide consumption could be reduced by nearly 80% (Falcon, 1980).

The vast potential of baculoviruses for pest control has hardly been exploited. Little more than a dozen baculoviruses are registered worldwide for use as biological insecticides in forestry and agriculture. In the past, most of these registrations were made by government agencies of the United States and Canada (Huber, 1986), but recently, some private companies have also engaged in the production and registration of virus preparations, particularly in Europe (Table 18.1).

Ecological aspects of the use of baculoviruses

Specificity

One of the most important characteristics of baculoviruses, besides their high virulence, is their outstanding specificity. As mentioned previously, no baculovirus so far has been found in an organism other than an arthropod. This lack of evidence for the occurrence of baculovirus in non-arthropod hosts is sustained by the absence of references to baculoviruses in the literature of medical, veterinary and phytosanitary science. In addition, several baculoviruses have been subjected to rigorous safety tests on many animals, including man, with no indication of possible risks for vertebrates or other non-target organisms (Burges et al., 1980).

Although baculoviruses have been isolated from several arthropod groups, the individual members of this virus family have a very narrow host range. In most cases, a given virus only infects a few closely related insect species, belonging to the same family, or even to the same genus (Gröner, 1986). The baculovirus with the widest host range known today is the NPV of Autographa californica (AcMNPV). It is reported to infect, in the laboratory, larvae from about a dozen insect families, but all belonging to the Lepidoptera (Payne, 1986). The other extreme with regard to the host range is baculoviruses as, e.g. NPVs from sawflies, for which no transmission to another host, not even from the same genus is known. In general, GVs seem to be more specific than NPVs. They have only been found in Lepidoptera and cross transmission between hosts from different Lepidopteran families has never been demonstrated.

The high selectivity makes baculoviruses ideal tools for integrated pest control. Their direct effect on non-target organisms, especially beneficial organisms, is minimal to non-existent. Of course, there is always the possibility of an indirect effect on the environment by the reduction of the

Table 18.1. Virus preparations commercialized in Europe

Target pest	Virus type	Product trade name	Company (R = Registration, P = Production)	Registered Country	Date
Adoxophyes orana	GV	CAPEX	Andermatt-Biocontrol A.G., Grossdietwil (R + P)	CH	02/89
Agrotis segetum	GV	AGROVIR	Saturnia, Copenhagen (R + P)	DK	[a]
Cydia pomonella	GV	MADEX	Andermatt-Biocontrol A.G., Grossdietwil (R + P)	CH	12/87
		GRANUPOM	Hoechst A.G., Frankfurt (R) PROBIS, Pforzheim (P)	D	03/89
		CARPOVIRUSINE	Calliope S.A., Beziers (R + P)	F	10/92
Mamestra brassicae	NPV	MAMESTRIN	Calliope S.A., Beziers (R + P)	F	07/93
Neodiprion sertifer	NPV	Monisärmiövirus (Kemira Sertifvirus)	Kemira Oy, Espoo (R + P)	SF	05/83
		VIROX	Microbial Resources Ltd (R + P)[c] Oxford Virology Ltd (P)	GB	1984
Spodoptera armigera	NPV	SPODOPTERIN	Calliope S.A., Beziers (R + P)	F	[b]

[a] Notification only, no registration necessary.
[b] Applied for registration.
[c] No longer in business.

target pest population, thus removing a competitor or reducing a common resource, namely the host insect. In most cases this effect is trivial and the only alternative would be not to control the pest at all.

Persistence

The behaviour and persistence of baculoviruses in the environment has been well studied (for review, see Fuxa, 1989). One of the known, and in many cases desired, effects is the increase of virus in the environment after a virus application. However, there are several reports that indicate that after release of virus for control of a pest population, the virus load accumulated in the environment is the same size or even smaller than after the collapse of the population through a natural epizootic, which occurs usually late in the season when the larvae are already much larger and more virus is being produced in them (Jaques, 1974; Podgwaite *et al.*, 1979).

By far the most important inactivating factor for viruses in the environment is the UV radiation of the sun. Half-lives of 1 or 2 days are common for foliar deposits of unprotected virus. Only when the sun is low, as for example in winter time, can they persist for a longer period (Entwistle and Adams, 1977).

In the soil, baculoviruses are retained within 10 cm of the surface. Protected from the sunlight, relatively massive viral populations can accumulate and persist there for several years. But since the virus cannot grow on the soil substrate, and viral proteins are not known to be toxic to any soil organisms, these virus loads do not represent a danger to the environment. They serve as a virus reservoir and seem to be the main source for the initiation of viral epizootics in nature (Olofsson, 1988). Baculoviruses adhere strongly to sand particles. Therefore, they migrate very slowly in the sand layers of the aquifers and are not a risk for contamination of the ground water, even if it is known from laboratory studies that they can persist for many years in aqueous suspensions (Lewis and Rollinson, 1978).

Genetic stability

Genetically, baculoviruses appear very stable. The granulosis virus of the codling moth for instance, has been found only once in a few diseased larvae collected nearly 30 years ago in Mexico. Since that time the virus has been multiplied inadvertently or deliberately in most laboratories that rear codling moth. Even so, the codling moth GV isolates found today in cultures all over the world still have the same DNA restriction enzyme pattern as the original Mexican isolate, and there are only one or two local strains known which differ in one band from the original (Crook et al., 1985). Though the genetic background for the high specificity of baculoviruses is not known, it has been noted that it is very difficult to change their host range. Many reports on successful adaptation of baculoviruses to new hosts, after examination by DNA analysis techniques, turned out to be mere induction or provocation of an indigenous virus, already present in the new host in an occluded form. To present a risk to vertebrates, the changes in the genome of baculoviruses would have to be of rather drastic nature, since the virus would virtually have to mutate into another virus family (see Crook and Winstanley, Chapter 21 for an account of using genetically engineered baculoviruses as insecticides).

To summarize, when all the available information is taken into account, baculoviruses seem to rate among the least environmentally disruptive pesticides currently available.

Economical aspects of the use of baculoviruses

Commercialization

It cannot be denied that the practical use of insect viruses in pest control is faced with some scepticism, in spite of their ecological benefits. As profitable and desirable specificity is from an ecological view point, it is negative with regard to the economics of the commercialization of baculovirus products. Due to their selectivity their market size and sales potential is very limited. Small, locally orientated companies could still live with that limitation, but they are confronted with another problem. In most countries, insect viruses have to be officially registered as insecticides and are subjected to the same regulations as chemical pesticides (Bode, 1988; and see Klingauf, Chapter 28). Therefore, the costs of their commercialization are of the same order of magnitude as for

conventional insecticides. Large companies on the other hand, which would have the necessary experience to handle registration, need a big market to make registration profitable (Table 18.2). One solution could be to divide the task. Whereas registration is administered by a big company, or by a government agency, production and marketing could be carried out by small companies.

Use

The use of highly selective pesticides, including baculoviruses, in the framework of integrated pest management programmes, requires sound knowledge of the biology of the pest species and their antagonists in the crop, and is therefore not as simple as when using broad spectrum chemicals. Since insect viruses have to be ingested to become effective, and are less persistent than chemical pesticides, more attention has to be given to correct timing and application of the sprays. All this makes the use of baculoviruses more cumbersome and more costly for the farmer.

Since most commercially produced viral pesticides are currently being used on a rather experimental basis, it is difficult to make price comparisons. The few preparations already sold on the market are substantially more expensive than their chemical alternatives. Of course, the use of baculoviruses has to be seen in the context of integrated control programmes. The direct costs of the control of codling moth with the granulosis virus, for instance, are about two or three times as high as the cost for chemical control. But by not using broad spectrum chemicals, the risk of spider mite infestation in orchards is greatly reduced (Dickler and Huber, 1984). As a consequence, less or no acaricides have to be applied, and the overall costs of plant protection measures can be the same or even lower than chemical treatments, in spite of the higher costs for the selective baculovirus preparations. There are cases where use of a baculovirus is substantially less expensive than chemical control. The application of VIROX for control of the European pine sawfly, Neodiprion sertifer, for example costs less than one-eighth of that of diflubenzuron, mainly because the virus is so efficient that very little of it is needed (Draeger, 1985).

The main advantages of using a selective insec-

Table 18.2. Different aspects in the commercialization of viral pesticides and their acceptance by industry

	Small, local industry	Big, multinational company
Production		
Small scale production, due to limited market size	+	−
Production technology simple	+	±
Registration		
Procurement of safety data expensive, requiring substantial financial investment	−	+
Experience in dealing with government authorities needed	−	+
Marketing		
Small sales potential	+	−
Targeted mostly at small consumers (home gardeners, organic growers)	+	−

ticide are environmental and social. It is mainly the general public who benefit from the use of ecologically safe and non-disruptive pesticides, in the form of an undisturbed environment. Usually, there is hardly any economic benefit for the farmer – the direct user of the product – and nobody can blame him for not being particularly keen to pay for somebody else's advantage. It is therefore, mandatory that the public pays its share of the costs, so that the use of environmentally safe pesticides, such as baculoviruses, becomes profitable for the farmer.

Relevance of baculoviruses for developing countries

Whereas baculoviruses, for reasons outlined above, face many problems in industrialized countries, they look very promising for use in third world nations. Little energy input and equipment is needed for their production, whereas the hand labour required is often abundant in these countries. Viral pesticides offer the possibility for developing nations to produce their own insecticides within the country, with the use of local resources and man power, thus reducing dependence on the pesticide products of industrialized countries.

Agricultural and forest ecosystems in developing countries are particularly sensitive to any disturbances. The use of broad spectrum insecticides can induce dramatic and unexpected secondary effects. Many potential tropical pests are

kept in check by natural enemies, but they can increase disastrously when their parasites and predators are reduced or eliminated by the inconsiderate use of broad spectrum insecticides. In this situation, the use of a highly specific baculovirus is attractive, because it provides the least possible interference to the environment. These facts have been recognized and the potential of baculoviruses has already been exploited in certain countries e.g. Brazil, Guatemala, Thailand, Columbia, Zimbabwe, and Malawi.

Conclusions

Baculoviruses represent environmentally very safe alternatives to chemical pesticides. In spite of their ecological benefits, their use in agriculture will only increase if accompanying economic problems are solved. As long as the farmer does not get a reward for the use of environmentally safer pesticides, he will, in most cases, choose the easier and cheaper chemical solution for his pest problems. A change in this attitude can hardly come from the farmer or from the plant protection industry. They are both subjected to the dictates of commerce. Only the consumers and governments can influence the situation. If the consumer prefers agricultural products that are produced with the help of environmentally safe pesticides, the farmer is inclined to abandon his traditional pest management practices. The politicians, however, have the possibility to change the market from the outside through regulations, taxes and subsidies. Governments could even ban the use of a given broad spectrum chemical if an environmentally safer product is available for the same purpose. We have, however, to accept that the preservation of nature has its price, and we have to be willing to pay it.

References

Bilimoria, S. L. (1986). Taxonomy and identification of baculoviruses. In *The Biology of Baculoviruses*, vol.I (ed. R. R. Granados and B. A. Federici), pp. 37–59. CRC Press, Boca Raton.

Bode, E. (ed.) (1988). Umwelthygienische Aspekte bei der Bewertung von Pflanzenschutzmitteln aus Mikroorganismen und Viren im Zulassungsverfahren. *Mitteilungen aus der Biologischen Bundesanstalt für Land-und Forstwirtschaft*, Berlin-Dahlem, Heft 243.

Burges, H. D., Croizier, G. and Huber, J. (1980). A review of safety tests on baculoviruses. *Entomophaga* 25, 329–340.

Crook, N. E., Spencer, R. A., Payne, C. C. and Leisy, D. J. (1985). Variation in *Cydia pomonella granulosis* virus isolates and physical maps of the DNA from three variants. *Journal of General Virology* 66, 2423–2430.

Dickler, E. and Huber, J. (1984). Mit Apfelwickler-Granulosevirus selektive, biologische Apfelwicklerbekämpfung möglich. *Besseres Obst (Wien)* 29, 301–303, 323–324.

Draeger, H. (1985). Überlegungen und Versuche zur Bekämpfung von *Neodiprion sertifer* im Nds. Privatwald. MD thesis, Fachhochschule Hildesheim Holzminden, Fachbereich Forstwirtschaft, Göttingen,

Entwistle, P. F. and Adams, P. H. W. (1977). Prolonged retention of infectivity in the nuclear polyhedrosis virus of *Gilpinia hercyniae* (Hymenoptera, Diprionidae) on foliage of spruce species. *Journal of Invertebrate Pathology* 29, 392–394.

Falcon, L. A. (1980). Economical and biological importance of baculoviruses as alternatives to chemical pesticides. In *Symposium Proceedings 'Safety Aspects of Baculoviruses as Biological Insecticides'* (Jülich, 13–15 November 1978) pp. 27–46. German Federal Ministry for Research and Technology, Bonn.

Federici, B. A. (1986). Ultrastructure of baculoviruses.

In *The Biology of Baculoviruses*, vol. I (ed. R. R. Granados and B. A. Federici), pp. 61–88. CRC Press, Boca Raton, FL.

Fuxa, J. R. (1989). Fate of released entomopathogens with reference to risk assessment of genetically engineered micro-organisms. *Bulletin of the Entomological Society of America* 35, 12–24.

Granados, R. R. and Williams, K. A. (1986). *In vivo* infection and replication of baculoviruses. In *The Biology of Baculoviruses* vol. I (ed. R. R. Granados and B. A. Federici), pp. 89–108. CRC Press, Boca Raton, FL.

Gröner, A. (1986). Specificity and safety of baculoviruses. In *The Biology of Baculoviruses* vol. I (ed. R. R. Granados and B. A. Federici), pp. 177–202. CRC Press, Boca Raton, FL.

Huber, J. (1986). Use of baculoviruses in pest management programs. In *The Biology of Baculoviruses*, vol. II (ed. R. R. Granados and B. A. Federici), pp. 181–202. CRC Press, Boca Raton.

Ignoffo, C. M. (1985). Manipulating enzootic-epizootic diseases of arthropods. In *Biological Control in Agricultural IPM Systems* (ed. M. A. Hoy and D. C. Herzog), pp. 243–262. Academic Press, Orlando, FL.

Jaques, R. P. (1974). Occurrence and accumulation of the granulosis virus of *Pieris rapae* in treated field plots. *Journal of Invertebrate Pathology* 23, 351–359.

Jaques, R. P. (1977). Stability of entomopathogenic viruses. *Miscellaneous Publication of the Entomological Society of America* 10, 99–116.

Lewis, F. B. and Rollinson, W. D. (1978). Effect of storage on the virulence of gypsy moth nucleopolyhedrosis inclusion bodies. *Journal of Economic Entomology* 71, 719–722.

Luckow, V. A. and Summers, M. D. (1988). Trends in the development of baculovirus expression vectors. *Bio/Technology* 6, 47–55.

Martignoni, M. E. and Iwai, P. J. (1986). A catalog of viral diseases of insects, mites, and ticks. *General Technical Report PNW-195*, USDA, Forest Service.

Olofsson, E. (1988). Environmental persistence of the nuclear polyhedrosis virus of the European pine sawfly in relation to epizootics in Swedish Scots pine forests. *Journal of Invertebrate Pathology* 52, 119–129.

Payne, C. C. (1986). Insect pathogenic viruses as pest control agents. In *Biological Plant and Health Protection* (ed. J. M. Franz), (*Progress in Zoology* 32), pp. 183–200. Gustav Fischer Verlag, Stuttgart.

Podgwaite J. D., Stone Shields, K., Zerillo, R. T. and Bruen, R. B. (1979). Environmental persistence of the nucleopolyhedrosis virus of the gypsy moth, *Lymantria dispar*. *Environmental Entomology* 8, 528–536.

Part IV USE OF GENETICALLY MODIFIED ORGANISMS

19

Assessing the potential benefits and risks of introducing natural and genetically manipulated bacteria for the control of soil-borne root diseases

Maarten H. Ryder and Raymond L. Correll

Introduction

The application of biological control agents is an established if under-exploited method for the control of agricultural pests and diseases. There are considerable benefits to be gained from this approach, however caution is necessary in the introduction of biocontrol agents because of possible effects on non-target organisms. In some situations the biocontrol organism itself may become a pest.

There are numerous examples of experimental biocontrol treatments for fungal and bacterial plant root diseases, though on a practical level there are less examples than for control of insect pests. Plant root diseases can often be controlled by management practices such as rotation with a non-host crop or by several years of crop monoculture, a practice which can lead to the build-up of a disease-suppressive soil (Hornby, 1983). Microbially mediated disease suppression can also occur in soil-less potting mixes (Hoitink and Fahy, 1986). Fungicides (e.g. Baytan for control of take-all) are used for control of disease, but success has been limited especially when applied to field crops. Biological control agents used to control soil-borne root diseases are most commonly soil bacteria or fungi.

On the whole, there are relatively few publications dealing with risk assessment in the literature on biocontrol. Recent papers include that by de Jong et al. (1990). The need for risk assessment is now much greater, particularly as genetic manipulation is beginning to be used more often in biological control (Jones and Kerr, 1989; Ryder and Jones, 1991). The use of recombinant DNA technology brings its own risks and benefits.

Given that there are both risks and benefits for pest control methods, procedures for their estimation are desirable. Risks and benefits are usually not assessed until a potential product is in its development stages, owing to the expense of the assessment process. However, where recombinant DNA technology is used, at least some assessment of risk should be done at an earlier stage. Methods for estimation of risk have been developed in various areas including the chemical industry (Kletz, 1984) and in the development of bioherbicides (Weidemann, 1991).

In this paper, we consider risks and benefits in relation to both natural and genetically manipulated micro-organisms for the control of plant root diseases, and also methods of assessment.

Natural (unmodified) organisms

Existing and potential benefits

Some benefits are clear. Decreased disease can lead to an increase in returns from saleable products, for example through greater yield of a field crop or reduced damage in the case of nursery seedlings.

Other benefits are indirect but no less important. These include:

1. Reduced use of chemical control measures (e.g. biocontrol agents could replace fungicides that are used extensively in plant nurseries to control damping-off).
2. For an integrated control programme (biological and chemical), the possibility of reduced rates of application of either the chemical or the biological agent, or of greater disease control with existing rates.
3. Control of disease where no other method exists. Biological control of crown gall is one example. Another example is the potential for replacement of chemical pesticides, particularly nematicides, for which registration has been withdrawn.
4. Reduced likelihood of development of resistance. In theory, resistance would be less likely to develop if the mechanism of biocontrol is direct competition for food or space. Where antibiotic production is part of the mechanism of disease control, only small amounts are needed in comparison to the application of a chemical agent. An example is the biosynthesis of phenazine-1-carboxylic acid (PCA) by *Pseudomonas* strain 2–79, where the production of only 100 to 250 mg of PCA per ha *in situ* in the plant root and rhizosphere (Thomashow *et al.*, 1990) could lead to disease suppression. The spread of resistance to control of crown gall by *Agrobacterium radiobacter* strain K84 has been pre-empted by the development of a new strain in which a plasmid that codes for production of the antibiotic agrocin 84 can no longer be transferred to pathogenic strains (Jones *et al.*, 1988).
5. Stimulation of plant growth in the absence of disease (Kloepper *et al.*, 1991). Soil micro-organisms commonly produce plant growth substances in culture and this could explain some of the observed increases in plant growth. As an example, *Bacillus subtilis* strain A.13 ('Quantum 4000') is sold as an 'inoculant', not a pesticide, as it promotes plant growth directly as well as reducing damping-off diseases on cotton and peanut (Turner and Backman, 1991). Thus, a grower may expect increased returns for employing a treatment even when there is little or no disease.

6. Greater flexibility in management. The use of a biocontrol measure would in many cases allow a grower more options in disease control. For example, crop rotation is a good control measure for take-all, a serious root disease of wheat. However, an effective biocontrol would give a farmer the option of applying a control in the current season, and allow wheat to be grown on land that had not been prepared by crop rotation.
7. Application of a biocontrol agent could hasten the development of disease suppressiveness in soil (Shipton *et al.*, 1975).

The basic information needed to assess benefits includes:

- Economic losses with and without biocontrol, consistency of biocontrol;
- Growth promotion: does it contribute to the final product? Consistency of growth promotion;
- Effectiveness compared to other control methods;
- Cost of control; estimate of environmental costs.

Existing and potential risks

Introductory remarks

A basic consideration is whether the introduced organism is endemic or new to the environment in which it is to be applied. If the organism is endemic, then the risks of unforeseen negative effects from applying the organism are very low. The introduction of a ubiquitous organism with pathogen-free status (e.g. *P. fluorescens*) may be quite simple. However, it would still be desirable to test the candidate biocontrol agent for effects on non-target plants. This process of host-range testing would probably form part of the normal procedure for registering a product. The taxonomy of the introduced organism should be considered, including the occurrence, within the species or genus, of other organisms that have known deleterious effects.

Most of the risks listed below for the introduction of a biocontrol agent are also encountered in the control of disease with the use of chemical agents. Some procedures used to estimate risks may be applicable in both situations.

1. Deleterious effects on the target crop (i.e. the crop to which the biocontrol agent is applied).
   This may occur through the application of a higher than optimal dose of the biocontrol agent, with possible phytotoxic effect, especially if the microbe produces a highly active antibiotic that can affect plant as well as microbial growth. Keel *et al.*,

(1990) list several bacterial antibiotics that are produced by biocontrol organisms and have negative effects on plant growth, in addition to their positive role in biological control.

2. Deleterious effects on non-target crops or organisms. Host-range testing would be appropriate in the case of organisms that either have negative effects or are related to known pathogens (Weidemann, 1991). A knowledge of the background population in the soil where the agent is to be applied is desirable. To determine this, we need a good selective medium or a species-specific DNA probe.

   (a) Direct risk to following crops, similar to carry-over effects of herbicides (Altman and Rovira, 1989).

   (b) Indirect risk from spread of the organism to susceptible crops in adjacent areas. This is comparable to spray drift during application of a chemical, but could be mediated by water as well as by wind, insects, animals and farm machinery.

3. Possible 'breakdown' of biological control, e.g. due to the development of resistance in the target pathogen.

4. Control of one disease can lead to worse effects from another disease ('disease trading').

5. Ineffective disease control in a particular season (lack of consistency).

Note that risks 4 and 5 especially, and also risk 3, are risks to the grower, i.e. a financial risk in applying the control measure.

The basic information needed to assess risks includes:

- Taxonomy of biocontrol agent;
- Host range for negative effects;
- Population dynamics of the biocontrol organism;
- Spread and persistence of the biocontrol organism.

Risk assessment

We address this topic in a later section, after consideration of genetically-manipulated organisms (GMOs). The procedures are likely to be similar for both an exotic natural organism and a GMO. Issues in risk assessment, including its scientific and regulatory aspects, relating to the introduction of GMOs have been placed in a broad ecologically-based framework by Tiedje *et al.* (1989).

Genetically manipulated organisms

Some potential benefits: what are the aims of genetic manipulation?

1. Increasing the effectiveness of biocontrol. Two examples are:

   (a) Over-production of antibiotics or other metabolites important in biocontrol (Shim *et al.*, 1987; Keel *et al.*, 1990). This approach could lead to development of a more effective biocontrol agent. However examples show that 'more is not necessarily better'.

   (b) Selectively increasing the population of an introduced strain in the rhizosphere (in association with plant genetic manipulation or by other means).

2. Ensuring the safety and future of biocontrol.

   (a) An example is the deletion of plasmid transfer functions to avoid the transfer of antibiotic resistance from a biocontrol organism, *Agrobacterium* strain K84, to pathogenic strains of *Agrobacterium* (reviewed by Farrand, 1990; Ryder and Jones, 1991).

   Agrobacterium strain K84 has been used commercially for the control of crown gall on stone fruits, roses and other cultivated plants in many countries for more than a decade (Kerr, 1980). The production of the antibiotic agrocin 84 by strain K84 is a major reason for the success of this strain in the control of disease caused by many types of pathogenic agrobacteria. Agrocin 84 biosynthesis is encoded on a plasmid, pAgK84, which can move to other agrobacteria, including pathogenic strains, at low but measurable frequency in some conditions in the soil. Pathogenic strains that have acquired the agrocin 84 biosynthesis plasmid are resistant to the antibiotic, and are no longer controlled by strain K84. To ensure the long-term effectiveness of the biological control of crown gall, the plasmid pAgK84 was engineered by deletion of 5.9 kb, including part of the transfer region, so that it could no longer be transferred to other agrobacteria. The altered strain, *Agrobacterium* K1026, has been sold in Australia for control of crown gall since 1988.

   (b) Where a biocontrol agent may have undesirable effects on alternate crops or other plants in a locality, a low long-term survival of the agent is desirable. This could be achieved by the incorporation of a suicide function into an organism so that it dies out at a certain signal. For example, when plant roots die, a particular compound from roots is no longer produced, that compound being needed to repress the suicide gene(s).

(c) Some types of manipulated organisms may be less able to survive in the environment in the long-term. An example is a biocontrol strain of *Pseudomonas corrugata* carrying the *lacZY* genes (constitutively expressed) from *E. coli* as a marker. The manipulated strain survived less well than its parent, during the first summer season (Fig. 19.1). Even though the populations were not significantly different at any one sampling time, when all the data were considered together in a two-way analysis, the difference was significant at P<0.05. The lower survival would be acceptable, from the point of view of biocontrol, as long as population of the biocontrol organism remained high enough during the growing season to give adequate disease control.

(d) Tracking organisms in the environment by use

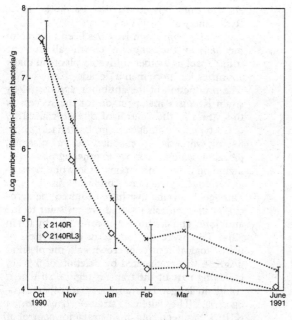

Fig. 19.1 Survival in the field of a biocontrol strain of *Pseudomonas corrugata*, with and without inserted *lacZY* marker genes (at Roseworthy, South Australia).

Pseudomonas corrugata strain 2140 (Ryder and Rovira, 1993) was recovered from roots and crowns of wheat at intervals after seed inoculation. The field test was sown in October 1990. Strain 2140R (parent) is resistant to rifampicin at 100 mg/l. Strain 2140RL3 contains the *lacZY* genes of *E. coli*, inserted in the chromosome by the method of Barry (1988). (Vertical lines show LSD (5%) for each sampling.)

of marker genes or unique DNA sequences will be very useful if not mandatory. Apart from the usefulness in scientific study, this can have additional benefits, for example where a biocontrol agent is suspected to have had deleterious effects on a particular non-target crop. The marker could be used to establish clearly the presence or absence of the control agent.

Potential risks

A genetically-manipulated organism will by definition be new to the environment. So, even for an endemic microbe, new risk factors will need to be considered. Some of these risks would be the same as for natural organisms (see above).

Potential risks include:

1. Greater or a wider range of negative effects. This may for example occur with a manipulated strain that produces more of an antibiotic or a greater range of antibiotics.
2. Establishment of the new organism in greater numbers than its parent, in certain environments, with possible harmful effects.
3. Effects as a result of gene transfer to other soil microorganisms. Genes for superior competitive ability could be transferred into pathogenic species.
4. Unforeseen deleterious effects.

Risk mitigation

For **natural (unmodified) organisms**, risks could be reduced by:

1. Selecting micro-organisms that do not cause negative effects. The 'safe' status of an organism could be established by performing a minimum set of toxicity and pathogenicity tests. Table 19.1 is a list of plants and other organisms that we have included in pathogenicity tests for *Pseudomonas corrugata* strain 2140, a biocontrol agent for take-all of wheat (Ryder and Rovira, 1993).
2. Recommendation of a minimum safe distance between the target crop and susceptible plants, where an organism causes concern due to its effects on non-target crops or native plants (de Jong *et al.*, 1990).
3. Use of endemic biocontrol agents, with the aim of specifically increasing the population of an organism that is native to the locality.
4. Ensuring that the correct amount of biocontrol agent is applied, where it is known that too high a dose has adverse effects.
5. Using genetic manipulation of the biocontrol agent

Table 19.1. Risk assessment for *Pseudomonas corrugata* used in biological control of take-all of wheat: suggested list for plant pathogenicity tests and for antagonism of organisms beneficial to plants

1	2	3	4	5
Wheat (target)	Tomato (host)	*Solanum* spp.	*Eucalyptus* spp.	*Rhizobium*
Barley	Potato	*Nicotiana* spp.	*Acacia* spp.	Mycorrhizal fungi
Rye-grass	Eggplant	*Anthocercis* sp.	*Allocasuarina* sp.	*Nitrobacter*
Medic	*Capsicum*	*Grammosolen* sp.	*Dodonea* sp.	*Azotobacter*
Sub. clover	Tobacco			Earthworms

1, Crops grown in rotation with target crop. 2, Known host and closely related crops (Solanaceae). 3, Native species closely related to known host (Solanaceae). 4, Native species not related to known host. 5, Organisms beneficial to plants.

to decrease risk factors where the characteristics involved are well understood and can be manipulated without incurring greater risks. As an example, it may be possible to remove the risk of pathogenic effects from an organism without affecting its biological control ability.

For **modified biocontrol agents** produced by genetic manipulation, risks could be minimized by:

1. Sufficient tests in the laboratory and glasshouse or controlled environment chamber before field tests are undertaken. The conditions should mimic the field as far as possible. Properties of the modified strain and its parent should be compared (for example growth rate, persistence in the soil environment, effects on plants and on disease).
2. Where attempts are made to increase the effectiveness of disease control, it will be important to test for potential adverse effects, particularly where the new organism has increased levels of expression of gene(s) involved. The effect of various doses of the parent and modified strain on both target and non-target plants should be ascertained. This concept is discussed in more detail in the next section, 'Methods of risk assessment'.
3. For genetic manipulations designed to make biocontrol safer, tests will still be necessary but may be qualitatively different from those that apply to 2, above.
4. It may be necessary to test organisms with DNA insertions more rigorously than those from which a DNA segment has been deleted.
5. Minimizing the possibility of transfer of foreign genes into other soil micro-organisms.

Methods of risk assessment

The assessment of risk in the release of an exotic biocontrol agent or GMO must recognize its potential ecological impact on the soil microflora, local economic crops, other economic crops, local flora, and animals. For the organism to pose any threat, it must be able to persist and to reproduce (Weidemann, 1991). Furthermore, it must be transported from the site of introduction to the site of potential damage (Alexander, 1985).

Since it will be impossible to 'recall' a microbial biological control agent after release, risk assessment should be carried out in advance, in the containment laboratory or glasshouse. In such tests, the detrimental effect of a biocontrol agent or GMO should be tested on a range of plant species. There should perhaps also be tests on other organisms, for example beneficial microbes and animals. Note that some of these tests may also be requirements for product registration. The range of tests undertaken should be appropriate to the case in question.

In tests on plants, detrimental effects may be pathogenic (acute), where plants show severe symptoms and may even die rapidly, or chronic. In the former case, the test should be a standard pathogenicity assay, where the organism under scrutiny is inoculated directly on to test plants, with and without wounding. The methodology for this type of test is not discussed in this paper. If the pathogenic effects are severe, the release of the organism should be questioned. In the latter (chronic) case, the effect will be a decline in vigour or production. Here, it is appropriate to examine the effect of a range of doses of the micro-organism on plant growth and/or production.

This concept is considered here in some detail. The range of doses should be similar to that which could be realistically observed in the field.

It will be impossible to characterize the effect of the test organism on all potential crop plants and native species. Some type of sampling scheme is then required. One potential scheme is the 'Centrifugal-Phylogenetic Host Range' approach as described in Weidemann (1991). This scheme was developed to assess the safety of biological weed control agents prior to release (see Greathead, Chapter 5; and Blossey, Chapter 8). The candidate biocontrol organism is first tested for deleterious effects on a series of plants (cultivars and species) that are taxonomically related to the host, and the range is subsequently widened to include plants from other taxa at progressively higher taxonomic levels. With any such sampling scheme there will always remain a small risk that a species not considered will be acutely susceptible to the test organism.

The centrifugal host-range method would be appropriate for use with biocontrol agents and GMOs where acute effects are found during pathogenicity testing. Where no effect or chronic effects are found intially, a combination of this approach and techniques developed in ecotoxicology for extrapolating between taxa, would result in a measure that could be used in risk assessment. Using the experimental and statistical approach outlined below, the probability of a plant having a response greater than a certain amount could be assessed.

In the case of a microbial agent released for disease control on a crop plant, we suggest that for safest possible use, the organism be tested for effects on crop plants commonly grown in rotation with the crop for which the disease control is intended. In addition, if the biocontrol agent is known to have a deleterious effect on a particular plant species, or is taxonomically related to a plant pathogen, we suggest that the centrifugal host-range approach be used.

Table 19.1 lists the range of plant species that we propose to use in order to assess the safety of a biocontrol isolate of *Pseudomonas corrugata*, an organism that is able to control the fungal root disease take-all of wheat, but which is also related to a pathogen of tomato. The list of over 20 species includes: crops grown in rotation with the crop to which the strain would be applied; the

known host (tomato) and taxonomically-related crops; native species closely related to the known host; some common native Australian plant species not related to the known host; finally, some soil organisms known to be beneficial to plants.

Where the plant response to the test biocontrol organism is not severe, its effect on growth and/or yield can be quantified on a selected list of species, such as that suggested in Table 19.1. It will be impractical to test all plant species, so some extrapolation from a sample will be required. There will be some risk of a detrimental effect being missed if we extrapolate from a sample of plants to all plants that occur in a particular region.

A similar problem occurs in ecotoxicology, where at times all the appropriate data are not available for all the target organisms or toxicants. These problems have been addressed by Suter and Rosen (1988) and Barnthouse *et al.* (1986). These authors consider relationships between lethal concentrations ($LC_{50}s$) and the maximum acceptable toxic concentration (MATC) and the relationship between MATC for one taxon and the MATC for another taxon. These relationships enable extrapolation of the existing toxicological data to new taxa, but at the risk of greater uncertainty.

One assumption discussed at length by Suter and Rosen (1988) is that of the representativeness of the selected set of test organisms. This problem must also be considered when selecting plant species for testing with a biocontrol agent or GMO.

A potential approach to quantifying the risks associated with the release of a natural biocontrol agent or GMO would be to perform a series of experiments in which the dose of the test microorganism is varied on a series of host plants. In doing this, we can build up confidence limits for the parameters of the regressions that we obtain for plant growth response against dose. Doses should be representative of those found or expected in the field. After such a series of trials, the probability of a plant having a response greater than a certain amount could be assessed. A high level of confidence would be required to ensure that none of the many crop and native plant species was likely to be susceptible.

The risk of detrimental effect must be tempered by the probability of the test organism surviving, being transported to the threatened plant, and then multiplying to numbers that can have a negative effect. If this information is also available, the risk posed can be quantified both for indigenous plant species and economic crops.

Conclusions

It is clear that we must weigh the two sides, risk and benefit, in the application of a biocontrol agent, especially with micro-organisms that are known or suspected to have deleterious side-effects. The benefit would be the estimated gain in production brought about by the introduction of the control agent and must include the savings made by replacing any chemical controls (together with their associated risks). The risks of biological control are a series of products of the likelihood of damage to a particular crop multiplied by the value of that damage. The detriment is the sum of these products. In economic terms, a release should only be carried out if the benefit exceeds the detriment by more than the financial cost of the release. After these considerations there may also be social and economic implications because the benefits and detriments can be felt by distinctly different groups of people. The potential for effects on indigenous plant and animal species should also be recognized.

Each example of genetic manipulation of a biocontrol agent should be treated on its merits. So far we have proceeded on a case-by-case basis in most countries. Both the benefits and the risks may be greater, in amount and variety, for the genetically manipulated biocontrol agents than for the parent strains from which they were derived. Initial testing of new organisms in realistic but controlled environments is essential. Nevertheless, behaviour in the field may not be as predicted from laboratory (microcosm or glasshouse) tests. When tested in the field, survival and spread of the organism should be monitored.

We suggest a way to quantify risks, that involves dose–response experiments where a biocontrol agent or GMO is tested against a selected range of organisms (mostly plants in this case). This should allow confidence limits to be placed on extrapolations from the group of test organisms used to other organisms.

Public perceptions of successes, or more particularly of failures, are important in these early stages of research and development in genetic manipulation and its application to biological control. We should be prepared to admit that there are risks and that there may be unforeseen risks involved in the introduction of biological control organisms and GMOs. Perhaps in future, after we have accumulated more information and experience, we may be able to make generalizations and take routes other than the case-by-case approach.

Acknowledgments

The authors thank B. M. Doube, R. Forrester, D. Pimentel, P. M. Stephens and G. W. Suter II for helpful discussions and comments on the manuscript. Monsanto Australia assisted with financial support for part of the experimental work reported here.

References

Alexander, M. (1985). Environmental release: reducing the uncertainties. *Issues in Science and Technology* 1, 57–68.

Altman, J. and Rovira, A. D. (1989). Herbicide-pathogen interactions in soil-borne root diseases. *Canadian Journal of Plant Pathology* 11, 166–172.

Barnthouse, L. W. and Suter II, G. W. (eds.) (1986). *User's Manual for Ecological Risk Assessment*. Oak Ridge National Laboratory, Environmental Sciences Division, Publication No. 2679, Oak Ridge, Tennessee.

Barry, G. F. (1988). A broad-host-range shuttle system for gene insertion into the chromosomes of Gram-negative bacteria. *Gene* 71, 75–84.

Farrand, S. K. (1990) *Agrobacterium radiobacter* strain K84: a model biocontrol system. In *New Directions in Biological Control: Alternatives for Suppressing Agricultural Pests and Diseases* (ed. R. R. Baker and P. E. Dunn), pp. 679–691. Alan R. Liss, Inc., New York.

Hoitink, H. A. J. and Fahy, P. C. (1986) Basis for the control of soilborne plant pathogens with composts. *Annual Review of Phytopathology* 24, 93–114.

Hornby, D. A. (1983) Suppressive soils. *Annual Review of Phytopathology* 21, 65–85.

Jones, D. A. and Kerr, A. (1989). *Agrobacterium radiobacter* K1026, a genetically-engineered derivative of strain K84, for biological control of crown gall. *Plant Disease* 73, 15–18.

Jones, D. A., Ryder, M. H., Clare, B. G., Farrand. S. K. and Kerr, A. (1988). Construction of a Tra⁻ deletion mutant of pAgK84 to safeguard the biological control of crown gall. *Molecular and General Genetics* **212**, 207–214.

de Jong, M. D., Scheepens, P. C. and Zadoks, J. C. (1990). Risk analysis for biological control: a Dutch case study in biocontrol of *Prunus serotina* by the fungus *Chondrostereum purpurea*. *Plant Disease* **74**, 189–194.

Keel, C., Wirthner, Ph., Oberhaensli, Th., Voisard, C., Burger, P., Haas, D. and Défago, G. (1990). Pseudomonads as antagonists of plant pathogens in the rhizosphere: role of the antibiotic 2,4-diacetylphloroglucinol in the suppression of black root rot of tobacco. *Symbiosis* **9**, 327–341.

Kerr, A. (1980) Biological control of crown gall through production of agrocin 84. *Plant Disease* **64**, 25–30.

Kletz, T. A. (1984). *Loss Prevention*. The Institute of Chemical Engineers, London.

Kloepper, J. W., Zablotowicz, R. M., Tipping, E. M. and Lifshitz, R. (1991). Plant growth promotion mediated by bacterial rhizosphere colonizers. In *The Rhizosphere and Plant Growth* (Beltsville Symposia in Agricultural Research 14), (ed. D. L. Keister and P. B. Cregan), pp. 315-26. Kluwer Academic Publishers, Dordrecht, the Netherlands.

Ryder, M. H. and Jones, D. A. (1991). Biological control of crown gall using *Agrobacterium* strains K84 and K1026. *Australian Journal of Plant Physiology* **18**, 571–579.

Ryder, M. H. and Rovira, A. D. (1993). Biological control of take-all of glasshouse-grown wheat using strains of *Pseudomonas corrugata* isolated from wheat field soil. *Soil Biology and Biochemistry* **25**, 311–320.

Shim, J.-S., Farrand, S. K. and Kerr, A. (1987). Biological control of crown gall: construction and testing of new biocontrol agents. *Phytopathology* **77**, 463–466.

Shipton, P. J., Cook, R. J. and Sitton, J. W. (1975). Occurrence and transfer of a biological factor in soil that suppresses take-all of wheat in eastern Washington. *Phytopathology* **65**, 511–517.

Suter, II G. W. and Rosen, A. E. (1988). Comparative toxicology for risk assessment of marine fishes and crustaceans. *Environmental Science and Technology* **22**, 548–556.

Thomashow, L. S., Weller, D. M., Bonsall, R. F. and Pierson, III L. S. (1990). Production of the antibiotic phenazine-1-carboxylic acid by fluorescent *Pseudomonas* species in the rhizosphere. *Applied and Environmental Microbiology* **56**, 908–912.

Tiedje, J. M., Colwell, R. K., Grossman, Y. L., Hodson, R. E., Lenski, R. E., Mack R. N. and Regal, P. J. (1989). The planned introduction of genetically engineered organisms: ecological considerations and recommendations. *Ecology* **70**, 298–315.

Turner, J. T. and Backman, P. A. (1991). Factors relating to peanut yield increases after seed treatment with *Bacillus subtilis*. *Plant Disease* **75**, 347–353.

Weidemann, G. J. (1991) Host range testing: safety and science. In *Microbial Control of Weeds* (ed. D. O. TeBeest), pp. 83–96. Chapman and Hall, New York.

20

Serodiagnostic methods for risk assessment of *Pseudomonas cepacia* as a biocontrol agent

Kenichi Tsuchiya

Introduction

To develop successful biocontrol of plant diseases by the introduction of naturally occurring (see Défago *et al.*, Chapter 12) or genetically engineered (see Dowling *et al.*, Chapter 14) antagonistic pseudomonads, careful research is needed to assess risks. This includes viability and translocation of the microbes in soil ecosystems, transfer of genetic material to and from the indigenous soil microbial communities, effects of introductions on biochemical functions of the soil microflora and so on. The development of detection methods is important, not only for risk assessment of survival or dispersal of biocontrol microorganisms, such as genetically engineered microorganisms (GEMs) in soil, but also to evaluate the efficacy of biocontrol agents against plant diseases.

Pseudomonas cepacia Palleroni & Holmes, was initially described as a phytopathogen (Burkholder, 1950) as well as a saprophyte in soil (Sinsabaugh and Howard, 1975). It has been reported to provide biological control of certain soil-borne diseases, such as *Fusarium*-wilt of onion (Kawamoto and Lorbeer, 1976), damping-off of radish, *Fusarium*-wilt of tomato and *Verticillium*-wilt of eggplant (Homma *et al.*, 1985). This bacterium is also known to produce various antibiotics, such as pyrrolnitrin (Elander *et al.*, 1968; Janisiewicz and Roitman, 1988; Homma *et al.*, 1989) and pseudane derivatives (Homma *et al.*,

1989), which are supposed to contribute to disease suppression. On the other hand, *P. cepacia* is extremely resistant to antimicrobial agents and is able to survive and multiply even in purified waters (Carson *et al.*, 1973). In addition, it is known to be an opportunistic pathogen and has been described as a synonym for *P. kingii* and *P. multivorans*, and as aetiological agents of clinical infection (Ederer and Matsen, 1972; Randall, 1980). There are a few reports on distinguishing between strains of clinical and environmental origin by differences in bacteriocin production, pectolytic activity and plasmid profile. However, no differences in toxicity to mice were detected (Gonzalez and Vidaver, 1979; Lennon and DeCicco, 1991).

The question remains as to whether *P. cepacia* strains of environmental origin are indentical to clinical strains, because neither the virulence factors nor ecology of these bacteria are sufficiently well known. For these reasons the practical use of *P. cepacia* as a biocontrol agent in agriculture has so far not been effected in Japan.

This paper deals with: (1) development and application of a specific and highly sensitive detection method of *P. cepacia* in soil by use of a selective medium and enzyme-linked immunosorbent assay (ELISA) with polyclonal and monoclonal antibodies (MABs); (2) comparison of *P. cepacia* strains of clinical and environmental origin by bioassay of antagonistic activity towards plant pathogens; and (3) serological diagnosis of *P.*

cepacia strains of clinical and environmental origin based on the reactivity to polyclonal and monoclonal antibodies.

Development and application of a combined method of selective medium and ELISA for risk assessment of *P. cepacia*

Although *P. cepacia* only rarely causes infection outside the hospital environment (Pallent *et al.*, 1983), natural sources of the bacterium appear to be ubiquitous in the soil and water. High versatility in utilizing nutrients and the ability to survive and multiply in nature are important in the risk assessment of this bacterium. Therefore, the development of a sensitive and specific detection method for this bacterium is necessary.

Development of a simple and highly sensitive ELISA for the specific detection of *P. cepacia* in soil

An antiserum highly specific to *P. cepacia* (*Pc*) was produced by immunization with glutaraldehyde-fixed (F) cells of a *Pc* strain RB425 (isolated from rhizosphere of lettuce) in rabbits. Three direct ELISA procedures that included a double-sandwich method, which uses alkaline phosphatase conjugated with γ-globulin, and three indirect ELISA methods using the avidin-biotin complex system (ABC-ELISA) were used.

When bacterial cells were applied directly to the microplates followed by reaction with F-conjugates, only 22 strains of *Pc* reacted specifically among a total of 151 bacterial strains that belonged to 9 genera.

By using this simple direct ELISA, *Pc* was detected at a concentration below 10^4 cfu/g soil in three types of soils that were artificially co-inoculated with either *P. solanacearum* or *Erwinia carotovora* at concentrations of approximately 10^5 to 10^9 cfu/g. Furthermore, both the reaction sensitivity and minimum detection level were increased by heat treatment (the sample tubes were placed in boiling water for 15 min) after which *Pc* in the soil could be detected at the level of approximately 10^2 cfu/g (Tsuchiya *et al.*, 1991).

Six monoclonal antibodies (MABs) were produced. They originated from an L-immunized (live cells of strain RB425) mouse (PCL-MABs) and from a F-immunized (fixed cells by glutar-

aldehyde) mouse (PCF-MABs). As reported previously (Takahashi *et al.*, 1990), these MABs recognized six distinguishable epitopes of three different types, thought to represent protein, lipoprotein and lipopolysaccharide (LPS) regions of the bacterial surface antigens. The tested strains of *Pc* could be divided into groups based on the reactivity to these six MABs (Takahashi *et al.*, 1990).

Two *Pc* strains, 356–2 and 356–5 with different specificity to MABs PCF-1A5 and PCF-2G1 were inoculated individually or simultaneously at 2–6×10^8 cfu/g soil into pots filled with field soil, which was infested with *P. solanacearum* (*ca* 10^6 cfu/g soil). Both *Pc* strains could be detected reliably in the presence and absence of the other *Pc* strain: the assay was by indirect-ABC ELISA using the corresponding MAB. The detection limit of Pc356–2 with PCF-1A5 was 10^3–10^4 cfu/g soil (Tsuchiya, 1991). The results showed that strain-specific detection and monitoring of a *Pc* strain, either artificially introduced into soil or naturally occurring, is possible by combining the most suitable MAB(s) with a highly sensitive direct ELISA protocol.

Detection of *P. cepacia* from the rhizosphere with a combination method of selective medium and ELISA

A new selective medium for *Pc*, S-PC-1, was developed. The medium consists of: $MgSO_4 \cdot 7H_2O$, 0.3 g; $(NH_2)_2CO$, 0.05 g; KH_2PO_4, 1.0 g; LiCl, 10 g; citraconic acid (as K-salt), 4 g; cycloheximide, 25 mg; ampicillin, 10 mg; L-cystine, 50 μg; agar, 15 g; and distilled water, 1000 ml. At 40 °C the medium was very selective for *Pc* strains among tested 236 strains belonging to 38 species in 8 genera of *Pseudomonas*, *Xanthomonas*, *Erwinia*, *Agrobacterium*, *Clavibacter*, *Bacillus*, *Rhizobium* and *Micrococcus* (Tsuchiya, submitted). A two-step incubation method, first at 40 °C for 4 days followed by 28 °C for 3 days, was effective in the elimination of contaminants during isolation of *Pc* from natural or inoculated soil.

The simple direct ELISA described above, in combination with S-PC-1 medium was used for specific detection and survey of *Pc* from roots and rhizosphere soil of various plants (Tsuchiya *et al.*, 1989). When the suspension prepared from root or rhizosphere soil was cultured by a two-step incubation method on S-PC-1 medium, very low concentrations (*ca* 10^3–10^4 cfu/g) of bacteria could be detected. Most of these colonies, presumably *Pc*, were isolated almost at the same concentration from all tested plants of soybean, barley, chinese cabbage, lettuce and tobacco. Samples from Welsh onion contained a larger number of *Pc* (*ca* 10^4–10^6 cfu/g). Heat treatment combined with the direct ELISA was useful for the detection of low concentrations of *Pc* (10^2 cfu/g). This is 10 to 100 times lower than the minimum detection level by direct colony counts on S-PC-1 medium.

The ELISA detection method has several advantages:

1. Rapid and simple assay saves labour, time and money and so is very useful for mass-screening of samples in routine field inspections.
2. Detection in low concentrations of the bacteria (*ca* 10^2 cfu/g) is equal or better in sensitivity and specificity compared with the other currently used methods, including culturing, PCR and so on.
3. Strain-specific detection and identification of the strains enables rapid evaluation and screening of large numbers of *P. cepacia* strains.

Comparison of the *P. cepacia* strains of clinical and environmental origin for antagonistic activity toward plant pathogens

Production of antibiotic substances of 37 *Pc* strains from infected plants, 44 strains from rhizosphere and 26 strains of clinical origin was tested by dual culture on potato dextrose agar (PDA) for fungal plant pathogens and by the plate-chloroform method on King's B medium for the bacterial plant pathogens (Tsuchiya, unpublished). About 80% of all strains and more than 90% of the rhizosphere and plant (environmental) strains showed antibiosis against four fungal plant pathogens, namely *Phytophthora capsici, Fusarium oxysporum* f. sp. *radicis-lycopersici, Pythium elongatum* and *Rhizoctonia solani* (AG-4). Activity was also detected against four bacterial plant pathogens: *Clavibacter michiganensis* subsp. *michiganensis; E. carotovora.* subsp. *carotovora;* and *P. solanacearum;* and *P. caryophylli*. On the other hand, although both antifungal and antibacterial activities of the clinical strains varied depending on the indicator pathogens tested, the antagonistic activities were lower than those shown by strains isolated from plant or soil (Fig. 20.1).

Seed bacterization with all strains from soil or plants suppressed the damping-off of radish seedlings, caused by *R. solani* (AG-4), although the level of suppression varied among the strains. On the other hand, most of the clinical strains, with few exceptions, showed no or very little biocontrol activity (Tsuchiya, unpublished).

It is noteworthy that this screening method could detect quantitatively the difference in the antagonistic ability between clinical strains and environmental strains. Therefore, this method can be used as an additional or alternative discrimination system for the evaluation of the safety of biocontrol agents, as well as in risk assessment concerning the origin of *P. cepacia* strains.

Serological diagnosis of *Pc* strains of clinical and environmental origin

To compare the serological relations between *Pc* strains of clinical and environmental origins, serodiagnosis by indirect ABC-ELISA technique was performed using ten polyclonal antisera, which classified Japanese clinical strains into ten serogroups A to J (Nakamura *et al.*, 1986). In addition, six monoclonal antibodies, which were different in their reactivity and specificity as described above were also applied for discrimination of the strains.

On the basis of specific reactivities 37 strains of plant origin were classified into three serogroups A (51.4%), C (43.2%) and D (5.4%). On the other hand, 44 strains of rhizosphere origin were grouped into three serogroups C (93.2%), E (4.5%) and A (2.3%). Almost 95% of environmental strains were classified into two major serogroups A and C within the 10 serogroups characterized (Tsuchiya, submitted).

None of the six MABs could specifically distinguish *Pc* strains of clinical origin from strains of plant and rhizosphere origin, although some differences in reactivity were found (Fig. 20.2). No direct relationship between reactivity to MABs and serogroups of the *Pc* strains was found, except in the case of PCF-1A5, which reacted specifically to the strains that belonged to serogroup A of both clinical and environmental origin.

Although antigen cross-reactions between strains of different serogroups were not totally eliminated with the six MABs currently available, this type of screening could provide a better discriminating 'fingerprint' system when improved MABs specific to beneficial or harmful properties of biocontrol agents become available.

Conclusion

When microbial biocontrol agents including GEMs are introduced into the environment, careful prior risk assessment must be made (see Ryder and Correll, Chapter 19). Although there are many areas that concern risk assessment, the development of effective detection and differentiation methods for the GEMs is essential, and will also contribute much to other research areas. The results can be summarized as follows:

1. By using simple direct ELISA the bacterium *P. cepacia* could be specifically detected at a concentration of 10^2 cfu/g from both articially inoculated soil and field soil, as well as from plant rhizosphere, regardless of the presence of other closely related bacteria.
2. Detection of a specific strain was possible from soil in which there were different strains of the same species by an indirect ABC-ELISA that used a specific monoclonal antibody corresponding to the target strain.
3. Most *P. cepacia* strains of environmental origin were antagonistic to various fungal and bacterial plant

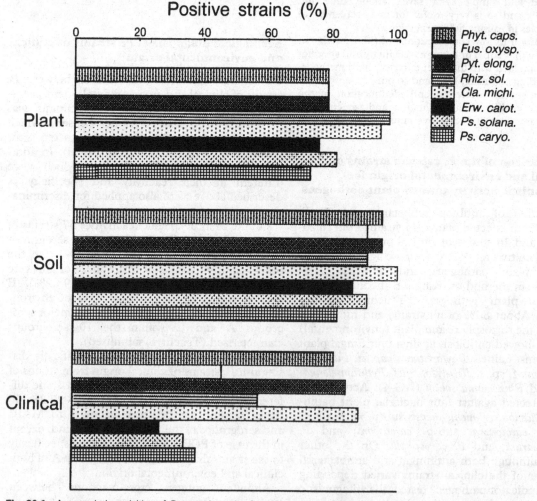

Fig. 20.1 Antagonistic activities of *P. cepacia* strains from plant, soil and clinical origins against fungal and bacterial plant pathogens.

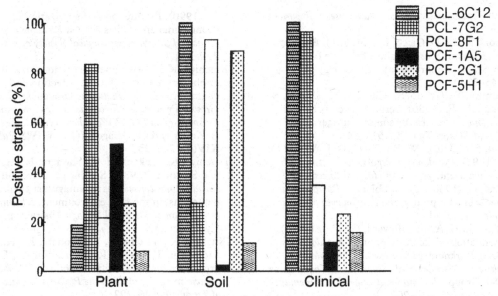

Fig. 20.2   Reactivity of *P. cepacia* strains from plant, soil and clinical origins with six monoclonal antibodies.

pathogens and suppressed disease. On the other hand, fewer strains of clinical origin showed suppressive activities, though some of them were as active as the environmental strains.

4. Ninety-five per cent of *P. cepacia* of environmental origin were classified into two major serogroups (A and C) among the serogroups of the clinical strains. None of the MABs enabled us to distinguish strains of clinical origin from environmental isolates. PCF-1A5, one of the six MABs used, showed specific reactivity to the strains that belonged to serogroup A.

I conclude by emphasizing the convenience of the serodiagnostic methods described here in the risk assessment of *P. cepacia* as a biocontrol agent, not only for discriminating strains of clinical and environmental origin but also for detecting or tracing specific strains of microbes present in low concentrations simply, rapidly and sensitively in routine field inspections. The results described here relate to *P. cepacia*, but these methods should also be applicable to risk assessment of other related micro-organisms.

References

Burkholder, W. H. (1950). Sour skin, a bacterial rot of onion bulbs. *Phytopathology* **40**, 115–117.

Carson, L. A., Favero, M. S., Bond, W. W. and Petersen, N. J. (1973). Morphological, biochemical, and growth characteristics of *Pseudomonas cepacia* from distilled water. *Applied Microbiology* **25**, 476–483.

Ederer, M. G. and Matsen, J. M. (1972). Colonization and infection with *Pseudomonas cepacia*. *Journal of Infectious Disease* **125**, 613–618.

Elander, R. P., Mabe, J. A., Hamill, R. H. and Gorman, M. (1968). Metabolism of tryptophans by *Pseudomonas aureofaciens*. VI. Production of pyrrolnitrin by selected *Pseudomonas* species. *Applied Microbiology* **16**, 753–758.

Gonzalez, C. F. and Vidaver, A. K. (1979). Bacteriocin, plasmid and pectolytic diversity in *Pseudomonas cepacia* of clinical and plant origin. *Journal of General Microbiology* **110**, 161–170.

Homma, Y., Kato, K. and Suzui, T. (1985). Biological control of soilborne root disease by *Pseudomonas cepacia* isolated from roots of lettuce and *Campanula* sp. *Annals of the Phytopathological Society of Japan* **51**, 349. (Abstract in Japanese.)

Homma, Y., Sato, Z., Hirayama, F., Konno, K., Sirahama, H. and Suzui, T. (1989). Production of antibiotics by *Pseudomonas cepacia* as an agent for biological control of soilborne plant pathogens. *Soil Biology and Biochemistry* **21**, 723–728.

Janisiewicz, W. J. and Roitman, J. (1988). Biological control of blue mold and gray mold on apple and pear with *Pseudomonas cepacia*. *Phytopathology* **78**, 1697–1700.

Kawamoto, S. O. and Lorbeer, J. W. (1976). Protection of onion seedlings from *Fusarium oxysporum* f. sp. *cepae* by seed and soil infestation

222 TSUCHIYA

with *Pseudomonas cepacia*. *Plant Disease Reporter* **60**, 189–191.

Lennon, E. and DeCicco, B. T. (1991). Plasmids of *Pseudomonas cepacia* strains of diverse origins. *Applied and Environmental Microbiology* **57**, 2345–2350.

Nakamura, Y., Hyodo, S., Chonan, E., Shigeta, S. and Yabuuchi, E. (1986). Serological classification of *Pseudomonas cepacia* by somatic antigen. *Journal of Clinical Microbiology* **24**, 152–154.

Pallent, L. J., Hugo, W. B., Grant, D. J. W. and Davis, A. (1983). *Pseudomonas cepacia* as contaminant and infective agent. *Journal of Hospital Infection* **4**, 9–13.

Randall, C. (1980). The problem of *Pseudomonas cepacia* in a hospital. *Canadian Journal of Public Health* **71**, 119–123.

Sinsabaugh, H. A. and Howard, G. W. Jr (1975). Emendation of the description of *Pseudomonas cepacia* Burkholder (Synonyms: *Pseudomonas multivorans* Stanier *et al.*, *Pseudomonas kingii* Jonsson; EO-1 Group). *International Journal of Systematic Bacteriology* **25**, 187–201.

Takahashi, Y., Tsuchiya, K., Shohara, K. and Suzui, T. (1990). Production and application of monoclonal antibodies against *Pseudomonas cepacia*. *Annals of the Phytopathological Society of Japan* **56**, 229–234.

Tsuchiya, K. (1991). The use of monoclonal antibodies as probes for the detection and characterization of *Pseudomonas cepacia*. In *Plant Growth-Promoting Rhizobacteria – Progress and Prospects (2nd PGPR Workshop)* (ed. C. Keel, B. Koller and G. Défago). *IOBS/WPRS Bulletin* **XIV**(8), 256–259.

Tsuchiya, K., Takahashi, Y., Shohara, K., Homma, Y. and Suzui, T. (1989). Serological detection of *Pseudomonas cepacia*, as an antagonist to plant pathogens, in the field environment. *5th International Symposium of Microbial Ecology* Abstracts p. 122. Kyoto, Japan.

Tsuchiya, K., Takahashi, Y., Shohara, K., Homma, Y. and Suzui, T. (1991). Rapid and simple ELISA procedure for the specific detection of *Pseudomonas cepacia* in soil. *Annals of the Phytopathological Society of Japan* **57**, 196–202.

21

Benefits and risks of using genetically engineered baculoviruses as insecticides

Norman E. Crook and Doreen Winstanley

Introduction

Insects are susceptible to a wide range of viral pathogens, but viruses of the family Baculoviridae, which occur only in arthropods, have received most attention because their high pathogenicity, narrow host range, and complete safety to vertebrates and plants, make them particularly promising candidates for biological pest control agents (see Huber, Chapter 18). Although natural strains of these viruses can in many cases provide highly satisfactory levels of control, the past 5 years has seen several attempts to produce genetically engineered baculoviruses with improved insecticidal properties. The chief target for these improvements has been reduction in the crop damage that continues to occur after infection of the insects.

A second reason for using genetically engineered baculovirus insecticides is to provide marked strains for ecological studies in a manner somewhat analogous to the way that animals are often marked physically in order to study in a more detailed way their behaviour and survival. Marked viruses may contain simply a short stretch of introduced DNA in their genome. This can be detected by various methods such as dot blotting, Southern blotting, restriction enzyme analysis (especially if the introduced DNA contains additional restriction sites), or, with much greater sensitivity, by use of the polymerase chain reaction. Alternatively, marked viruses may contain a reporter gene that expresses a protein, usually an enzyme, which can be detected easily and specifically. The *E. coli lacZ* gene, which codes for β-galactosidase, has been found particularly useful. In many cases, there has been a need for more precise and accurate ecological studies to provide reliable data for assessment of the risk of releasing genetically engineered baculoviruses which contain genes designed to enhance the insecticidal properties of the virus.

The vast majority of investigations undertaken on the genetic improvement of baculoviruses has been done with *Autographa californica* nuclear polyhedrosis virus (AcMNPV), which is the best characterized baculovirus, and techniques for genetic manipulation of this virus are well documented (King and Possee, 1992; O'Reilly *et al.*, 1992).

Recombinant baculoviruses for improved pest control

Several genes have now been introduced into baculoviruses in attempts to produce a virus that kills, or inhibits feeding, more rapidly than the wild-type. These genes fall into two broad groups: those that code for toxins, which will kill or paralyse the insect; and those that code either for hormones or for hormone regulatory proteins, which will alter the level of hormones essential for the normal development of the insect.

The *Bacillus thuringiensis* δ-endotoxin gene,

which has been widely used to produce transformed plants with resistance to insects, has also been introduced into baculoviruses (Martens et al., 1990; Merryweather et al., 1990). These recombinant viruses expressed large amounts of the toxin in insect cells but there was no reduction in LD_{50} values with susceptible larvae. Recently, Ribeiro and Crook (1993) showed that massive levels of δ-endotoxin can be expressed in larvae, either from full length or truncated genes in AcMNPV without altering the pathogenicity of the virus. However, a single insect killed by this virus contained the equivalent of 3000 LD_{50} doses of toxin when administered orally. This demonstrates that although the toxin was still highly active by the oral route, it had no significant effect when expressed within cells in the insect. Since insect neurotoxins are active by injection, it might be expected that their expression within insect tissues would be more likely to be effective. However, the first attempt to produce a neurotoxin-expressing recombinant baculovirus also met with little success (Carbonell et al., 1988). In that case, a synthetic gene, designed to code for insect toxin-1 from the scorpion, Buthus eupeus, was cloned into AcMNPV. High levels of the gene transcripts, but only low levels of the toxin, were detected and the recombinant virus did not cause any paralytic effect. More success has been obtained with recombinants that contained the insect-specific neurotoxin gene from the scorpion, Androctonus australis. Stewart et al. (1991) produced two recombinants, one of which included a copy of the AcMNPV gp67 signal peptide sequence at the 5' end of the coding region of the scorpion toxin, to facilitate protein secretion. This virus performed better than the one without the signal peptide and, using second instar Trichoplusia ni larvae, gave a 25% reduction in ST_{50} value, a slight reduction (30%) in LD_{50}, and up to 50% reduction in leaf damage, compared with wild-type virus. McCutchen et al. (1991) used the same gene, unmodified, and obtained a 30% reduction in ST_{50} and a 40% reduction in LD_{50} with second instar Heliothis virescens larvae. In both cases the scorpion toxin was expressed using the p10 promoter adjacent to the polyhedrin locus so that the recombinant viruses were occluded. Another insect-specific toxin gene that has been found to improve greatly the effectiveness of a baculovirus insecticide was obtained from the mite, Pyemotes

tritici (Tomalski and Miller, 1991). A recombinant virus that expressed this gene caused paralysis of 100% of larvae within 2 days after injection of non-occluded virus. An occluded recombinant virus fed to larvae also caused paralysis or death much earlier than wild-type virus.

Expression of hormones or hormone regulatory proteins at high levels and at times when they may not normally be present in the insect might be expected to have a deleterious effect on insect development. Juvenile hormone esterase inactivates juvenile hormone in larvae and results in cessation of feeding and the initiation of metamorphosis. However, expression of the enzyme by a non-occluded recombinant virus reduced feeding and growth only in first instar larvae and failed to have any significant effect on later instars (Hammock et al., 1990). Eldridge et al. (1991, 1992) have produced various recombinant viruses that express eclosion hormone, a neuropeptide which is associated with a number of physiological and behavioural functions, particularly ecdysis. Although high levels of active hormone were produced, feeding rates and survival times of insects infected with the recombinant viruses were no different to those infected with control viruses that lacked the hormone gene. This was also true for a recombinant, which in addition to expressing eclosion hormone also had the egt gene deleted (see below).

An alternative strategy to expression of a toxic protein has been explored by O'Reilly and Miller (1989; 1991). They discovered a baculovirus gene (egt), which codes for ecdysteroid UDP-glucosyltransferase. Expression of egt enables the virus to inhibit moulting of its infected host. Larvae normally stop feeding prior to moulting but expression of egt prevents this. Larvae infected with an egt deletion mutant displayed considerably reduced feeding and a 22% reduction in ST_{50} value compared with larvae infected with wild-type virus.

Strategies for release of genetically engineered baculoviruses

Many baculoviruses have been extensively field tested in many countries and a number have been fully registered for commercial use. No unforeseen effects have occurred as a result of using these viruses. Release of genetically engineered viruses, however, requires much more exhaustive safety testing and involves many more restrictions on the way in which the release is carried out. One of the main concerns about releasing a genetically engineered organism is that it will spread outside the area in which it was released and become permanently established in the environment. If the organism is specific for the pest that it is intended to control and if it has no deleterious side effects, then this may be acceptable. However, at the moment some form of containment is required. This can be either physical, to block all mechanisms by which virus could be transmitted beyond the release site, or it can be genetic, so that the virus is weakened in some way to reduce its persistence. Many insect viruses, including most baculoviruses, are able to persist in the environment by virtue of their occlusion body (OB) – a thick protein matrix that surrounds and protects the virus particles. Modification of the way in which the virus is occluded can therefore reduce its environmental persistence. The simplest and most extreme method is to remove the OB all together by deletion of the matrix protein gene. However, this causes such a drastic reduction in persistence that it is unlikely that non-occluded virus could be applied in the field in a conventional manner and still be effective as an insecticide. A more subtle modification of the OB has been achieved by deletion of the p10 gene (Williams *et al.*, 1989). The p10 protein is expressed at very high levels late in the replication cycle. It is a non-essential gene but OBs produced by p10-negative mutants are more susceptible to physical damage than normal OBs and it is predicted than they would be less persistent in the environment (Vlak *et al.*, 1988). A third strategy combines use of a non-occluded recombinant with the ability to apply it as normally occluded virus (Miller, 1988). This is achieved by co-infection with both the recombinant virus and wild-type virus. Cells that are infected with both genotypes produce progeny virus of both genotypes. Since these are physically indistinguishable, both become occluded within polyhedra. Such co-occluded virus can contain up to 80% recombinant virus particles but this percentage falls quite rapidly on passage through insects so that after 3–4 passages (depending upon inoculum level)

less than 5% of virus particles are recombinants (Miller, 1988).

Field-release testing

The first release of a genetically engineered baculovirus was in 1986 in the UK (Bishop, 1986). This virus was identical to wild-type AcMNPV except that it contained an 80bp non-coding oligonucleotide, inserted in a non-regulatory intergenic region, to allow the fate and persistence of the virus to be studied in a 'cabbage-patch' ecosystem. Prior to release, extensive host-range testing of this and all subsequently released viruses was carried out. In 1987, a release was made in the same system of a marked (with a different oligonucleotide) virus which also had the polyhedrin gene deleted, to study the persistence of a non-occluded virus (Bishop *et al.*, 1988). These tests confirmed the greatly reduced persistence of non-occluded virus – following the deaths of the released larvae (which had been infected in the laboratory) no infectious virus was detected at the test site and even the corpses of virus-killed larvae did not yield any infectious virus. These releases of genetically engineered viruses, together with further releases at the same site in subsequent years (Bishop *et al.*, 1989), have provided detailed data on the transmission and persistence of these pathogens within this ecosystem. A high degree of physical containment that severely limited movement of insects, birds and other animals into or out of the site was used throughout these studies. In addition, at the end of each experiment the site was disinfected and shown to be free from virus. In the USA, a genetically engineered baculovirus was released without physical containment in 1989 using the co-occlusion strategy (Wood *et al.*, 1990). A polyhedrin-negative AcMNPV, which contained no additional inserted DNA, was co-occluded with wild-type virus, such that polyhedra contained equal numbers of both viruses, and sprayed on three successive occasions onto a 2 acre cabbage plot in New York State. Prior to each application, 2700 third instar *T. ni* larvae were introduced into the plot. All the test larvae died of virus and it was estimated that at the end of the growing season the test site contained 10^{14} progeny polyhedra per acre containing on average 42% polyhedrin-negative virus (Wood and Granados,

1991). In subsequent years the amount of this genetically engineered virus in the soil and on plants cultivated on the site has been monitored to obtain an accurate assessment of its long-term persistence. Results from bioassays of soil samples in 1990 and 1991 showed that progeny polyhedra contained less than 10% engineered virus indicating that polyhedrin-minus virus is lost as it cycles from one insect to another in the natural environment (H. A. Wood, personal communication).

The first release of a genetically engineered baculovirus other than AcMNPV was due to take place in the USA in 1993 (H. A. Wood, personal communication) and was also the first release of a recombinant virus in a forest ecosystem. *Lymantria dispar* MNPV has been modified by removal of the polyhedrin gene and insertion of a *lacZ* gene to facilitate monitoring. The polyhedrin-minus virus was co-occluded with wild-type LdMNPV.

Risks of using genetically engineered baculoviruses

Most of the potential risks from the use of genetically engineered baculoviruses are similar to those of naturally occurring baculoviruses and are dealt with elsewhere (see Huber, Chapter 18). In general, the high specificity and low persistence of these viruses severely limit the likelihood of unforeseen effects occurring.

Replication and gene expression in non-target hosts

The introduction of other genes, even broad host range toxins, expressed under the control of very late baculovirus promoters, such as the polyhedrin or p10 promoters, is unlikely to modify the specificity of the virus since these promoters are active only at very low levels, if at all, in non-permissive host cells *in vitro* (Morris and Miller, 1992) and it is likely that this will also be the case *in vivo*. In addition, the long-term persistence of viruses that kill their host more rapidly is likely to be reduced since less virus is produced by the insect and released into the environment. So far, all releases of recombinant viruses that express a foreign gene have utilized a very late baculovirus promoter to express the gene. Expression from early, and to some extent, late promoters has been shown to be less host specific than very late promoters (Morris and Miller, 1992). Thus, genes could be expressed from early promoters that are active prior to expression of any viral proteins, in cells which the virus can enter even though it may be unable to replicate. However, if there was no replication of virus or viral DNA, expression would be possible only at primary sites of infection in the gut and only from DNA present in the inoculum virus. This would indicate that high levels of virus inoculum and highly active gene products would be needed to produce any effect in an insect.

The possibility of any risk to vertebrates appears to be very small, although low levels of replication in mammalian cells have been reported. Himeno et al., (1967) observed polyhedral bodies in a small proportion of FL cells (a human amnion cell line) after transfecting them with DNA from *Bombyx mori* NPV. However, these authors were unable to infect the cells using native virus. McIntosh and Shamy (1980) observed replication of AcMNPV in a small proportion (<5%) of Chinese hamster cells at very high multiplicities of infection, but other workers using the same virus and cell line have been unable to obtain any evidence of replication (Volkman and Goldsmith, 1983; Gröner et al., 1984) and have questioned the validity of the earlier data. Volkman and Goldsmith (1983) exposed a total of 35 vertebrate cell lines, including 23 of human origin, to AcMNPV and used a sensitive immunoassay to detect any virus replication. Cells were incubated at both 27 °C (normal optimum for insect viruses) and 37 °C (optimum for mammalian cells) and assayed for the presence of virus at four time points up to 7 days after infection. No evidence of frank viral replication was obtained in any of these cell lines although virus uptake appeared to be quite common. Even though mammalian cells may be able to take up baculoviruses, there appear to be other blocks that prevent the expression of genes on the viral genome even when the gene is under the control of a promoter that is normally active in the cell. For example, Carbonell and Miller (1987) showed that a recombinant AcMNPV was taken up by both a human and a mouse cell line but that there was no expression of a sensitive reporter gene (chloramphenicol acetyltransferase) under the control of a mammalian active promoter from Rous sarcoma virus.

Recombination between baculoviruses

One of the possible risks involved in the use of genetically engineered viruses in the environment is that an introduced gene could be transferred to another naturally occurring virus. Although the genetically engineered virus may have been extensively tested and may have been modified to reduce its persistence, transfer of an introduced gene into another virus with perhaps a different host range could have unintended consequences. Recombination events that involve baculovirus DNA are known to be common in some situations. For example, foreign genes are introduced into baculovirus expression vectors by recombination between DNA from a plasmid transfer vector that contains the foreign gene and baculovirus DNA (Luckow and Summers, 1988), and host cell DNA frequently becomes inserted into a baculovirus genome following repeated passage in cell culture (Beames & Summers, 1988). Examples of recombination between two different baculoviruses have only rarely been recorded. Recombination of AcMNPV and *Rachiplusia ou* NPV was obtained *in vitro* in *T. ni* cells (Summers *et al.*, 1980). Croizier and Quiot (1981) obtained a high level of recombinants after mixed infection of *Galleria mellonella* larvae with *G. mellonella* NPV and AcMNPV but mixed infection with AcMNPV and *R. ou* NPV in *G. mellonella* which is susceptible to both viruses, always resulted in AcMNPV progeny virus.

Kondo and Maeda (1991) showed that recombination between two viruses could result in new viruses with an expanded host range. Recombinants between AcMNPV and *B. mori* NPV were obtained after co-infection of the two viruses in a cell line susceptible only to AcMNPV and then selection in another cell line susceptible only to *B. mori* NPV. Some of these recombinants, which were able to infect both cell lines, were further characterized and showed that recombination occurred at many regions along the genome and frequently involved exchange of the polyhedrin gene. The viruses had an expanded host range in insects as well as in cell culture. A similar procedure was used to isolate recombinants between *Spodoptera exigua* NPV and *S. frugiperda* NPV and, once again, recombination was found to take place at many points along the genome (Maruniak, 1992).

Many different genotypic variants were isolated from *Pieris brassicae* larvae following infection with a single genotype of *Pieris rapae* GV (Smith and Crook, 1993). It appeared that *P. rapae* GV was unable to undergo complete replication in *P. brassicae* but initiated an infection by activation of an inapparent, perhaps latent, virus in the larvae. Restriction enzyme mapping of the progeny virus genomes suggested that the genotypes resulted from recombination between *P. rapae* GV and *P. brassicae* GV. In a somewhat similar study, in which larvae of *Trichoplusia ni* were inoculated with *Choristoneura fumiferana* NPV, the progeny viruses were found to be mutants of AcMNPV composed of three distinct plaque phenotypes (Fraser and Wang, 1986). The origin of these was unknown but it was speculated that they could have arisen by recombination between the inoculum and either contaminant or latent AcMNPV.

Most of these examples involved recombination between closely related viruses but there is some recent evidence that recombination may occur between very different baculoviruses (Crook *et al.*, 1993). In this case, recombination took place between *Cydia pomonella* granulosis virus and a mutant AcMNPV when the two viral genomes were cotransfected into *Spodoptera frugiperda* cells. It was possible to detect a very low recombination frequency because the mutant AcMNPV lacked the p35 gene and replicated very poorly in *S. frugiperda* cells (Clem *et al.*, 1991) whereas the recombinant was able to replicate normally. It is unlikely that this would occur in insects since *C. pomonella* granulosis virus and AcMNPV do not share a common host. However, AcMNPV does replicate in many lepidopterous species that are susceptible to a range of other baculoviruses and it would be of interest to determine whether recombination occurs in these cases.

Many of the above recombination events have been detected when the recombinant had a selection advantage over the inoculum viruses. When there was very little selection pressure, as was the case when two distinct genotypes of *Anticarsia gemmatalis* NPV were used to co-infect *Spodoptera frugiperda* cells, recombination appeared to occur frequently and resulted in many different genotypes (Croizier and Ribeiro, 1992). It was suggested that recombination is probably an active process in wild populations of *A. gemmatalis*

Table 21.1. Host range of baculoviruses

Virus	Maximum recorded range of species infected by one virus	Orders with susceptible species
A. californica MNPV	11 families (some only at very high doses)	Lepidoptera
Mamestra brassicae MNPV	4 families (some only at very high doses)	Lepidoptera
Other MNPVs	Usually 1, maximally 3 families	Lepidoptera
SNPVs	1 family	Lepidoptera, Hymenoptera, Diptera, Decapoda
GVs	1 family	Lepidoptera

NPV and accounts for the large number of genotypes which occur in these populations. Miller (1988) attempted to quantify the extent of recombination when there was no selective advantage. He co-infected cells with wild-type AcMNPV (occlusion positive, clear plaque) and a genetically engineered virus in which the polyhedrin gene was replaced with the *E. coli lacZ* gene. Expression of β-galactosidase was detected using X-gal and thus resulted in an occlusion negative, blue plaque phenotype. Progeny virus from mixed infections were plaqued and screened for potential recombinants (occlusion positive, blue plaque). Although quite a high percentage of these were observed, further analysis failed to isolate a stable recombinant and it was argued that the observed result was not due to recombination. However, it was suggested that further work was needed to clarify these issues.

Results from studies done in the laboratory therefore suggest that recombination between closely related viruses may occur when two viruses co-infect an insect. A very low frequency of recombination may also be possible between less closely related viruses, but the narrow host range of most baculoviruses greatly reduces the likelihood of two such viruses replicating in the same insect. Whether recombination might pose a risk in the release of genetically engineered baculoviruses could depend also on the construction of the virus. Transfer of an introduced gene from the genetically engineered virus to a wild-type strain of the same virus should not cause a problem unless the genetically engineered strain has other modifications, for example, to limit its persistence. Field studies are therefore needed to determine whether there are detectable rates of recombination in the natural environment and whether this is likely to have any adverse consequences.

Conclusion

Genetically engineered baculoviruses have now been produced which, in laboratory bioassays, show significant improvements in insecticidal activity over wild-type virus. Several field releases of genetically marked baculoviruses have provided accurate measurements of field persistence under defined conditions and provide an informed basis on which to assess the risk from the release of viruses with improved insecticidal activity. Some concern has been expressed about the potential risk of an uncontrolled release of a genetically improved AcMNPV virus due to its relatively broad host range (Williamson, 1991). All baculoviruses, including AcMNPV, have an extremely narrow host range relative to any chemical insecticide. The host range of AcMNPV is restricted to the Lepidoptera and infects only a small proportion of these insects. This host range is considered broad only by comparison with other baculoviruses, many of which are restricted to one or a small number of closely related species (Table 21.1). Using a baculovirus with a narrower host range would further reduce any possible risk but few attempts have been made to modify other viruses (except *B. mori* NPV, which infects the silkworm). Granulosis viruses (GVs), in particular are considered to have a narrow host range

(Crook, 1991) and infect only a few closely related species. Genetic engineering of GVs has not been possible due largely to the lack of adequate cell culture systems for these viruses. However, these problems are now being overcome (Winstanley and Crook, 1993) and before long we can expect to see genetically engineered GVs. Another strategy that would lessen any unforeseen risk in an uncontrolled release would be to use a deletion mutant that contained no additional DNA. AcMNPV lacking the *egt* gene has been shown to kill insects more quickly and inhibit feeding relative to wild-type virus (O'Reilly and Miller, 1991). This could provide improved control without any risk of gene transfer in the environment.

References

Beames, B. and Summers, M. D. (1988). Comparisons of host cell DNA insertions and altered transcription at the site of insertions in few polyhedra baculovirus mutants. *Virology* 162, 206–220.

Bishop, D. H. L. (1986). UK release of genetically marked virus. *Nature* 323, 496.

Bishop, D. H. L., Entwistle, P. F., Cameron, I. R., Allen, C. J. and Possee, R. D. (1988). Genetically engineered baculovirus insecticides. *Aspects of Applied Biology* 17, 385–395.

Bishop, D. H. L., Harris, M. P. G., Hirst, M., Merryweather, A. T. and Possee, R. D. (1989). The control of insect pests by viruses; opportunities for the future using genetically engineered virus insecticides. In *Progress and Prospects in Insect Control* (ed. N. R. McFarlane), pp. 145–155. British Crop Protection Council, London.

Carbonell, L. F. and Miller, L. K. (1987). Baculovirus interaction with non-target organisms: a virus-borne reporter gene is not expressed in two mammalian cell lines. *Applied and Environmental Microbiology* 53, 1412–1417.

Carbonell, L. F., Hodge, M. R., Tomalski, M. D. and Miller, L. K. (1988). Synthesis of a gene coding for an insect-specific scorpion neurotoxin and attempts to express it using baculovirus vectors. *Gene* 73, 409–418.

Clem, R. J., Fechheimer, M. and Miller, L. K. (1991). Prevention of apoptosis by a baculovirus gene during infection of insect cells. *Science* 254, 1388–1390.

Croizier, G. and Quiot, J. M. (1981). Obtention and analysis of two genetic recombinants of baculoviruses of Lepidoptera, *Autographa californica*

Speyer and *Galleria mellonella* L. *Annales de Virologie* 132, 3–18.

Croizier, G. and Ribeiro, H. C. T. (1992). Recombination as a possible major cause of genetic heterogeneity in *Anticarsia gemmatalis* nuclear polyhedrosis virus wild populations. *Virus Research* 26, 183–196.

Crook, N. E. (1991). Baculoviridae: subgroup B – Comparative aspects of granulosis viruses. In *Viruses of Invertebrates* (ed. E. Kurstak), pp. 73–110. Marcel Dekker, New York.

Crook, N. E., Clem, R. J. and Miller, L. K. (1993). An apoptosis-inhibiting baculovirus gene with a zinc finger-like motif. *Journal of Virology* 67, 2168–2174.

Eldridge, R., Horodyski, F. M., Morton, D. B., O'Reilly, D. R., Truman, J. W., Riddiford, L. M. and Miller, L. K. (1991). Expression of an eclosion hormone gene in insect cells using baculovirus vectors. *Insect Biochemistry* 21, 341–351.

Eldridge, R., O'Reilly, D. R. and Miller, L. K. (1992). Efficacy of a baculovirus pesticide expressing an eclosion hormone gene. *Biological Control* 2, 104–110.

Fraser, M. J. and Wang, H. (1986). Molecular approaches in the analysis of cross infections with baculoviruses. In *Fundamental and Applied Aspects of Invertebrate Pathology* (ed. R. A. Samson, J. M. Vlak and D. Peters) pp. 55–56. Wageningen, Foundation of the Fourth International Colloquium of Invertebrate Pathology, Veldhoven, The Netherlands.

Gröner, A., Granados, R. R. and Burand, J. P. (1984). Interaction of *Autographa californica* nuclear polyhedrosis virus with two nonpermissive cell lines. *Intervirology* 21, 203–209.

Hammock, B. D., Bonning, B. C., Possee, R. D., Hanzlik, T. N. and Maeda, S. (1990). Expression and effects of the juvenile hormone esterase in a baculovirus vector. *Nature* 344, 458–461.

Himeno, M., Sakai, F., Onodera, K., Nakai, H., Fukada, T. and Kawade, Y. (1967). Formation of nuclear polyhedral bodies and nuclear polyhedrosis virus of silkworm in mammalian cells infected with viral DNA. *Virology* 33, 507–512.

King, L. A. and Possee, R. D. (1992). *The Baculovirus Expression System*. Chapman and Hall, London.

Kondo, A. and Maeda, S. (1991). Host range expansion by recombination of the baculoviruses *Bombyx mori* nuclear polyhedrosis virus and *Autographa californica* nuclear polyhedrosis virus. *Journal of Virology* 65, 3625–3632.

Luckow, V. A. and Summers, M. D. (1988). Trends in the development of baculovirus expression vectors. *Bio/Technology* 6, 47–55.

Martens, J. W. M., Honée, G., Zuidema, D., van Lent, J. W. M., Visser, B. and Vlak, J. M. (1990).

Insecticidal activity of a bacterial crystal protein expressed by a recombinant baculovirus in insect cells. *Applied and Environmental Microbiology* **56**, 2764–2770.

McCutchen, B. F., Choudary, P. V., Crenshaw, R., Maddox, D., Kamita, S. G., Palekar, N., Volrath, S., Fowler, E., Hammock, B. D. and Maeda, S. (1991). Development of a recombinant baculovirus expressing an insect-selective neurotoxin: potential for pest control. *Bio/Technology* **9**, 848–852.

McIntosh, A. H. and Shamy, R. (1980). Biological studies of a baculovirus in a mammalian cell line. *Intervirology* **13**, 331–341.

Maruniak, J. E. (1992). Contribution of molecular biology to the improvement of insect viruses as biological control products. *Pesquisa Agropecuária Brasileira* **27**, 142–150.

Merryweather, A. T., Weyer, U., Harris, M. P. G., Hirst, M., Booth, T. and Possee, R. D. (1990). Construction of genetically engineered baculovirus insecticides containing the *Bacillus thuringiensis* subsp. *kurstaki* HD-73 delta endotoxin. *Journal of General Virology* **71**, 1535–1544.

Miller, D. W. (1988). Genetically engineered viral insecticides. In *Biotechnology For Crop Protection* (ed. P. A. Hedin, J. J. Menn and R. M. Hollingworth), pp. 405–421. American Chemical Society, Washington, DC.

Morris, T. D. and Miller, L. K. (1992). Promoter influence on baculovirus-mediated gene expression in permissive and nonpermissive insect cell lines. *Journal of Virology* **66**, 7397–7405.

O'Reilly, D. R. and Miller, L. K. (1989). A baculovirus blocks insect molting by producing ecdysteroid UDP-glucosyl transferase. *Science* **245**, 1110–1112.

O'Reilly, D. R. and Miller, L. K. (1991). Improvement of a baculovirus pesticide by deletion of the egt gene. *Bio/Technology* **9**, 1086–1089.

O'Reilly, D. R., Miller, L. K. and Luckow, V. A. (1992). *Baculovirus Expression Vectors – A Laboratory Manual*. W. H. Freeman, New York.

Ribeiro, B. M. and Crook, N. E. (1993). Expression of full-length and truncated forms of crystal protein genes from *Bacillus thuringiensis* subsp. *kurstaki* in a baculovirus and pathogenicity of the recombinant viruses. *Journal of Invertebrate Pathology* **62**, 121–130.

Smith, I. R. L. and Crook, N. E. (1993).

Characterization of new baculovirus genotypes arising from inoculation of *Pieris brassicae* with granulosis viruses. *Journal of General Virology* **74**, 415–424.

Stewart, L. M. D., Hirst, M., Ferber, M. L., Merryweather, A. T., Cayley, P. J. and Possee, R. D. (1991). Construction of an improved baculovirus insecticide containing an insect-specific toxin gene. *Nature* **352**, 85–88.

Summers, M. D., Smith, G. E., Knell, J. D. and Burand, J. P. (1980). Physical maps of *Autographa californica* and *Rachiplusia ou* nuclear polyhedrosis virus recombinants. *Journal of Virology* **34**, 693–703.

Tomalski, M. D. and Miller, L. K. (1991). Insect paralysis by baculovirus-mediated expression of a mite neurotoxin gene. *Nature* **352**, 82–85.

Vlak, J. M., Klinkenberg, F. A., Zaal, K. J. M., Usmany, M., Klingeroode, E. C., Geervliet, J. B. F., Roosien, J. and Van Lent, J. W. M. (1988). Functional studies on the p10 gene of *Autographa californica* nuclear polyhedrosis virus using a recombinant expressing a p10-beta-galactosidase fusion gene. *Journal of General Virology* **69**, 765–776.

Volkman, L. E. and Goldsmith, P. A. (1983). *In vitro* survey of *Autographa californica* nuclear polyhedrosis virus interaction with nontarget vertebrate host cells. *Applied and Environmental Microbiology* **45**, 1085–1093.

Williams, G. V., Rohel, D. Z., Kuzio, J. and Faulkner, P. (1989). A cytopathological investigation of *Autographa californica* nuclear polyhedrosis virus p10 gene function using insertion/deletion mutants. *Journal of General Virology* **70**, 187–202.

Williamson, M. (1991). Biocontrol risks. *Nature* **353**, 394.

Winstanley, D. and Crook, N. E. (1993). Replication of *Cydia pomonella* granulosis virus in cell cultures. *Journal of General Virology* **74**, 1599–1609.

Wood, H. A. and Granados, R. R. (1991). Baculoviruses as pest control agents. *Annual Reviews in Microbiology* **45**, 69–87.

Wood, H. A., Hughes, P. R., Van Beek, N. and Hamblin, N. (1990). An ecologically acceptable strategy for the use of genetically engineered baculovirus pesticides. In *Insect Neurochemistry and Neurophysiology*, 1989 Edited (ed. A. B. Borkovec, and E. P. Masler), pp. 285–288. Humana, Clifton, NJ.

22

Mathematical modelling of gene exchange in soil

James M. Lynch, M. J. Bazin and J. Choi

Introduction

The Organization for Economic Co-operation and Development (OECD) has been active in developing procedures for safely carrying out experiments in genetic engineering (OECD, 1986) and in developing procedures for good developmental practices to evaluate the products of genetic manipulation in the field (OECD, 1990). Most member countries, as well as the Commission of the European Communities (CEC), have accepted the guidelines to develop regulatory frameworks.

OECD funded a workshop at King's College London, under the Theme III of the Biological Resources Management Project (Bazin and Lynch, 1994). The aim was to bring together biologists and mathematicians to attempt to work towards an understanding of risk assessment of gene exchange in terrestrial ecosystems.

The purpose of this paper is to give a very brief overview of the workshop and to build on this by a report on the construction of a recent model of gene exchange between soil bacteria.

Risk analysis

The first step in risk analysis is to identify hazards. In Britain the Royal Commission on Environmental Pollution (1991) suggested the GENHAZ procedure based on HAZOP (Hazard and Operational Study), which had proved very successful in identifying possible hazards in the

chemical industry. The main elements of this system are:

INTENTION	What is intended to happen during, or as a result of, the release
DEVIATION	A departure from the intention uncovered by systematically applying guide words
CONSEQUENCE	A result of deviation
CAUSE	Means by which a consequence could occur
ACTION	A step to be taken as a result of identifying a serious consequence with a realistic cause

Thus, as with most mathematical models, the initial step is a word model. The important next step from identifying such hazards is to move towards a risk assessment that can be addressed biometrically.

Several trials involving the contained use of and field release of genetically engineered micro-organisms (GEMs) have now been undertaken by biologists. Mathematical models can help to develop the application of the results to a range of soils, crops and environments, and can also help in the design of better experimental procedures.

Considerable discussion of risk assessment followed the presentation at the workshop by Dr J. Dunster. For example, it was discussed how the risk of contracting cancer as a result of radioactivity in the human population was estimated (Dunster, 1994). The basis of the estimation was a dose–response curve in which the frequency of carcinomas in a population was plotted against dose. The presentation identified clearly a major reservation that concerned the assessment of risk

231

associated with the release of GEMs. That is, unless a specific risk is identified, a quantitative responce (except a possibly trivial one) is not possible. If GEMs are released *something* untoward *will* happen, even if the untoward event is an operator dropping a sack of GEMs on his toes! In the trivial case, then, the probability of risk is unity. Trials and tribulations of risk assessment need to be distinguished.

Trials (Hazards)
1. Ecosystem disruption
2. Harm to non-target organisms
3. Sociological harm

Tribulations (Complications)
1. Dispersal
2. Survival and growth
3. Horizontal gene transfer

Trials, which lead to hazard assessment, need dose–response models, whereas *tribulations*, which involve exposure assessment, should involve transport models (dispersal) and fate models (reproduction and interaction).

If specific risks associated with the release of GEMs are identified, it might be possible to take each of these in turn and associate a probability of occurrence with them. However, it might have to be accepted that statistically-sound trends be established rather than attempting to quantify risks *per se*. In this respect, it might be possible to balance specific risks against specific benefits.

Population dynamics

In order to achieve better models, more work in basic microbial population dynamics and interactions is required. Controlled laboratory experimentation with chemostat cultures, column reactors and microcosms are of particular importance. Also, where possible, such investigations should be undertaken in concert with field studies. Of particular importance are field and laboratory measurements of gene exchange, particularly chromosomal genes. The kinetics of this process were identified as being of fundamental importance for the development of a quantitative predictive model.

As is often emphasized by microbiologists, the soil is a heterogeneous environment; physical, chemical and biological properties are not distributed evenly and may be discontinuous. But even in a well-mixed soil composed of equally-sized particles, biological activity will result in the formation of chemical gradients which, in turn, affect the distribution of microbial cells. In root-free soil these result chiefly from the uni-directional flow of nutrients down the soil column by leaching. Even a simple process such as the application of a fertilizer that contains ammonium salts to the surface of the soil, and its subsequent nitrification after a rainfall event, leads to gradients in the concentrations of ammonium, nitrite and nitrate ions (McLaren, 1969; Saunders and Bazin, 1973). Thus, such systems change with respect to distance down the soil column as well as time and, as they are thermodynamically open (there is an input and output of both matter and energy), they may come to a steady state. When a plant root is present there is also an exchange of materials between this organ and the surrounding soil through the rhizosphere. This imposes an additional change with respect to radial distance from the root.

Two basic modelling approaches have been used to describe solute leaching in the soil. Both are summarized by Addiscott and Wagenet (1985); one describes solute movement in terms of partial differential equations and the other, the 'layer' approach, subdivides the soil column into theoretical compartments, each of which is considered to be well mixed. In the latter case, sets of ordinary rather than partial differential equations can be employed. Both approaches can be used as a basis for the modelling of microbial activity. For example, Hutson *et al.* (1994) adapted a chemical fate model (LEACHB), which numerically solves a set of partial differential equations, to model single-species microbial growth and predator–prey interaction in soil using a programme the authors called LEACHM. Scott *et al.* (1994) employed a simple layer model to describe the growth of a root and its colonization by genetically modified micro-organisms. We have modified this programme in an attempt to model plasmid transfer dynamics in an experimental soil microcosm.

Sun *et al.* (1993) grew donor strains of *Pseudomonas cepacia* and *Enterobacter cloacae*; both contained the transmissible plasmid *R388* : : Tn1721, together with recipient strains in separate continuous-flow packed-bed reactors. The densities of donor, recipient and transconjugant population were estimated in the effluents from the columns. The results obtained are reproduced in Fig. 22.1. In columns containing *E. cloacae*, all three populations survived at steady-state densities. On the other hand, only the recipient strain of *P. cepacia* maintained a stable density while the transconjugants and donors decreased in what appears to be an exponential fashion. We have used the kinetic description of plasmid transfer suggested by Levin *et. al.* (1979) to reproduce these results qualitatively.

Following the methodology described by Scott *et al.* (1994), we consider a column reactor to be subdivided into a series of horizontal, well-mixed theoretical compartments, each receiving effluent from the compartment above and expelling effluent to the compartment below. The specific

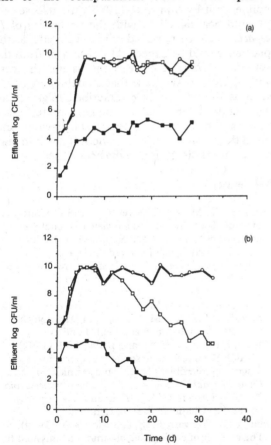

Fig. 22.1 Data of Sun *et al.* (1993) showing the population densities of donor (□), recipient (○) and transconjugant (■) strains in the effluent from continuous packed-bed column reactors containing (a) *E. cloacae* and (b) *P. cepacia*.

growth rate, μ, of each strain is described by the Monod (1942) function. Thus:

$$\mu = \frac{\mu_{mi}S}{K_{si} + S} \tag{22.1}$$

where, S is the concentration of growth-limiting nutrient and the subscript, i, refers to the particular strain represented, i.e. donor (d), recipient (r) or transconjugant (t). Some part of the biomass in each compartment was assumed to be retained while the fraction leaching was assumed to be constant and designated α. The mass action model of Levin *et al.* (1979) describes transconjugant formation by plasmid transfer as:

$$\frac{dT}{dt} = \gamma(D + T)R \tag{22.2}$$

where D, R and T are the densities of the donor, recipient and transconjugant populations and γ is the conjugational transfer rate, assumed to be constant. Thus, for the jth theoretical compartment of a continuous-flow column reactor, the rate of change in population density of the three strains due to growth, leaching and plasmid transfer is:

$$\frac{dD_j}{dt} = \mu_{dj}D_j + \frac{F}{V}(D_{j-1} - D_j)\,\alpha \tag{22.3}$$

$$\frac{dR_j}{dt} = \mu_{rj}R_j + \frac{F}{V}(R_{j-1} - R_j)\,\alpha - \gamma(D_j + T_j)R_j \tag{22.4}$$

$$\frac{dT_j}{dt} = \mu_{tj}T_j + \frac{F}{V}(T_{j-1} - T_j)\,\alpha - \gamma(D_j + T_j)R_j \tag{22.5}$$

where F is the leaching rate through the system and V is the volume of each theoretical compartment.

This model was not fitted quantitatively to the data of Sun *et al.* (1993) but it has successfully differentiated qualitatively between the behaviour of *E. cloacae* and *P. cepacia* when different parameter values were used. A theoretical construct was employed that divided the column into ten compartments, each with an arbitrary volume of unity and solved equations 22.3 to 22.5 numerically for

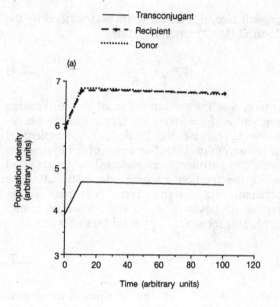

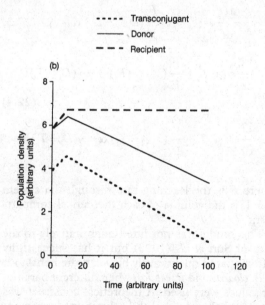

Fig. 22.2 Qualitative simulation of the experimental results shown in Fig. 22.1: (a) *E. cloacae* and (b) *P. cepacia*.

each of the compartments using a Runge–Kutta routine. Arbitrary values for the parameters were employed and a range of values tested. The simulated results, shown in Fig. 22.2, were particularly sensitive to the value assigned to α, the fraction of cells leaving each compartment, and the differing responses of *P. cepacia* and *E. cloacae* could be mimicked by manipulating just this parameter. The system was also sensitive to the maximum specific growth rate: saturation-constant ratios but this was expected as this quantity determines how quickly a population grows and utilizes nutrient.

The results reported above should be regarded as preliminary but they do represent one approach to the mathematical modelling of gene exchange in the soil that might be of value for understanding and predicting the dynamics of the process. As pointed out by Sun *et al.* (1993), the release of a plasmid-bearing GEM with the properties of *P. cepacia* might be regarded as relatively safe, as the plasmid would be expected to disappear from the system naturally. However, this is not the case with *E. cloacae*. We hope that modelling exercises such as the one we have described might be of value in determining what ecological strategies are necessary to ensure *P. cepacia*-type behaviour, and avoid the *E. cloacae*-type behaviour if it would lead to long-term disruption to the ecosystem.

References

Addiscott, T. M. and Wagenet R. J. (1985). Concepts of solute leaching in soils: a review of modelling approaches. *Journal of Soil Science* **36**, 411–424.

Bazin, M. J. and Lynch J. M. (eds.) (1994). *Environmental Gene Release: Models, Experiments and Risk Assessment*. Chapman & Hall, London.

Dunster, H. J. (1994). Risk assessment. In *Environmental Gene Release: Models, Experiments and Risk Assessment* (ed. M. J. Bazin and J. M. Lynch), pp. 139–148. Chapman and Hall, London.

Hutson, J. L., Scott, E. M. and Bazin, M. J. (1994). LEACHB – a root-zone solute transport model describing microbial population dynamics. In *Environmental Gene Release: Models, Experiments and Risk Assessment* (ed. M. J. Bazin and J. M. Lynch), pp. 13–28. Chapman & Hall, London.

Levin, B. R., Stewart, F. M. and Rice, V.A. (1979). The kinetics of conjugative plasmid transmission: fit of a simple mass action mode. *Plasmid* **2**, 247–260.

McLaren, A. D. (1969). Nitrification in soil: systems approaching steady state: *Proceedings of the Soil Society of America* **33**, 551–556.

Monod, J. (1942). *Recherches sûr la Croissance Bacteriennes*. Herman et Cie, Paris.

OECD (1986). *Recombinant DNA Safety Considerations*. OECD, Paris.

OECD (1990). *Good Developmental Practices for Small Scale Field Research with Genetically Modified Plants and Micro-organisms*. OECD, Paris.

Royal Commission on Environmental Pollution (1991). *GENHAZ. A System for the Critical Appraisal of Proposals to Release Genetically Modified Organisms into the Environment*. Fourteenth Report Cm 1557. HMSO, London.

Saunders, P. T. and Bazin, M. J. (1973). Non-steady studies of nitrification in soil: theoretical considerations. *Soil Biology and Biochemistry* **5**, 545–557.

Scott, E. M., Rattray, E. A. S., Prosser, J. I., Glover, L. A., Lynch, J. M. and Bazin, M. J. (1994). A mathematical model for dispersal of bacterial inoculants colonising the wheat rhizosphere. *Soil Biology and Biochemistry* (in the press).

Sun, L., Bazin, M. J. and Lynch, J. M. (1993). Plasmid dynamics in a model soil column. *Molecular Ecology* **2**, 9–15.

23

Pest resistance to *Bacillus thuringiensis*: ecological crop assessment for *Bt* gene incorporation and strategies of management

C. Howard Wearing and Heikki M. T. Hokkanen

Introduction

While integrated pest management research has sought to reduce the use of pesticides in agriculture over the past 30 years, current public and political pressure has increased dramatically the urgency of the search for alternatives. Few are immediately available, and this places particular reliance on some key biological control agents. One such agent is *Bacillus thuringiensis* Berliner (*Bt*), which is the most widely used biopesticide. Sales have doubled in the past 4 years and are conservatively forecast to rise by 20% per year in the future, leading to a market of US$ 300–500 million by 1995 or 5–10% of global insecticide sales by the year 2000 (Anonymous, 1991*a,b*). In addition to the expanding use of *Bt* as a microbial pesticide, the *Bt* genes responsible for its efficacy as a pesticide have become a primary source of insect toxins for those who seek to produce transgenic crops resistant to insects. It was reported by Watkinson (1992) that at least 50 plant species have been transformed with *Bt* genes and the list is being constantly added to. This brings closer the prospect of *Bt* toxins becoming widely and persistently distributed in the environment.

There is a growing criticism and concern about this genetic approach to pest control (Harris, 1991; van Emden, 1991; Anderson, 1992).

Insects have exceeded most projections of their ability to develop resistance to pest control methods applied against them, and this now includes *Bt*. Despite the complexity and polygenic basis of the mode of action of *Bt* when used as a microbial pesticide, both diamondback moth, *Plutella xylostella* (L.) (Tabashnik *et al.*, 1990; Shelton *et al.*, 1993) and Indian meal moth, *Plodia interpunctella* (Hübner) (McGaughey, 1985), have developed resistance following repeated exposure to sprays in the 'field' (Fig. 23.1). A similar recessive genetic mechanism appears to be involved in both species, controlled by one or few genes (Tabashnik *et al.*, 1992*a*; see also Van Rie, 1991). The presence of *Bt* genes in crops (*Bt*-crops) adds a new dimension to the potential for *Bt* resistance and early indications of rapid resistance development have been reported (van Emden, 1991). McGaughey and Whalon (1992) have provided a comprehensive review of the potential for insect resistance to *Bt* and resistance has also featured strongly in recent workshops on the development of *Bt* (e.g. Milner and Chandler, 1992). Many pest control specialists share a common concern that great care should be exercised in the deployment of *Bt* genes in crops, notably the δ-endotoxins, to ensure the continued susceptibility of the target insects (Tabashnik *et al.*, 1991; Llewellyn *et al.*, 1992; McGaughey and Whalon, 1992; Wigley

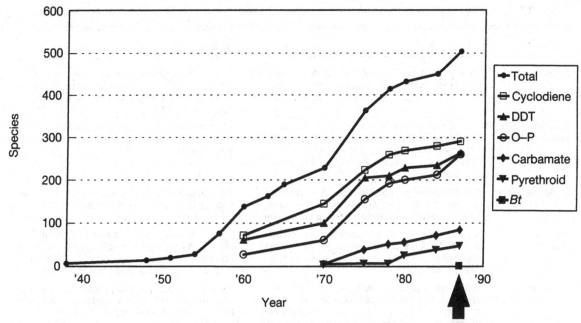

Fig. 23.1 Timing of occurrence of *Bt* resistance (arrow) in the chronological sequence of the number of insect and mite species resistant to at least one type of insecticide (total), and species resistant to each of the five principal classes of insecticides (after Georghiou, 1990).

and Chilcott, 1992). In the past, plant breeders have encountered similar problems in the management of resistant germplasm, such as when using cultivars that carry single-gene resistances to the Hessian fly (Gould, 1986). Their strategy was also to retain susceptibility of the insects if possible, though this was not always successful (Gould, 1986, 1990), rather than face the more difficult problem of the management of insect resistance after it has arisen, as has so often occurred with insecticides (Hoy, 1992).

Insect resistance

Based on past experience with insecticides, there are four key factors that govern selection for resistance: selection intensity; degree of dominance of resistance; size of refugia; and gene flow (e.g. Caprio and Tabashnik, 1992a). In this context, it is useful to compare and contrast the features of insecticides and transgenic *Bt* toxins that would influence resistance development (Table 23.1).

Bt toxins used alone appear especially vulnerable to resistance development by the target insects. Their persistence is a special problem and

it is important that research is intensified to develop and deploy tissue- and time-specific promoters of *Bt* gene expression (see also Harris, 1991; Van Rie, 1991; Llewellyn *et al.*, 1992). *Bt* toxins share with insecticides the major drawback that resistance requires the selection of only single genes. If *Bt*-crops are to be durable, methods for the inclusion of multiple *Bt* and other resistance genes in plants are essential (Brattsten, 1991). In the long-term, the potential flow of *Bt* genes into wild and feral host plants is especially disturbing for future resistance development in insects.

These problems with *Bt* toxins are balanced in part by some positive features. In particular, the initial limitation of the *Bt* toxin to the transgenic crop is especially important in allowing the survival of susceptible insects in refugia. The species and stage selectivity of *Bt* toxins should also assist in delaying resistance by the restriction of selection pressure and by encouraging the survival of natural enemies. Tissue- and time-specific promoters appear to be a promising way of controlling the persistence of the *Bt* toxin dose in the future. However, McGaughey and Whalon (1992) point out that promoters may not prevent

Table 23.1. Comparative features of insecticides and transgenic *Bt* toxins that contribute to resistance development

Feature	Insecticide	*Bt* toxins
Persistence	Very variable, but reduced in most new products. Not as persistent as potential *Bt* gene expression, at least not since organochlorines were withdrawn	Potentially continuous, including year-round and through the life of annual, biennial or perennial crops. Could be reduced by tissue- and time-specific promoters
Species selectivity	Low to high and variable with use, including variable safety to natural enemies. Traditionally selectivity is low but has been enhanced by improved screening and by formulation, timing and placement	High, including safety to natural enemies. Could be further enhanced by tissue- and time-specific promoters
Stage specificity	Variable, depending on the product, formulation and application. Insecticides may be toxic to several stages, including adults, eggs and larvae	Toxic to plant-feeding stages, usually larvae, and especially young larvae (dose-related). Not toxic to many adult insects
Distribution in/on plant	Very variable, including inability to reach some sites and tissues. Dependent on formulation and application methods, and highly dependent on external conditions	Potentially very consistent and throughout the plant, including parts very difficult to reach with conventional chemical applications. Could be restricted with tissue- and time-specific promoters
Distribution in crop ecosystem	With most application methods, there is widespread distribution of insecticide on the target crop, on non-target plants (e.g. weeds and other alternative host plants of the pests), and on other parts of the crop ecosystem	Initially confined to the tissues of the transgenic crop. Not present in other plants, except longer term through spread in pollen and seed
Distribution beyond the crop ecosystem (spread/drift)	Initial distribution controlled during application but dependent on application method and condition (e.g. wind). Drift to non-target areas common, including alternative host plants of pests and toxicity to non-target species	Initial distribution strongly restricted to the transgenic crop(s). Thereafter, spread primarily through seed and pollen, with potential for out-crossing with related plants. In the long-term, very widespread distribution is possible, including exposure to non-target segments of the pest populations and non-target species
Dose	Low to high at time of application depending on the rate selected and the efficiency of the applicator. Thereafter, often very variable and unknown doses depending on biotic and abiotic factors, formulation etc.	Low to high depending on the manipulation of gene expression, but normal distribution of dose can be expected in different tissues. Could be further controlled with tissue- and time-specific promoters
Genetic basis	Single toxins. Insect resistance is frequently monogenic, but may be more complex	Single gene, but with potential for deployment of multiple genes in transgenic crops
Compatibility with IPM	Low to high depending on selectivity (see above)	High but not universal (e.g. mating disruption – see text)

some lower expression of *Bt* toxin in other tissues and could potentially add to resistance risk.

These features of *Bt*-crops give a guide to the operational procedures available for resistance management as described by McGaughey and Whalon (1992) and summarized in Table 23.2.

Strategies to retain susceptibility to *Bt* genes are already being proposed before *Bt*-crops become available (McGaughey and Whalon, 1992; Wigley and Chilcott, 1992). Many of these strategies are based largely on the existing methods of insecticide resistance management, and Gould (1988) and McGaughey and Whalon (1992) have also drawn on the experience of conventional plant breeders when confronted with new insect biotypes (see also Brattsten, 1991). Wigley and Chilcott (1992) focused on the fundamental decision to incorporate *Bt* genes into plants, particularly in relation to polyphagous (cosmopolitan) pests. They advocated procedures to minimize the risk

Table 23.2. Operational procedures available for the management of resistance to *Bt* in transgenic crops

1. *Retention of susceptible insects for dilution of selection for resistance*
   1.1. Refugia – Non-transgenic (susceptible) crop cultivars within the *Bt*-crop
          – Other non-transgenic (susceptible) host plants of crop pests (e.g. weeds, ground cover plants, shelter trees) within and around the *Bt*-crop
          – Use of tissue- and time-specific promoters to leave non-essential crop tissues unprotected with *Bt* toxin
   1.2. Immigration
          – Distribution and density of non-transformed crops and alternative host plants of the crop pests, in relation to the distribution and density of *Bt*-crops
          – Release of susceptible insects

2. *Dose and its persistence and distribution*
   2.1. Dose control through *Bt* gene expression
          – Low dose option
          – High dose option
          – Ultra high dose option
   2.2. Duration and distribution of dose
          – Tissue-specific promoters (e.g. buds, fruit, seeds)
          – Time specific promoters (e.g. fruit ripening, winter buds)
          – Wound-specific promoters
          – Elicitor-specific promoters

3. *Gene management*
   3.1. Gene options
          – Single *Bt* gene/multiple *Bt* genes
          – Different *Bt* genes in different crops in relation to target pest and to *Bt* use as a microbial pesticide (see Wigley and Chilcott 1992)
          – *Bt* genes combined with other genes
   3.2. Mixtures
          – Pyramids
          – Multilines and mosaics
   3.3. Rotations or alternations

4. *Use in integrated pest management programmes*
   Numerous options exist for the combination of *Bt*-crops with other pest management procedures, such as the use of natural enemies

---

of cross-resistance in polyphagous pests that attack several *Bt*-crops. Because there is strong evidence that differences in *Bt* toxin activity and resistance are linked to the affinity of toxins for binding to specific sites in the insect gut (Van Rie, 1991), they proposed that genes with different binding sites be deployed in different hosts of the polyphagous pests. Similarly, they proposed that CryIA proteins, which are the principal toxins in *Bt* bioinsecticides, should not be used to transform plants where the risk of cross resistance is present.

The special concern of these authors with polyphagous pests appears well justified. Georghiou and Taylor (1986) considered that polyphagous pests have developed resistance to insecticides more slowly than monophagous pests because polyphagy permits greater reservoirs of untreated host plants. If this advantage is nullified by the incorporation of *Bt* genes in many hosts, the greater biochemical ability of polyphagous pests to develop resistance may be given opportunity to develop to its full potential.

Wigley and Chilcott (1992) proposed that crops and their pest complex be assessed for their suitability for *Bt* gene incorporation *before* the genes are deployed, with special regard to the risk of resistance in polyphagous pests that may attack several *Bt*-crops. In New Zealand for example, lightbrown apple moth (*Epiphyas postvittana* (Walker)) feeds on apple, kiwifruit, and white clover, all of which are current targets for *Bt* gene incorporation. This same pest attacks lucerne, lupin, citrus, grape, poplar, willow, *Pinus radiata*, stone fruits (various *Prunus* spp.), berryfruits (various *Rubus* and *Ribes* spp.) and a number of other crops. Some of these are already being discussed for transformation with *Bt*, and others may become candidates.

It is this aspect of the potential for resistance to *Bt* toxins to develop in insects that we focus on in this paper, with special attention being given to the biological/ecological factors (see Georghiou and Taylor, 1986) that influence it. Our objective is to provide procedures that could be followed to determine the suitability of a crop for *Bt* gene incorporation and, using case studies, show how these procedures could assist in the derivation of ecological operational strategies to retain *Bt* susceptibility in the pest complex. The whole pest complex of the crop needs to be addressed, this includes both polyphagous and oligophagous species, because each presents unique problems. The strategies developed could be combined with those based on dose and genetic considerations (Table 23.2) to provide a resistance management programme for the crop. This approach is less ambitious but similar to that proposed by Gould

(1988), to conduct life table and behavioural studies of pests to help design genetically modified crops.

Bt-crops are a new phenomenon in pest control, which raise new questions about the distribution and ecology of pest species. However, we agree with Roush (1989) that valuable recommendations to delay resistance development in insects can be made with a modest amount of information about the pesticides (*Bt* toxins in this case) used and the existing knowledge of population dynamics (and behaviour) of each pest (in each *Bt*-crop). This empirical process also assists in the assessment of both the suitability of a crop for *Bt* gene incorporation and the research needed to improve this assessment.

Principles of crop assessment for *Bt* gene incorporation

To assess the suitability of a crop for *Bt* gene incorporation, it is necessary to analyse the crop and associated pest ecosystem in relation to the primary determinants of the evolution of resistance viz. selection intensity, degree of dominance of the resistance, size of refugia, and gene flow. The degree of dominance that may arise is unknown, but selection intensity, refugia and gene flow can be modified through operational procedures (Table 23.2). We agree with many authors (Georghiou and Taylor, 1986; Roush and Croft, 1986; Gould, 1988, 1990; Roush, 1989; and see review of McGaughey and Whalon, 1992) that refugia and the immigration of susceptible insects are critical in the delay of resistance development. Reduced selection intensity and the provision or availability of susceptible insects are closely linked. We therefore place special emphasis in this assessment on low selection intensity, this includes the identification of sources of susceptible insects that can mate successfully with those surviving on the *Bt*-crop.

The assessment process has three parts: assessment of the crop itself for features that could impact on selection for resistance in its pest complex; assessment of each pest in the complex for its propensity to develop resistance to *Bt* toxins, including its host range which may include other *Bt*-crops; and finally, assessment of the combined crop and pest data in relation to the

specific features of transgenic crops (Table 23.1) and the operational management procedures available (Table 23.2). This last stage aims to provide guidance for the decision on whether *Bt* genes should be deployed in the crop, and if so, strategies to retain pest susceptibility. The key elements in the crop decision and strategy development are:

1. Pest host range and distribution in relation to the *Bt*-crop.
2. Pest mobility in relation to the *Bt*-crop (emigration and immigration).
3. Availability of refugia and potential for crop manipulation to create them.
4. The potential for the use of tissue- and time-specific promoters of *Bt* gene expression.
5. Natural enemies of the pests and compatibility of the *Bt*-crop with them and with other components of IPM programmmes.

The decision to transform a crop with *Bt* genes and the strategies of resistance management will be specific to each crop and location (e.g. country), and will be illustrated by the case studies. However, the principles of the crop and pest stages of the assessment will first be described.

The crop

Selection intensity and refugia
Because *Bt* genes are present in the crop tissues, various aspects of crop production will have a major impact on selection intensity.

Crop duration
The duration of the crop (annual, perennial) and the cycles of crop plantings (e.g. series of sowings of the same crop in different areas of a farm) will contribute in the determination of the persistence of the *Bt* toxin in the ecosystem of the pests. These factors will require evaluation in relation to the life cycle and mobility of each pest susceptible to *Bt*.

Crop rotation
The impact of crop rotation on selection intensity is similar. However, it is especially important to identify other crops or potential crops in the rotation that could contain the *Bt* gene(s), especially those that share polyphagous *Bt*-susceptible pests. The scale and frequency of rotation will also be important in relation to the life cycle and mobility of each *Bt*-susceptible pest.

Crop area

The hectarage and distribution of a *Bt*-crop may be critical in the determination of selection intensity when compared with

1. the availability of other non-transformed host plants, and
2. the hectarage and distribution of other *Bt*-crops which share the same polyphagous pests.

Size of crop unit

The size of cropping units (e.g. fields, orchards, pasture) and their contiguity is a further component of selection intensity. In particular, it may affect the ability of immigrant susceptible insects to reach all parts of the *Bt*-crop, this depends on the mobility of each pest species. Very large cropping areas and crop units within a farm, compared to pest mobility and gene flow, may be sufficient reason to reject the use of a *Bt*-crop, unless there is a mechanism for the provision of refugia within it.

Crop uniformity

Some crop stands have to be highly uniform with respect to cultivar, because of production and harvesting requirements. In addition, weed control may be critical and eliminate many of the alternative non-transformed host plants of the crop pests. Selection intensity would be high in these conditions, whereas other crops may comfortably accommodate a proportion of *Bt*-free plants. The presence of non-transgenic crop cultivars as a component within the *Bt*-crop can provide a refuge for susceptible insects. Crop management procedures need to be assessed to determine the applicability of this technique in each case. These refugia may function for both susceptible pests and natural enemies.

Other host plants

Other potential sources of reduced selection intensity and refugia in the crop ecosystem should be identified. This could include shelter or windbreak trees, weeds and ground-cover plants that are hosts of polyphagous pests in the crop. This is discussed in detail for specific pests. These refugia may harbour both susceptible pests and natural enemies. Some natural enemies have habitat requirements that hedgerows and shelter provide.

Tissue- and time-specific promoters

The prospects of using these promoters for control of *Bt*-expression in the crop should be investigated at the earliest opportunity to minimize selection intensity and provide refugia. The priority for their use should be determined by the particular tissue distribution characteristics of key pests in the *Bt*-crop. Wound-induced *Bt*-expression could reduce selection pressure if triggered after damage has reached a threshold (Gould, 1988).

Bt as a biopesticide

Bt is already used on many crops as a biopesticide. Even as *Bt*-crops are being introduced, this practice may need to continue in neighbouring areas of the crop or it may become necessary in the future. And even if not used on the crop in question, it may be used on another non-transgenic crop that is host to a common polyphagous pest species.

Gene flow

There are two major aspects of gene flow that affect resistance development to *Bt*-crops, gene flow in the plant community and gene flow in the insect community.

Bt gene flow

Non-transgenic host plants provide the refugia and sources of immigrant susceptible insects that will delay resistance development to *Bt*-crops. It is important, therefore, to identify the potential for spread of the *Bt* gene from the crop to related host plants within and outside the cropping area through pollen transfer and seed dispersal. The extent and significance of this spread will vary greatly. For a pest with a restricted host range on a crop bred from wild relatives, this gene flow could be important in resistance development (see Ellstrand, 1988), especially if the climate is suitable for the rapid spread of seedlings. Potential rapid *Bt* gene flow out of the cropping system to closely related plants may be a reason to reject the development of a *Bt*-crop attacked by oligophagous pests.

Gene flow in insect populations

The crop features that affect gene flow in the insect community will be specific to each pest and will require evaluation in the assessment process (see below). However, crop-related barriers to immigration of susceptible species could include the size of the cropping unit or the presence of shelters and wind-breaks. In contrast, shelter trees may themselves provide a valuable source of susceptible immigrants if they are attacked by polyphagous pests which also feed on the *Bt*-crop.

Pest complex

The pests within the complex attacking the crop will include those susceptible to existing *Bt* toxins (e.g. Lepidoptera), those potentially susceptible to new *Bt* toxins (e.g. Coleoptera), and pests unlikely to be controlled in this way (e.g. Aphidoidea). The last category require assessment only to determine their possible impact on those controlled with the *Bt* gene(s). For instance, *Bt* resistance in a lepidopterous pest may be accelerated by toxicity of an aphicide to its natural enemies. Similar incompatibilities may arise with fungicide and herbicide use, including the destruction of important non-transgenic alternative host plants of the *Bt*-crop pests.

The pests

Many factors in the ecology of pests influence their propensity for development of resistance to insecticides. This has been extensively reviewed and documented (e.g. US National Research Council, 1986) and those pests with special ability to develop resistance have been termed 'resistance recidivists' (Roush, 1989). The two species that have so far developed field resistance to *Bt* as a biopesticide (*P. xylostella*, *P. interpunctella*) fall into this category. However, the historical background of a pest in the development of resistance to insecticides may be only a poor guide to its potential for resistance to *Bt*-crops. Certainly, assumptions that pests are unlikely to develop resistance to *Bt*-crops based on their past performance (with insecticides and/or resistant plants), are unwarranted.

Each pest on a *Bt*-crop needs to be assessed in relation to the selection intensity from feeding on crop tissues that contain the *Bt* gene(s), the potential for refugia, the gene flow between insects in the *Bt*-crop and susceptible insects on other hosts, and its reproductive rate.

Selection intensity and the potential for refugia
Host range

The host range of *Bt*-susceptible insects is of fundamental importance in their selection for resistance to *Bt*-crops. Whereas insecticides may indiscriminately affect insects from any host plant in the ecosystem of the crop (Table 23.1), *Bt*-crops will exert no selection pressure on insects that feed on non-transformed host plants. This greatly increases the potential for refugia of polyphagous pests, which frequently feed on weeds, ground-cover plants or shelter trees. Similar benefits may accrue for oligophagous pests if another untransformed host plant (e.g. a related weed species) is in the crop.

The most critical aspect of host range in relation to selection intensity by *Bt*-crops is the proportions of the pest population that develop on different hosts, both within the crop ecosystem on refugia and on external hosts that can supply susceptible immigrants. While this question has been of some interest for insecticidal control (Roush, 1989) and the development of IPM programmes, it has not been perceived as critical, and there is little published information on this subject for most pests. For resistance management of *Bt*-crops, data on the composition of the pest population in relation to host-plant origin is urgently needed. Only rough approximations are likely to be possible in the present assessment process, through knowledge of the key major host plants of pests and their distribution. It is equally critical to identify the likely contribution of other host *Bt*-crops to the immigrant population of polyphagous pests that arrive in the *Bt*-crop that is under assessment. These pre-adapted migrants could seriously aggravate resistance development.

Situations may occur in which the *Bt*-crop or crops would provide the vast majority of the host plants of some key pests, with little potential for refugia or immigration of susceptible insects (e.g. Colorado potato beetle, *Leptinotarsa decemlineata* (Say), Ferro 1993). *Bt* gene incorporation would appear most unwise in these circumstances, at least until highly effective tissue specific promoters are available.

Distribution of the pest in the crop
In addition to host plant range and proportions, pest assessment must include information on the distribution of each *Bt*-susceptible pest within the

crop plant and between the crop plants in the field. Discontinuous distribution of an insect pest in a *Bt*-crop over space and time provides opportunities for the reduction of selection pressure and encouraging the establishment of refugia. Feeding preferences for particular parts of the plant or periods in the plant development may enable the use of tissue- and time-specific promoters that target the *Bt* toxin to the major part of the population while allowing a proportion to feed on *Bt*-free tissues. This could be especially appropriate where damage to the unprotected tissue results in no economic loss (Llewellyn *et al.*, 1992). Similarly, the *Bt* toxin may be expressed only at a critical time in crop development when maximum economic damage is incurred, such as during seed production, flowering, or the seedling stage. These critical periods should be identified for each pest.

Stages of the insect not affected by *Bt* are usually known. Information on their distribution in the crop ecosystem may assist in resistance management, such as for the development of techniques to improve the mating of adults with susceptible immigrants.

Natural enemies
There are conflicting views on whether natural enemies will speed up or delay pest adaptation to plant resistance (McGaughey and Whalon, 1992). Gould *et al.* (1991) have shown that both may occur and that site- or area-specific research is needed to determine the impact of natural enemies. Nevertheless, their results indicated that, in general, selection for adaptation to a resistant plant type (e.g. *Bt*-crop) is lower at a given level of pest population suppression when that suppression is achieved by the combined action of plant resistance and natural enemies than by strong resistance alone. Therefore, pest assessment should always include identification of the important natural enemies, the stages which they attack, and where known, their role in pest regulation. They may be of sufficient importance that *Bt* gene expression should be maintained at a dose level that permits some natural mortality from parasites and/or predators. The use of partial resistance has been strongly advocated by some authors (van Emden, 1991) to exploit these interactions. If they are to be used successfully, the identification of important natural enemies may also affect the choice of chemicals used to control other pests in the pest complex.

Other control methods
Unilateral control of pests with *Bt*-crops would be repeating the mistakes of the past. Durability of control will be improved by combining *Bt*-crops with other compatible pest management methods. These should be identified for each *Bt*-susceptible pest to facilitate development of an IPM programme. The availability and use of a range of control methods in combination with *Bt*-crops may be critical for the decision to proceed with *Bt* gene incorporation. This approach is already advocated for the use of *Bt*-biopesticides (Shelton *et al.*, 1993).

Gene flow
As with refugia providing susceptible insects, the immigration of susceptible insects into the *Bt*-crop is critical in delaying resistance development. This usually refers to adult insects presumed to be unaffected by the presence of the toxin gene and able to mate with adults that have survived on the *Bt*-crop. Emigration rates will be important in the spread of resistance.

Pest mobility
Information on adult pest mobility and dispersal is needed for both males and females to assess the potential for immigration of susceptibles to contribute to the dilution of resistance and for emigration to contribute to its spread. The host range and distribution of the pest, including its pre-exposure to *Bt* genes in other *Bt*-crops, will be a critical part of this assessment (see above).

Pest density in the Bt-crop
Another essential step in this stage of the assessment is to estimate the likely population density of adults surviving on the *Bt*-crop. This is necessary for evaluation of both the potential dilution of resistance by susceptible immigrants and the possible spread of resistance through emigration. All these factors interact with the crop characteristics identified in the first stage of the crop assessment e.g. area and crop unit size.

Mating behaviour
Mating behaviour may be important in determining the ability of immigrants to mate with adults

that have developed on the *Bt*-crop. The timing and frequency of mating by both males and females could also be important and these features should be considered for each pest, if known.

Actual measurements of gene flow in pest populations are not undertaken frequently but where this information is available, it will greatly assist with the assessment of the likely impact of immigration and emigration on resistance development. Caprio and Tabashnik (1992*a*) reported that measurements of gene flow between insect populations have seldom been as high as 10% (e.g. Pashley *et al.*, 1985); in their own research on diamondback moth, they estimated that the 5–10 individuals per generation that exchange between fields was well below 2% of the field populations and unlikely to retard resistance development. However, in a more recent paper (Caprio and Tabashnik, 1992*b*), allozyme studies indicate high gene flow in this species. They conclude that local variation in insecticide resistance among Hawaiian diamondback moth populations is not an indication of restricted gene flow and probably results from local variation in selection.

Models of insecticide resistance have generally indicated that gene flow of susceptible insects retards the rate of resistance development by pests in the treated crop (e.g. Comins 1977*a,b*). However, Caprio and Tabashnik (1992*a*) concluded that gene flow can accelerate adaptation among finite populations, such as through the spread of advantageous rare alleles. A different assessment of the benefits/risks of gene flow would be required if a crop is considered to contain finite populations.

Where gene flow is known to be low, this may be sufficient reason not to deploy *Bt*-crops against a specific pest. Sufficient refugia would need to be assured for this situation.

Rate of reproduction

The rate at which pests develop resistance to insecticides has previously been shown to be positively correlated with their numbers of generations per year (Tabashnik and Croft, 1985; Georghiou and Taylor, 1986). However, recent analyses challenge this conclusion (Tabashnik *et al.*, 1992*b*). Georghiou and Taylor (1986) also speculated that higher fertility should be associated with higher risk of resistance. The role of reproductive rate in resistance development is therefore uncer-

tain. Nevertheless, it is considered advisable to assess the rate of reproduction of each *Bt*-susceptible pest in terms of generation turnover and fertility. Present evidence suggests that it would be unwise to use *Bt*-crops to control pests of high reproductive rate unless other aspects of their ecology compensate for the associated risks.

Case studies

These assessment principles were applied to three crops in specific geographical locations to determine their suitability for *Bt* gene incorporation. This required detailed analysis of the crops and their pests (Hokkanen and Wearing, 1995; Wearing and Hokkanen, 1994) and only a brief summary of the findings can be reported here.

Apples – New Zealand

The development of IPM for apples in New Zealand over the last 30 years has sought to reconcile the needs of exporting damage- and pest-free produce (because of quarantine requirements) with environmental protection and minimum pesticide residues (Wearing *et al.*, 1993). The primary obstacles to more selective and biological control methods are the fruit-feeding tortricid larvae, which are susceptible to *Bt* as a biopesticide but not sufficiently to meet export requirements. *Bt* gene(s) incorporation in existing apple cultivars (*Bt*-apple) is being pursued as a possible solution to this problem within the IPM programme. The key crop and pest features of this case study are presented in Table 23.3.

General strategies for the management of *Bt*-apple were derived from the analysis and included: (i) avoiding the use of pesticides toxic to *Bt*-susceptible codling moths and leafrollers in refugia or arrival by immigration; (ii) avoiding the use of pesticides toxic to natural enemies of these pests; (iii) not using *Bt* genes primarily active in the *Bt*-biopesticide or with potential for cross-resistance; (iv) using mating disruption in combination with *Bt*-apple wherever there is doubt about the ability of susceptible immigrants and refugia to prevent resistance development; and (v) ensuring the expression of *Bt* genes in both the fruit and leaves. Problems are foreseen in confining *Bt*-expression to either fruit or leaves. For

Table 23.3. Apples – New Zealand. Summary of important crop and pest features influencing the decision on *Bt* gene incorporation

The crop	The pests
Market: Export, low pest tolerance **Duration:** Perennial, >20 years **Rotation:** Not applicable **Area:** 11 000 ha, regionally concentrated **Size of crop unit:** 5–40 ha,. contiguous, sometimes up to several kilometres but broken by shelter, pasture and scrub **Crop uniformity:** Mix of cultivars, single rows to >1 ha. Blocks preferred **Other host plants of pests:** Shelter trees and sward plants **Main other potential *Bt*-crops in the environment:** Kiwifruit and grass/clover pasture **Tissue-specific promoters:** Numerous possibilities – flowering, fruits, shoots, leaves, ripening. ***Bt* as a biopesticide:** Used by organic growers only but important potential for all growers ***Bt* gene flow from crop:** Slow by human distribution of fruit and seed, or by pollen crossing with wild apples.	**Key pests for *Bt* genes:** Codling moth, tortricid leafrollers, and noctuids (*Graphania* sp.) **Codling moth:** Oligophagous fruit feeder; main sources of population are neglected apple and walnut trees. No other *Bt* host plants likely. Highly mobile males, and females able to invade orchards in most locations. High ratio of susceptible: *Bt*-selected insects expected in most *Bt*-apple orchards but refugia needed in some locations. Natural enemies limited. Compatible IPM methods available – e.g. mating disruption, granulosis virus. **Leafrollers:** Highly polyphagous fruit and leaf feeders. *Bt*-crops very minor part of host plant range. Numerous host plants in sward and shelter to provide refugia of susceptibles and natural enemies. Highly mobile adults. Great potential for high ratios of susceptible: *Bt*-selected insects due to refugia and immigration. Other potential *Bt* host crops are kiwifruit and clover. Latter is extensive with potential for *Bt* resistance. Many natural enemies and some compatible IPM methods – e.g. mating disruption ***Graphania*:** Highly polyphagous. Apple is a minor temporary host. Main host plants are in pasture. Greatest risk of resistance is in *Bt*-pasture **Other pests, diseases, weeds:** Potential problems from insecticides against scales, aphids, and mealy bugs, which could be toxic to leafroller refugia and natural enemies. Herbicides could also eliminate refugia of leafrollers and noctuids

instance, if the *Bt* toxin is expressed in fruit only, a very high expression level would be needed to control larger larvae, which initially fed on non-expressing leaves but moved later to feed on fruit (e.g. leafrollers). On the other hand, absence of *Bt* toxin in fruit, but expression in leaves could select rapidly for fruit-feeding behaviour in the larvae.

Specific strategies for codling moth, *Cydia pomonella* L., include the retention of infested neglected apple and walnut trees close to orchards, planting a small ratio of *Bt*-free trees in the orchard as refugia, using mating disruption with *Bt*-apple in large or isolated orchards, the introduction of predators of eggs and larvae from North America and Europe, and using codling moth granulosis virus in combination with *Bt*-apple.

Specific strategies for leafrollers include not using shelter tree cultivars that have been transformed with *Bt*, avoiding the planting of *Bt*-clover in orchards or nearby pasture, not transforming white clover with *Bt* genes (which could induce cross-resistance to *Bt* genes in apple), encouraging the establishment and/or survival of non-transformed hosts of leafrollers in and around orchards, and using *Bt*-apple in combination with resistant apple cultivars.

Kiwifruit – New Zealand

Pests of kiwifruit were reviewed by Steven (1990, 1992). As for apples, IPM has had to reconcile the needs of pest-free fruit for export and the environmental and health requirements of minimal pesticide use. The primary obstacles to progress are the armoured scale pests and the leaf- and fruit-feeding tortricids, which include those leaf-rollers that attack apple and also some other species. *Bt* provides effective control of these tortricids (and oecophorids of the genus *Stathmopoda*) when used as a biopesticide on kiwifruit (Sale *et al.*, 1985) but the cost is prohibitive using *Bt* alone compared

Table 23.4. Kiwifruit – New Zealand. Summary of important crop and pest features influencing the decision on *Bt* gene incorporation

The crop	The pests
Market: Export, low pest tolerance **Duration:** Perennial, >20 years **Rotation:** Not applicable **Area:** 17 500 ha, regionally concentrated **Size of crop unit:** 5–60 canopy ha, contiguous up to several kilometres, but comprising fields of 1–2 ha surrounded by shelter **Crop uniformity:** Extreme. One female cultivar 'Hayward', but 1 in 9 plants male (hermaphrodite plant under development) **Other host plants of pests:** Intensive shelter trees and sward plants. Shelter is both a source of refugia and possible barrier to immigration/emigration **Main other potential *Bt*-crops in the environment:** Apple and clover/ryegrass pasture **Tissue- and time-specific promoters:** Many possibilities – flowers, fruits, shoots, leaves, ripening **Bt as a biopesticide:** Used by organic and conventional growers, especially to reduce bee toxicity at flowering and to reduce chemical residues at harvest **Bt gene flow from crop:** Potential extremely limited because the male vines producing pollen need not be transformed with *Bt*	**Key pests for *Bt* genes:** Tortricid leafrollers and oecophorids (*Stathmopoda* spp.) **Leafrollers:** Highly polyphagous fruit and leaf feeders. *Bt*-crops are a very minor part of the host range. Numerous host plants in sward and shelter to provide refugia of susceptibles and natural enemies. Highly mobile adults. Great potential for high ratios of susceptible: *Bt*-selected insects due to refugia and immigration. Other *Bt* host crops are kiwifruit and clover. Latter is extensive with potential for resistance to *Bt*. Many natural enemies and some compatible IPM methods – e.g. mating disruption under development. **Stathmopoda:** Native fruit-feeding larvae which also feed on decaying plant tissues of kiwifruit and many other host plants. Ecology is little understood and mobility unknown, but probably many refugia in the crop ecosystem. **Other pests, diseases, and weeds:** Potential problems from insecticide use against scales and leafhoppers which could be toxic to leafroller and *Stathmopoda* in refugia and to their natural enemies

with conventional insecticides, unless fruit is grown specifically for the organic market. These costs have restricted the use of *Bt* by conventional growers, although there has been a recent increase in use to assist reduction of spray residues. *Bt* gene incorporation is being pursued as a possible solution to this problem within the IPM programme. The key crop and pest features of this case study are presented in Table 23.4.

This analysis provided some general strategies for the deployment of *Bt*-kiwifruit which were similar to those for *Bt*-apple. In addition, it is important that the male kiwifruit vines are not transformed with *Bt* and retain their role as unusual but valuable refugia of susceptible insects and natural enemies. Conversely, some shelter tree species at present used in kiwifruit orchards could be avoided because they are not hosts for leaf-rollers (e.g. *Casuarina* spp., *Bambusa* spp.) and would not therefore provide refugia for susceptible insects. This is the reverse of current insecticide-based policy in which non-host shelter trees are preferred.

One group of pests common to both apple and kiwifruit, the tortricid leafrollers, also attack legumes, this includes clovers in pasture. White clover is a high priority in the research programme to develop *Bt*-crops. White clover/rye-grass pasture dominates the landscape of New Zealand over many hundreds of square kilometres. Although it may take many years for *Bt*-clover to become widespread in the pastures, the potential for development of resistance and emigration of preselected leafrollers to *Bt*-apple and *Bt*-kiwifruit crops is clear. This could also have implications for all horticultural crops on which *Bt* is used as a biopesticide.

Fortunately, the target of transformation of white clover with *Bt* is not leafrollers but primarily the hepialid complex of *Wiseana* spp. Further benefits of *Bt*-clover could be obtained through *Bt* toxicity to several noctuid species and the sod webworms (Pyralidae); two clover casebearer species (*Coleophora* spp.) could be targets for *Bt* genes in white clover seed crops. The key issue here is that all these lepidopterous pests, and especially the hepialids, are in families very different to the Tortricidae. The selection of *Bt* genes for use in white clover should be done with extreme care to avoid the use of genes with potential for cross resistance in leafrollers exposed to *Bt*-apple and *Bt*-kiwifruit, or for cross-resistance to genes used in the biopesticide (see Van Rie, 1991; Wigley and Chilcott, 1992).

The question of whether *Bt* genes should be put into white clover at all, when it has potential for expression over such vast areas and can spread through pollen and seed, should also be subject to full evaluation along the lines described in this chapter.

The decision on *Bt* gene incorporation for New Zealand apple and kiwifruit

Evaluation of the biology and ecology of apple and kiwifruit pests indicates that under a strictly controlled IPM programme, the incorporation of *Bt* genes in these crops is acceptable from the view of *Bt* resistance potential. However, the same genes, or genes with potential for cross resistance in leafrollers, should not be used in white clover; and the same genes, or genes with potential for cross resistance in Lepidoptera, should not be those used in *Bt*-biopesticides on these crops.

The most important features of *Bt* are its human and environmental safety while being selectively toxic to key pests. None of these attributes should be wasted. *Bt*-apple and *Bt*-kiwifruit should never be grown commercially in New Zealand in combination with broad-spectrum pesticides that are toxic to adult Lepidoptera and their natural enemies. Provision of refugia, susceptiable immigrants and natural enemies are the key elements that must be protected. This strategy alone appears likely to ensure the durability of *Bt*-apple and *Bt*-kiwifruit for insect pest control. Tissue-specific promoters may be able to augment this management programme in the future but will be only a supplement to the recommended strategy.

Oilseed crucifers – Canada

Oilseed crucifers ('canola') are grown on over 4 million ha in Canada. One very important pest species, the diamondback moth *Plutella xylostella* is usually susceptible to certain *Bt* toxins and could possibly be controlled by *Bt*-transgenic canola plants. Another important pest, the bertha armyworm *Mamestra configurata*, is likely to be susceptible to the recently discovered strain and

the new toxin (CryIH) of *Bt* (Bart *et al.*, 1993), but not to the other strains. Flea beetles – of equal or greater importance as pests than the two Lepidoptera – are not susceptiable to any known *Bt* toxins and must be controlled by chemical seed treatments. Fortunately, seed treatments appear compatible with the use of *Bt*-crops in controlling pests in most cases. The growing system and pest features for oilseed crucifers in Canada are outlined in Table 23.5. Hokkanen and Wearing (1995) assessed the system in greater detail, and concluded that the control of the two susceptible key pests appears possible through the use of *Bt*-transformed crops, provided that strict resistance management is practised. The resistance evolution in *Plutella xylostella*, however, to conventional *Bt* sprays is disturbing, particularly as at least in some of the reported areas the inflow of susceptible genotypes should have been adequate (wild crucifers and untreated crops within the range of dispersal). It thus appears that we cannot rely on adequate immigration of susceptible individuals for the diamondback moth, but refugia must be provided within the crop. Through the use of seed mixtures, as suggested by Hokkanen and Wearing (1995), the instant, continuous supply of susceptible individuals should take place, possibly prohibiting the evolution of resistance. If necessary, the proportion of non-transformed crop genotype could easily be raised over the suggested 10% of the seed (e.g. up to 20%). Resistance management is crucial if *Bt*-transformed crops are to be used for diamondback moth control in Canada (or elsewhere), because there is a good possibility that some of the annually immigrating moths may arise from areas where *Bt*-crops (e.g. cabbages) are grown. Thus, the moths may already be pre-adapted to the *Bt* toxins.

General discussion and conclusions

The detailed case studies (Hokkanen and Wearing, 1995; Wearing and Hokkanen, 1994) provided a valuable insight into the unique features in each crop and its ecosystem that will affect the management of *Bt* genes in a given location. Geographical and regional differences in crop management and the pest complex are likely to be extremely important. For instance, the large apple orchards of parts of North America may severely

Table 23.5. Rapeseed-canola in Canada. Summary of important crop and pest features influencing the decision on *Bt* gene incorporation (adapted from Hokkanen and Wearing, 1995)

The crop	The pests
Pest tolerance: High – great capacity for crop plant compensation **Duration:** Annual **Rotation:** Break crop, 3–4 year rotation with cereals **Area:** About 4 million ha (1993), concentrated in the Central Plains area **Size of crop unit:** Typically 10–100 ha **Crop uniformity:** Regionally many cultivars of three *Brassica* and one *Sinapis* species are grown; single fields of one cultivar **Other host plants of pests:** Many weed and cultivated crucifers; *Sinapis* and *Thlaspi* species are abundant and provide potentially large refugia for crucifer specialists **Main other potential *Bt* crops in the environment:** Cabbages **Tissue- and time-specific promoters:** Expression only in the leaves is useful but must be continuous ***Bt* as a biopesticide:** Not on canola but is used in cabbages ***Bt* gene flow from the crop:** Crossing with wild relatives shown not to occur, but rapeseed itself may be a weed	**Key pests for *Bt* genes:** Diamondback moth, bertha armyworm **Diamondback moth:** Oligophagous, multivoltine (2–3 in canola), *Brassica* specialist. Populations in Canada originate mainly from cabbage fields in the USA. The moths migrate annually in the spring; many could already carry genes for *Bt* resistance. Most of progeny in Canada presumably die in the winter, but a part may migrate south and survive. Highly mobile. Cabbage also likely to become *Bt* transformed. Refugia management essential, but easy to create through seed mix with nontransformed cultivars. Natural enemies likely to be important if toxic insecticides avoided. Some compatible IPM methods available **Bertha armyworm:** Highly polyphagous, univoltine. Many alternative, nontransformed host plants likely to remain close to transformed crops **Other pests, diseases, weeds:** Seed treatments against flea beetles and *Lygus* bugs appear compatible with the *Bt*-crop. Occasional sprays against aphids or red turnip beetle are a risk to *Bt*-susceptibles in refugia, as well as to natural enemies. Herbicides may eliminate alternative cruciferous weed hosts (effectively controlled in canola, but may remain in other crops and roadsides)

limit immigration of susceptible insects; the many wild fruit trees of Europe may provide good sources of susceptible pest insects for immigration into the smaller European orchards; both areas have different pest complexes with different host ranges and potential for refugia; the tolerance of pests is usually a little higher for domestic than export production and this may permit different strategies of *Bt* gene deployment, such as with tissue-specific promoters and damage thresholds.

The important questions surrounding the decision to produce and release a *Bt*-crop into the environment primarily concern selection intensity and are biological and ecological, not the operational factors that are so much the focus of insecticide resistance management (e.g. application methods, dose, mixtures, rotations). If the biological and ecological characteristics of a crop, its pests, and their management indicate very high probability of rapid resistance development to *Bt* genes, it would be arrogant to assume that our mastery of the conventional operational factors is such that resistance will be prevented. Practical experience with insecticide resistance management shows otherwise.

However, there is much that can be done in analysing and then managing the crop ecosystem to reduce selection intensity and the potential for resistance to *Bt*-crops. This is the priority in determining crop suitability for transformation and the management of pest susceptibility. Manipulation of dose, mixtures, rotations and application often create an illusion of control of resistance while the debate about their real benefits rages on (Mallett, 1993).

A fundamental ecological principle of the management of susceptibility to *Bt*-crops, that has come from this and other analyses, is the need to reduce selection intensity by the protection of susceptible immigrants and refugia. It would be foolish to plant a transgenic crop and then apply a pesticide (e.g. for another pest or weed) that is toxic to the susceptible adult insects or the host plant refugia. Similarly, while mating disruption has assisted management of resistance after the event (Suckling *et al.*, 1990), it may be inappropriate for susceptibility management when mating between susceptible and selected insects is part of the programme. Mating disruption would also minimize the presence of susceptible insects in refugia within the crop. The gene flows arising from high levels of susceptible immigrants and refugia may be adequate in themselves to prevent the development of resistance to *Bt* toxins in transgenic plants, particularly if care is taken to encourage this process. There are many different ways of protecting refugia and susceptible immigrants, as shown by the case studies. In these circumstances, the need for, and cost of, mating disruption may not be justified. On the other hand, mating disruption has the advantage of both reducing the population density in the crop ecosystem and minimizing the risk of selected insects mating with each other; it also probably increases the likelihood that those selected insects that mate, will do so with susceptible insects on the edge of, or outside, the disrupted area. Therefore, mating disruption offers a powerful tool for resistance management, particularly if: (i) there is uncertainty surrounding the ratio of susceptible selected insects; (ii) there is dependence on immigrant susceptible insects to delay resistance rather than refugia; and/or (iii) there is a need to manage insect pests at very low levels in the crop (e.g. export apple crop).

Bt-crops, because of their highly selective toxicity, are also very well suited for the creation of refugia. This could influence the decision on whether to transform a crop with *Bt* because it can be a powerful means of overcoming ecological characteristics of pests, such as poor mobility, which would encourage resistance. Gould (1990) and Roush (1991) have proposed a general strategy of high *Bt* dose and the provision of refugia for resistance management of *Bt*-crops. We would support the latter but are uncertain about the high dose. McGaughey and Whalon (1992) have questioned whether this dose can be expressed so consistently with or without tissue-specific promoters. It is probable that *Bt* genes will have the wide variation in expression which characterizes other plant genes influenced by plant nutrition and health, and other internal and external factors. High dose in *Bt*-crops may be another illusion of operational control. The decision on whether to use *Bt* genes in a crop should not be influenced by the high-dose option whereas refugia will be of benefit whatever the dose.

Roush and Croft (1986) identified the need for increased research on refugia and immigration for insecticide resistance management. This is still true today and includes *Bt*-crops.

A different strategy for the deployment of resistant plants has been proposed by van Emden (1991), who advocated partial resistance to take advantage of interactions with other controls, especially natural enemies. The protection of natural enemies is another ecological priority for the assessment and use of *Bt*-crops which takes advantage of their selectivity. The same procedures proposed to protect refugia should be extended to ensure the survival of natural enemies. Choice of a low dose may not need to be as precise or controlled as for the high-dose option, which must control all heterozygotes to be effective, but the dose strategy would again be lower in the decision hierarchy than refugia or immigration of susceptible insects.

For all these foregoing reasons we oppose the use of *Bt*-crops with conventional broad-spectrum pesticides. This approach would also be a waste of the important environmental and health advantages of *Bt*.

The increase in numbers of crops being transformed with *Bt* is a special concern for polyphagous pests. Assessment of the leafroller complex in New Zealand (Wearing and Hokkanen, 1994) confirms the view of Wigley and Chilcott (1992) that special attention should be devoted to these pests when assessing a crop for transformation. The degree of polyphagy, the proportions of the pest population on different hosts, and pest mobility will all be important in the assessment. These are all areas where information is often lacking and research is urgently required. This problem shows the need for coordination of the production and release of *Bt*-crops rather than the current rush to produce and market them. It should be noted that high mobility and polyphagy are often correlated (Dennill and Moran, 1990) and this may generally assist in a delay to resistance development in these pests. The concept of avoiding resistance development by the use of different *Bt* genes in the different host plants of polyphagous pests has been challenged by recent results with a resistant strain of *Heliothis virescens* (Gould *et al.*, 1992). This strain shows cross-resistance to a range of *Bt* toxins differing in structure and activity.

Gould (1988) suggested that polyphagous pests with many hosts may, under selection pressure from the *Bt*-crop, move increasingly to other host plants. This could be a highly desirable result if the other hosts are weeds or plants without economic or other importance. Removal of toxic insecticides from the crop ecosystem would be important for permitting such a process. Gould (1988) also proposed that a combination of plant resistance and repellency would be especially useful for the management of resistance in polyphagous pests; this may be achievable through the anti-feedant properties of *Bt*.

It seems probable that the first transgenic crops with *Bt* genes will express the *Bt* δ-endotoxin(s) throughout their tissues and throughout the life of the crop. Tissue- and time-specific promoters appear to offer a means of reducing the persistence of toxin expression for the future. The case studies revealed complexities in the use of these promoters (Wearing and Hokkanen, 1994), such as confining *Bt* gene expression to the fruits or to the leaves of apple and kiwifruit. Insect behaviour and its interactions with resistance selection require very careful consideration before using such options. Some crops, such as cotton (Van Rie, 1991; Llewellyn *et al.*, 1992), may be more suited to this approach than apples and kiwifruit in New Zealand. Limiting the damage to non-essential tissues will be assisted by the deployment of *Bt*-crops within IPM programmes in which natural enemies and other compatible control methods can play a role. McGaughey and Whalon (1992) have cautioned against assuming that promoters will enable *Bt* to be confined absolutely to the target tissues. If this is not possible and *Bt* is expressed at variable levels in a variety of other tissues, the use of promoters may be undesirable for resistance development.

Another aspect of the persistence of *Bt* gene expression is its potential for spread from the *Bt*-crops to related plants in the environment. This may happen very slowly for some plants in some locations (e.g. kiwifruit in New Zealand). However, the problem of spread could be serious for resistance development in many pests, particularly oligophagous pests that could then have an increasing source of pre-adapted immigrants entering the *Bt*-crop. Research to develop means of limiting spread is a priority for resistance management as well as other aspects of environmental risk (Ellstrand, 1988; Regal, 1988).

The history of the development of resistance to insecticides contains frequent episodes in which resistance has occured through mis-management, and often over-use of insecticides by a few individuals. This may have contributed to the resistance of diamondback moth to *Bt*-biopesticide in Hawaii (Gibbons, 1991). The application and doses of insecticides have been (and still are) in the hands of numerous different operators with varying skills. The insecticides are only as good as those who use them and the costs of resistance are high (Kazmierczak *et al.*, 1992). The situation of *Bt*-genes is similar but the application and dose of the genes will be determined by those who develop the *Bt*-crops. Now is the time to consider very carefully which crops are appropriate for this technology and which are not; and also to determine how the genes will be presented to the target insects, as mixtures, high doses or low doses. But this is only the start. The management of *Bt*-crops, as with insecticides, will be in the hands of numerous operators with variable skills. The durability of *Bt*-crops will be most dependent on the education of end users about practices which will ensure the continued susceptibility of the target pests – refugia, immigration of susceptible insects, natural enemies. This will only be possible in a well constructed and presented IPM programme.

Acknowledgments

We thank our colleagues Drs Max Suckling, David Steven and Peter Wigley for valuable criticism and correction of an earlier draft, and the Organization for Economic Co-operation and Development (OECD) Cooperative Research Project 'Benefits and Risks of Introducing New Organisms in Agricultural Practice', for a fellowship to the first author to carry out this study.

References

Anderson, C. (1992). Researchers ask for help to save key biopesticide. *Nature* 355, 661.

Anonymous, (1991*a*). *Bt* Report. *Biocontrol News and Information* 12, 236.

Anonymous, (1991*b*). Conference reports. *Bt* '91. *Biocontrol News and Information* 12, 351–352.

Bart, L., Jansens, S., van Audenhove, K., Buysse, L., Decock, C., Piens, C., Saey, B., Seurinck, J. and Peferoen, M. (1993). Novel *Bacillus thuringiensis* insecticidal crystal protein with a superior activity against noctuid larvae. *Society for Invertebrate Pathology XXVI Annual Meeting, Program and Abstracts* Abstract No. 53, p. 48.

Brattsten, L. B. (1991). Bioengineering of crop plants and resistant biotype evolution in insects: counteracting coevolution. *Archives of Insect Biochemistry and Physiology* 17, 253–267.

Caprio, M. A. and Tabashnik, B. E. (1992a). Gene flow accelerates local adaptation among finite populations: simulating the evolution of insecticide resistance. *Journal of Economic Entomology* 85, 611–620.

Caprio, M. A. and Tabashnik, B. E. (1992b). Allozymes used to estimate gene flow among populations of diamondback moth (Lepidoptera: Plutellidae) in Hawaii. *Environmental Entomology* 21, 808–816.

Comins, H. N. (1977a). The development of resistance in the presence of migration. *Journal of Theoretical Biology* 64, 177–197.

Comins, H. N. (1977b). The management of pesticide resistance. *Journal of Theoretical Biology* 65, 399–420.

Dennill, G. B. and Moran, V. C. (1990). The possible role of directed mobility in the recruitment onto novel plant species and host plant specificity of herbivorous insects and mites. *South African Journal of Science* 86, 116–118.

Ellstrand, N. C. (1988). Pollen as a vehicle for the escape of engineered genes? *Trends in Biotechnology* 6, S30–S32.

Ferro, D. N. (1993). Potential for resistance to *Bacillus thuringiensis*: Colorado potato beetle (Coleoptera: Chrysomelidae) – a model system. *American Entomologist*, Spring 1993, 38–44.

Georghiou, G. P. (1990). Overview of insecticide resistance. In *Managing Resistance to Agrochemicals*, Chapter 2 (ed. M. B. Green, H. M. LeBaron and W. K. Moberg). ACS Symposium Series 421, American Chemical Society, Washington, DC.

Georghiou, G. P. and Taylor, C. E. (1986). Factors influencing the evolution of resistance. In *Pesticide Resistance: Strategies and Tactics for Management*, pp. 157–169. National Academic Press, Washington, DC.

Gibbons, A. (1991). Moths take the field against biopesticide. *Science* 254, 646.

Gould, F. (1986). Simulation models for predicting durability of insect-resistant germplasm: Hessian fly (Diptera: Cecidomyiidae)-resistant winter wheat. *Environmental Entomology* 15, 11–23.

Gould, F. (1988). Evolutionary biology and genetically engineered crops. *Bioscience* 38, 26–33.

Gould, F. (1990). Ecological genetics and integrated pest management. In *Agroecology* (ed. C. R. Carroll,

J. H. Vandermeer and P. Rosset), pp. 441–458. McGraw Hill, New York.

Gould, F., Kennedy, G. G. and Johnson, M. T. (1991). Effects of natural enemies on the rate of herbivore adaptation to resistant host plants. *Entomologia Experimentalis et Applicata* 58, 1–14.

Gould, F., Martinez-Ramirez, A., Anderson, A., Ferre, J., Silva, F. J. and Moar, W. J. (1992). Broad-spectrum resistance to *Bacillus thuringiensis* toxins in *Heliothis virescens*. *Proceedings of National Academy of Sciences of the USA* 89, 7986–7990.

Harris, M. K. (1991). *Bt* and pest control. *Science* 253, 1075.

Hokkanen, H. M. T. and Wearing, C. H. (1995). Assessing the risk of pest resistance evolution to *Bacillus thuringiensis* engineered into crop plants: a case study of oilseed rape. *Field Crops Research* (in the press).

Hoy, M. A. W. (1992). Proactive management of pesticide resistance in agricultural pests. *Phytoparasitica* 20, 93–97.

Kazmierczak, R. F. Jr, Norton, G. W., Knight, A. L. and Rajotte, E. G. (1992). Economic effects of resistance and withdrawal of organophosphate pesticides on an apple production system. *Journal of Economic Entomology* 86, 684–696.

Llewellyn, D., Cousins, Y., Mathews, A., Hartweck, L., and Lyon, B. (1992). Expression of *Bacillus thuringiensis* protein genes in transgenic crop plants. *Proceedings of a Workshop on Bacillus thuringiensis*, pp. 69–75. Canberra, Australia, 24–26 September 1991.

McGaughey, W. H. (1985). Insect resistance to the biological insecticide *Bacillus thuringiensis*. *Science* 229, 193–195.

McGaughey, W. H. and Whalon, M. E. (1992). Managing insect resistance to *Bacillus thuringiensis* toxins. *Science* 258, 1451–1455.

Mallett, J. 1993. Possible problems with strategies against insect adaptation to transgenic insecticidal crops. *Resistant Pest Management Newsletter* 5, 24.

Milner, R. and Chandler, C. (eds) (1992). *Proceedings of a Workshop on Bacillus thuringiensis*. CSIRO Division of Entomology, Canberra, Australia, 24–26 September 1991.

Pashley, D. P., Johnson, S. J. and Sparks, A. N. (1985). Genetic population structure of migratory moths: the fall armyworm (Lepidoptera: Noctuidae). *Annals of the Entomological Society of America* 78, 756–762.

Regal, P. J. (1988). The adaptive potential of genetically engineered organisms in nature. *Trends in Biotechnology* 6, S36–S38.

Roush, R. T. (1989). Designing resistance management programs: How can you choose? *Pesticide Science* 26, 423–441.

Roush, R. T. (1991). Strategies to manage resistance to Bt: defense of transformed plants in seed mixtures. *Abstracts of the 1st International Conference on* Bacillus thuringiensis. *St Catherine's College, Oxford, 28–31 July 1991.*

Roush, R. T. and Croft, B. A. (1986). Experimental population genetics and ecological studies of pesticide resistance in insects and mites. In *Pesticide Resistance: Strategies and Tactics for Management*, pp. 257–270. National Academic Press, Washington, DC.

Sale, P. R., Sawden, D. and Steven, D. (1985). Trials with *Bacillus thuringiensis* on kiwifruit 1982–1984. *Proceedings of the New Zealand Weed and Pest Control Society* 38, 162–164.

Shelton, A. M., Robertson, J. L., Tang, J. D., Perez, C., Eigenbrode, S. D., Preisler, H. K., Wilsey, W. T. and Cooley, R. J. (1993). Resistance of diamondback moth (Lepidoptera: Plutellidae) to *Bacillus thuringiensis* subspecies in the field. *Journal of Economic Entomology* 86, 697–705.

Steven, D. (1990). Entomology and kiwifruit. In *Kiwifruit Science and Management* (ed. I. J. Warrington and G. C. Weston), pp. 362–412. New Zealand Society for Horticultural Science. Palmerston North, New Zealand.

Steven, D. (1992). Pest Lepidoptera of kiwifruit in New Zealand. *Acta Horticulturae* 297, 531–536.

Suckling, D. M., Khoo, J. G. I. and Rogers, D. J. (1990). Resistance management of *Epiphyas postvittana* (Lepidoptera: Tortricidae) using mating disruption. *New Zealand Journal of Crop and Horticultural Science* 18, 89–98.

Tabashnik, B. E. and Croft, B. A. (1985). Evolution of pesticide resistance in apple pests and their natural enemies. *Entomophaga* 30, 37–49.

Tabashnik, B. E., Cushing, N. L., Finson, N. and Johnson, M. W. (1990). Field development of resistance to *Bacillus thuringiensis* in diamondback moth (Lepidoptera: Plutellidae). *Journal of Economic Entomology* 83, 1671–1676.

Tabashnik, B. E., Finson, N. and Johnson, M. W. (1991). Managing resistance to *Bacillus thuringiensis*: lessons from the diamondback moth (Lepidoptera: Plutellidae). *Journal of Economic Entomology* 84, 49–55.

Tabashnik, B. E., Schwartz, J. M., Finson, N. and Johnson, M. W. (1992a). Inheritance of resistance to *Bacillus thuringiensis* in diamondback moth (Lepidoptera: Plutellidae). *Journal of Economic Entomology* 85, 1046–1055.

Tabashnik, B. E., Rosenheim, J. A. and Caprio, M. A. (1992b). Do we really understand evolution of pesticide resistance? *Resistant Pest Management Newsletter* 4, 30.

US National Research Council (1986). *Pesticide Resistance: Strategies and Tactics for Management.* National Academy Press, Washington, DC.

Van Emden, H. F. (1991). The role of host plant resistance in insect pest mismanagement. *Bulletin of Entomological Research* 81, 123–126.

Van Rie, J. (1991). Insect control with transgenic plants: resistance proof? *Trends in Biotechnology* 9, 177–179.

Watkinson, I. (1992). Global view of present and future markets for Bt products. In *Proceedings of a Workshop on* Bacillus thuringiensis (ed. R. Milner and C. Chandler), pp. 3–8. Canberra, Australia, 24–26 September 1991.

Wearing, C. H. and Hokkanen, H. M. T. (1995). Pest resistance to *Bacillus thuringiensis*: case studies of ecological crop assessment for Bt gene incorporation and strategies of management. *Biocontrol Science and Technology* 4, in press.

Wearing, C. H., Walker, J. T. S. and Suckling, D. M. (1993). Development of IPM for New Zealand apple production. *Proceedings on the 2nd ISHS Conference Integrated Fruit Production. Acta Horticulturae* 347, 277–284.

Wigley, P. W. and Chilcott, C. N. (1992). Present use of, and problems with, *Bacillus thuringiensis* in New Zealand. *Proceedings of a Workshop on* Bacillus thuringiensis (ed. R. Milner and C. Chandler), pp. 34–37. Canberra, Australia, 24–26 September 1991.

24

An international perspective for the release of genetically engineered organisms for biological control

Max J. Whitten

Introduction

During the 1970s a set of powerful biochemical tools became available for the characterization and manipulation of nucleic acids and proteins. These included the ability to sequence DNA segments rapidly, to excise specific segments, to execute changes to the nucleotide sequences, if and as desired, and then to insert the modified segment elsewhere into the genome. During this period it was also demonstrated that genetic material could be transferred into the genomes of foreign organisms and, under appropriate conditions, be expressed thus leading to synthesis of RNA and protein products. For example 'expression systems' using eukaryote genes coding for proteins such as interferon, insulin, growth regulators, etc. could be constructed in bacteria, yeast, insect or other eukaryote tissue cultures. Expression of these foreign genes in the host represented a valuable source of such products.

In the latter part of the 1970s and for much of the 1980s, recombinant-DNA technology was largely used as a powerful and creative tool in molecular biology and genetic analysis. It was principally directed towards fundamental questions such as the improvement of our understanding of gene structure and function and for unravelling chromosome organization.

The notion that novel genetic combinations could be created with recombinant-DNA (r-DNA) technology, which would have been unlikely or impossible to arise in nature, caused molecular biologists to pause and consider the risks from inadvertent or deliberate construction of novel organisms that might cause environmental or health problems should they enter the environment by accident or design. In response to these concerns, the scientific community in some countries developed guidelines for the physical and biological containment of cells, viruses or organisms that had been constructed using r-DNA technology. Generally these guidelines were voluntary, but in many Western countries, government control over research funds, or 'duty-of-care' considerations by private enterprise, meant that the guidelines effectively had a binding influence on the conduct of research. The research community led the field in these developments. The issues of concern with r-DNA manipulation were first identified at a Cold Spring Harbour meeting in 1973. This was followed by an international meeting at Asilomar in 1975, which called for the US National Institute of Health to develop guidelines for conducting research and to examine possible risks. The NIH established a Recombinant Advisory Committee and published the first guidelines in 1976.

Concern over the consequences of the inadvertent or deliberate release of novel genotypes into the environment initially focused on microorganisms, where the techniques were first developed. However, with the ability to transform or introduce foreign genes into model eukaryotes such as tobacco, *Drosophila melanogaster* and mouse, the objectives of the research and the nature of the concern shifted. New phenotypes, not easily achievable by more conventional genetic methods, could be constructed. In many instances, it was envisaged that these novel forms might improve plant or animal performance or provide resistance to pests and diseases. It was, therefore, envisaged that such transformed organisms would eventually take their place and survive in agriculture or the natural environment. Indeed, genetically engineered pest insects such as the sheep blowfly (*Lucilia cuprina*), mosquito species, and a wide range of potential natural enemies, from viruses, bacteria, fungi, nematodes through to insect parasitoids and predators have been viewed as possible early candidates for genetic manipulation with the new technologies (Whitten, 1991; Oakeshott and Whitten, 1993).

Although r-DNA technology was widely applied to both prokaryotes and eukaryotes during the 1980s, two factors tended to reduce scientific and eventually community concern over the perceived dangers of genetic engineering. First, a lengthening list of transformed plants, animals and microbes had been constructed and studied under contained conditions without misadventure; but, as I suggested earlier, the object of these manipulations was the study of fundamental questions rather than some exercise in phenotypic manipulation *per se*. Accordingly, these transformed organisms were not destined for deliberate release into the environment but were investigated under physical containment. Often the researcher ensured, as an added precaution, some measure of biological containment. This involved inclusion of genes that debilitated the phenotype so that sustained survival under field conditions was unlikely or not possible. Thus, the second factor that caused a lull in the debate about genetic engineering was the effective biological, as well as physical, containment of most genetically engineered or transformed organisms during the 1980s.

So far some 600 planned releases of genetically engineered plants have taken place world-wide in supervised field trials without misadventure. This experience has given added cause for optimism. Similarly, the experience with engineered human and animal vaccines (dead and alive) and microbes as bio-reactors for human and veterinary pharmaceuticals, has shifted the pendulum, perhaps temporarily, away from the risk and towards the benefits side of the debate.

During the late 1980s a significant change in research emphasis took place. Biologists strengthened efforts to use the tools of r-DNA technology to modify the phenotypes of a whole range of plants, animals and microbes so that the novel organisms could provide some economic, social or environmental advantage to humans subsequent to their deliberate release into the environment, or at least after their use in non-containment situations. The phenotypic modifications of special interest to the present debate would include herbicide resistance, and host resistance to pests and diseases in crop, pasture and forest plants; host resistance to pests and diseases in domestic animals; and genetic engineering of a wide range of organisms (bacteria, fungi, viruses, nematodes, arthropods, etc.) that can be used as natural enemies for the control of weeds, pests and diseases; and finally, veterinary and human vaccines (dead and alive), microbes in bioremediation and as bio-reactors for veterinary and human pharmaceuticals.

Examples to illustrate applications of genetic engineering for such purposes are legion, but let me refer to some that are a reality (or very nearly so), which more closely pertain to biological control. A commercial product of a genetically engineered *Agrobacterium radiobacter* which can be used to 'vaccinate' stone fruit species against crown gall (Jones *et al.*, 1988; Jones and Kerr, 1989) caused by a pathogenic form of *A. radiobacter*, has been available in New South Wales since December 1988 and elsewhere in Australia since 1991 under the trade name 'NoGall'. To my knowledge at the time of writing, this remains the only instance of a transformed organism commercially available for deliberate release into the environment, anywhere in the world; but certainly that situation will quickly change. A similar type of plant protection, which has now become a 'cause celebre' in modern agriculture, rekindling much debate on genetic engineering, has been the construction and experimental release of the 'iceminus' strains of *Pseudomonas syringae*. This type of biological control relies on ecological displacement of the natural pathogenic strain of *P. syringae*, which causes frost damage to certain crops with the engineered ice-minus strain that lacks the genetic information for initiating ice nucleation (see Pimentel, Chapter 2). The status of much of genetic engineering technology concerned with the release of genetically manipulated organisms (GMOs) in North America has been influenced by this particular case. This point is well illustrated elsewhere in this volume.

Some dozens of crop plants have been transformed to express a toxin coding gene from *Bacillus thuringiensis*, and some of these have been field tested, though none are yet commercially available (Feitelson *et al.*, 1992). Other plants have been engineered with genes coding for viral proteins, antisense RNA or ribozyme activity, designed to protect the host against viral diseases (Michael and Wilson, 1992). Similarly, herbicide resistance genes have been incorporated into a wide range of crop plants, and currently are in various stages of evaluation (Mullineaux, 1992). Moving to biological control agents, the actual engineering of natural enemies can encompass changes to either host specificity or efficacy. For example *Bt* strains have been modified to alter specificity while entomopox viruses (Christian *et al.*, 1992) and baculoviruses (Vlak, 1992; and see Crook and Winstanley, Chapter 21) are being manipulated to enhance virulence or reduce the time interval between infection and death of the host. Perhaps a more obvious example is the transformation of pseudomonads with the gene for *Bt* toxin (Feitelson *et al.*, 1992). Thus, genetic engineering of a wide range of organisms promises to provide important alternative options to biological control for plant and animal protection, or to improve the efficacy of biological control agents.

The links between recombinant-DNA technology and biological control

It is useful to explore the ramifications for conventional biological control now that the release of genetically engineered pathogens and other natural enemies comes close to reality. This linkage is well illustrated by reference to the commercialization and registration of the genetically engineered

versions of NoGall which was surrounded by controversy. The case featured prominently in the parliamentary enquiry into genetic engineering by the Australian Government in 1992 (Report of the House of Representatives Standing Committee on Industry, Science and Technology: Anonymous, 1992). NoGall was first registered in New South Wales in 1976 and its active ingredient was a naturally occurring strain, K84, of *A. radiobacter*. However, K84 which acted as a 'natural enemy' to the pathogenic strain *A. r. var. tumefaciens*, had the annoying capacity to transfer genetic material to *A.r. var. tumefaciens*. This transfer conferred immunity to the pathogen, thereby destroying the effectiveness of the natural enemy. Kerr and colleagues at the Waite Research Institute used elegant genetic engineering tools to remove a specific gene from the K84 strain. The resulting K1026 strain was equally effective as K84 as a biological agent but it had lost the capacity to transfer immunity to the pathogenic *A.r. var. tumefaciens*. When K84 was initially registered in 1976 as a pesticide, it was regarded as a biological control agent. With the development of a genetically improved strain K1026, it was described during the 1988 NSW registration process as a 'mutant strain' or 'the more modern K1026 strain' and its registration was expedited in NSW because the NSW Department of Agriculture did not 'require data on pesticides which are the same as, or very similar to, products which are already registered' (Anonymous, 1992). However, it was a further 3 years before NoGall received federal registration and became available elsewhere in Australia.

It is interesting to read how the Australian parliamentary committee grappled with the evidence and views from the molecular biologists, the regulatory authorities, and the public interests which were represented by the Australian Conservation Foundation and other researchers concerned with the release of NoGall and other GMOs (Anonymous, 1992). Pertinent conclusions drawn by the enquiry included:

- NoGall strain K1026 was safe, despite being a GMO;
- registration procedures and criteria for chemicals may be inappropriate for living organisms;
- the fast-tracking of NSW registration was implicated in the three year process of federal clearance for other States which the Committee regarded as grossly excessive;

- the value of voluntary guidelines were questioned in the face of 'the commercial imperative'.

An underlying concern throughout the enquiry was the perceived lack of vigorous assessment of conventional biological control agents. One witness stated:

'Another question which we have debated within CSIRO is whether, following the British model, it is appropriate to bring under the umbrella of an authority like the ACRE (Advisory Committee for Releases into the Environment) committee a much broader range of things, perhaps including biological control agents. Scientific logic certainly says that you should, but I have heard other arguments which say that would make the process too unwieldy.'

On this question the Committee concluded:

'The Committee has not examined the production of organisms by methods other than genetic manipulation or whether there may be special dangers involved in their release. The Quarantine Service already provides a mechanism for examining the risks of importing exotic organisms. The Government should consider whether the production within Australia of organisms by means other than genetic manipulation and the release of such organisms requires special clearance procedures.'

In Australia the Biological Control Act was enacted in 1985 to provide the basis for an orderly debate over the introduction and release of conventional biological control agents (see Cullen and Whitten, Chapter 26). This Act was without precedent elsewhere in the world. Biological control programmes during the 1980s in Australia had reached an impasse because common law made it possible for a minority group to secure an injunction to halt the release of natural enemies destined to serve a greater public benefit. Such an injunction was granted against the biological control of Paterson's Curse because beekeepers and some graziers considered the species to be a useful nectar source and fodder plant. The broad basis of the Biological Control Act could allow for consideration of GMOs as natural enemies. However, it is feared that such a broadening of the usage of the Biological Control Act to include GMOs as biocontrol agents could attract close scrutiny to an Act that has demonstrated its usefulness for conventional biological control since 1985.

In some countries uncertainty in the community and among regulators over the environmental impact of releasing novel organisms into the environment has focused attention on the evaluation processes for introduction and release of classical biological control agents. For example, in New Zealand, the draft legislation on the release of GMOs embraces the importation and release of all organisms for the purpose of biological control, again showing the potential for these two issues to become linked. The draft New Zealand Bill and accompanying discussion paper was released late in 1992.

Similarly, in the UK, the ACRE is concerned with the release of all non-indigenous organisms in the UK, regardless of whether they have been genetically manipulated or occur naturally outside the UK. Western Australia is also contemplating amendments to the relevant wildlife conservation legislation to encompass all novel organisms, whether GMOs or simply exotics.

In some other countries this conceptual linkage between conventional biological control and the release of GMOs appears to exist and suggests that the fate of both technologies can no longer be regarded as entirely separate. It seems probable that the future of biological control will be influenced by the progress of the debate on environmental release of GMOs whether or not for purposes of biological control.

Consequences of considering the product, not the process

It is interesting to note how the linkage between genetic engineering generally and biological control has come about, especially in the USA, UK, New Zealand and Australia. Initial attempts to regulate GMOs placed emphasis on the techniques applied in the construction of the organism. This had the obvious advantage that it helped regulators define the class of activities under consideration and therefore the scope of the legislation; but it placed undue emphasis on the process, not the product. It bestowed an unwelcome degree of mystique on genetic engineering technology *per se*. Researchers stressed that attention should focus on the product rather than the means of its generation. After all, it hardly matters whether a gene function is lost by way of precise deletion through r-DNA techniques as with NoGall or *P. syringae*, or by classical selection for a null mutant. Indeed, the former technology can be more precise. However, the significant operational disadvantage for regulatory authorities is how to define the scope of introductions or releases that should be covered by legislation or regulations. In the process of de-mystification of r-DNA technology, attention necessarily focuses on any novel release, regardless of its prior manipulation. Consequently, the introduction or release of any exotic biological control agents tend to get captured in the debate. In New Zealand the Hazardous Chemical Legislation, which embraces the release of GMOs, also covers classical biological control. However, the administrative arrangements will attempt to speed up the evaluation and approval procedures where there is no obvious genetic manipulation involved. Similar devices will need to be sought in other countries, or the process will become totally unworkable.

The global connection for GMOs

If we accept that GMOs represent an important option for future biological control programmes, then there is a further point that needs to be recognized. It is that the r-DNA debate has a global dimension and events on one continent will have repercussions elsewhere. Governments, transnational companies and even smaller commercial enterprises have become acutely aware of events on other continents. For example, the stinging criticism of certain US commercial interests for apparently using Latin America to field trial genetically engineered vaccines rather than the USA has had ramifications on the other side of the globe. One country in Africa sought to import an engineered vaccine from the USA but the relevant company, aware of the Latin American experience, refused to make the strain available without an environmental impact statement. This commercially understandable position by the US company has been viewed as an intolerable imposition by the Africans and the issue remains unresolved to date. Similarly, the careful educational strategy of David Bishop and his colleagues at the Institute of Virology in Oxford prior to seeking and being refused approval for the release of an engineered insect virus will clearly influence investors, research workers and regulators in other countries, regardless of the reasons for the set back.

These examples illustrate that application of r-DNA technology offers certain important advantages; it can reduce the need for biological control by providing alternative options or it can increase the range and efficacy of natural enemies. However, the disadvantage is that it is placing undue emphasis on conventional biological control practices and risk evaluation procedures. Where actual GMOs are being proposed as natural enemies, biological control itself becomes central to the debate. Further, the prospect of releasing GMOs regardless of their intended purpose has led, in some countries, to classical biological control being included in draft legislation. Finally, the global perspective of the r-DNA debate means that events in one country or continent, whether in the developed or developing world, will have significant repercussions in virtually every other continent.

Towards a global strategy for introducing GMOs into the environment

Following the above considerations, it seems desirable that a global strategy be defined for the development and release of GMOs regardless of the purpose of release. Such a strategy might improve the prospects of an orderly debate and raise the probability that this valuable new technology will play a role in improving the economic and social position of people in both the developing and developed world, and in making a substantial contribution to sustainable development. The balance of this chapter suggests criteria that might be applied to ranking, in a temporal sense, the situations in which r-DNA technology is applied to agriculture. It concentrates on principles rather than on addressing the real problems of strategies and tactics for implementing such concepts.

The need for a global strategy suggested itself at the August 1991 International Congress on Plant Protection in Rio, Brazil. Some research reported at that Congress suggested that the direction of r-DNA technology for plant protection was science and technology driven rather than responding to market forces. For example, some impressive examples of US research were given at the Rio Congress of tomato and potato crops being protected against herbivorous pests by the expression of *Bt* toxin genes in the tissues of these vegetables destined for human consumption. Strong consumer rejection of such commodities would have a negative impact well beyond the individual companies seeking a prompt return on their research investment. Clearly, we would be inclined to place quite low in a priority listing any research that entailed expression of foreign genes, coding for toxic chemicals in animal or plant tissue that was destined for human consumption. It should be noted that the toxins referred to here have no vertebrate toxicity; neither do they persist in vertebrate systems. However, we are talking about perceptions of the wider community, not a rational and informed assessment of the issues. Even the concept of the differences between vertebrates and invertebrates, or toxin specificity will be vague notions in the minds of many. Simple concerns about 'unknowns' and 'black boxes' will take precedence in the debate. How would such a list look and what criteria are applicable? Criteria might include:

- environmentally safe;
- limited impact on non-target organisms;
- not directly linked to food supplies for humans;
- removal or inactivation of a gene affecting a physiological pathway (e.g. over-ripening);
- effectiveness is sustainable, i.e. resistance unlikely to develop;
- can ultimately be withdrawn from production if such action is deemed desirable.

The preferred temporal order for implementing genetic engineering technology might be:

1. Synthesis of molecules using GMOs, e.g. insulin, interferon, growth regulators, *Bt* toxins.
2. Synthesis and release of killed GMOs, e.g. dead vaccines.
3. Synthesis and release of GMOs as live vaccines (perhaps this option could be placed further down the list).
4. Modification to host-plant resistance by the removal of specific genes, preferably in plants not directly in the food chain. Examples could include: removal of a gene whose product was an essential cue to a herbivore, e.g. the caterpillar of *Acraea horta* has a specific relationship with a cyanogenic plant species (*Kiggelaria africana*). The disruption of the pathway which synthesises the specific cyclopentenyl cyanogenic glycoside would free this plant from a major herbivore. There are undoubtedly many cases where a plant is recognized as a suitable host for a herbivore by the presence of one or more specific chemicals present in the plant tissue.

5. Modification to host-plant resistance by the addition of one or more genes that would identify the host plant as a likely source of herbivores for the relevant natural enemies. For example, the cotton boll weevil evolved on *Hampea* spp. in Mesoamerica where a diverse and complex array of its parasitoids are known to exist. Some of these do not recognize cotton as a host plant for the weevil and therefore have not made the shift in host habitat. Genetic engineering of cotton to incorporate the gene(s) which code for the specific plant substance(s) in *Hampea*, if practicable, should score high, compared with transformation with *Bt* toxin genes, both environmentally and scientifically, as a goal for a GMO to be released into the environment.
6. The genetic engineering of microbial pathogens or arthropod natural enemies to improve efficacy or modify specificity. There are many possibilities under this heading and these could in turn be prioritized according to the criteria listed above.
7. Modification to plants to enhance pathogen resistance especially to viruses, bacteria or fungi with preference being given to plants not directly in the human food chain.
8. Genetic engineering of a crop plant not directly used for human consumption using a gene(s) coding for a biological toxin, preferably with tissue specificity or inducibility by wounding.
9. Genetic engineering of a crop plant used in providing food for human consumption with gene expression restricted to tissues not used for consumption, e.g. potato leaves and shoots but not tubers, etc.
10. As for 9, but where the gene is expressed in the plant component used for human consumption. Here the gene might be the absence of a cue for herbivores whether physical (colour) or chemical. A similar line of reasoning might apply to domestic animals. The use of an animal growth regulator coupled with a human promoter sequence say in pigs or dairy cattle would seem to be an undesirable combination. (Indeed, it was the unauthorized release into the human food chain of some transformed pigs using a combined human/pig growth regulator construct that helped trigger the current parliamentary enquiry into the need to legislate regulatory control of genetic engineering in Australia.)

The final set of questions we might wish to address is where should the release of genetically engineered organisms take place – in developed countries or the developing world. From what has already been said, it is desirable that any strategy not be seen as testing a risky proposition in someone else's backyard – especially a developing country. One could argue that a collaborative venture should be preferable – that would suggest including a country that is both impoverished but which has a substantial scientific capability, such as Brazil or China. Field trials should be conducted simultaneously in both collaborating countries though it may be desirable, once the environmental or other external concerns have been dealt with, for the larger field or commercial application to be conducted in the partner country which has the cheaper labour force and which has more to gain in the short term from improved plant or animal production.

Some specific examples which could be considered as good early candidates might be:

1. Engineered baculoviruses for pest control in soybean involving collaboration between the USA and Brazil.
2. Engineered baculoviruses for pest control in cotton involving collaboration between Australia and China.
3. A similar project to 2 between the EEC and a north African country.

While it is difficult to envisage some international authority dictating what should be done, where and when, and by whom, the commercial judgment by those governments and transnational companies who have made substantial investments in the field of genetic engineering might dictate some strategy similar to that outlined in this paper. A difficulty for the larger operators could be actions by smaller operators who need a quick return on their investment and are prepared to jeopardize the orderly development of an exciting new field that represents an important advance to modern agriculture and the discipline of biological control.

Implementation of a global strategy

There is no international authority that could police a policy aimed at the orderly application of genetic engineering technology. Therefore, legislation or regulation at the global level is not an option. Instead, two broad approaches might be considered. First, peer pressure following broad discussion and a general consensus about the ideal sequence of applications might influence invest-

ors, scientists and regulators. Secondly, international agencies such as FAO, OECD, WHO, World Bank, etc., might develop in-house policies about preference areas of usage of GMOs. Having done so, these organizations could then lobby funding agencies to support applications that demonstrate safety, efficacy, sustainability and economic soundness of the use of GMOs in the biological control context. By the accumulation of examples that indicate the value and safety of the new technology, we could optimize the application of this valuable technology for the maximum benefit of people in both the developing and developed world.

References

Anonymous (1992). Genetic manipulation: the threat or the glory? In *Report of the House of Representatives Standing Committee on Industry, Science and Technology*, Australian Government Publishing Service, Canberra.

Christian, P. D., Hanzlik, T. N., Dall, D. J. and Gordon, K. H. (1992). In *Molecular Approaches to Fundamental and Applied Entomology* (ed. J. Oakeshott and M. J. Whitten), pp. 128–163. Springer-Verlag, New York.

Feitelson, J. S., Payne, J. and Kim, L. (1992). *Bacillus thuringiensis*: insects and beyond. *Biotechnology* 10, 271–275.

Jones, D. A. and Kerr, A. (1989). The efficacy of *Agrobacterium radiobacter* strain K 1026, a genetically-engineered derivative of strain K84, in the biological control of crown gall. *Plant Disease* 73, 15–18.

Jones, D. A., Ryder, M. H., Clare, B. G., Farrand, S. K. and Kerr, A. (1988). Construction of a Tra-deletion mutant of pAgK84 to safeguard the biological control of crown gall. *Molecular and General Genetics* 212, 207–214.

Michael, T. and Wilson, A. (ed.) (1992). *Genetic Engineering with Plant Viruses*. CRC Press, Boca Raton, FL.

Mullineaux, P. M. (1992). Genetically engineered plants for herbicide resistance. In *Plant Genetic Manipulation for Crop Protection* (ed. A. M. R. Gatehouse, V. A. Hilder and D. Boulter), pp. 75–107, CABI, Wallingford.

Oakeshott, J. G. and Whitten, M. J. (eds) (1993). *Molecular Approaches to Fundamental and Applied Entomology*. Springer Series in Experimental Entomology, Springer-Verlag, New York.

Vlak, J. M. (1992). Genetic engineering of

baculoviruses for insect control. In *Molecular Approaches to Fundamental and Applied Entomology* (ed. J. Oakeshott and M. J. Whitten), pp. 90–127. Springer-Verlag, New York.

Whitten, M. J. (1991). Pest management in 2000: what we might learn from the 20th century. In *Pest Management and the Environment in the Year 2000*, pp. 9–44. Kuala Lumpur, Malaysia.

Part V ECONOMICS AND REGISTRATION

25

Development of the biocontrol fungus *Gliocladium virens*: risk assessment and approval for horticultural use

Robert D. Lumsden and J. F. Walter

Introduction

The fungus *Gliocladium virens* Miller, Giddens and Foster is an important biological control agent (Papavizas, 1985). A formulation of this fungus (strain Gl–21) was recently registered with the US Environmental Protection Agency (EPA) by W. R. Grace & Co. Conn. The formulation was developed in cooperation with the Biocontrol of Plant Diseases Laboratory (BPDL), US Department of Agriculture (USDA) (Lumsden *et al.*, 1991). It is intended for use against damping-off diseases of vegetable and ornamental seedlings caused by the soil-borne plant pathogens, *Pythium ultimum* and *Rhizoctonia solani* in glasshouse operations (Lumsden and Locke, 1989). This fungus is one of the first to be registered for biocontrol of plant diseases and will soon be available in the US for commercial use in glasshouse applications under the trade name, Gliogard[TM]

Certain criteria were considered important in the early stages of development of biocontrol agents (Lumsden and Lewis, 1989). In the development of a screening method for the selection of an appropriate micro-organism, the following points were considered. The screening would involve: (1) the use of a relatively uniform, commercially available soil-less medium that is used extensively in commercial glasshouses where the disease problem occurs; (2) targeted pathogens were selected that are important in the confines of

a glasshouse where use of a biological control agent would probably be most successful because of a relatively uniform environment; (3) micro-organisms indigenous to the US were selected because non-indigenous micro-organisms might be conceived as more likely problems for the US environment; (4) a single isolate of a biocontrol agent for control of both pathogens was preferred over a mixture of isolates; and (5) a high value crop, important in the ornamental production industry, was selected to defray the cost of development and registration. On the basis of these factors, consideration of which would make the process for registration and commercialization of an agent for control of plant diseases easier, a screening programme was initiated and over 100 isolates of fungi, bacteria and actinomycetes were tested (Lumsden and Locke, 1989). Most of these isolates had some prior history of biocontrol potential or were isolated from survival structures of plant pathogens, and thus were likely to have antagonistic capabilities. From these screenings, *Gliocladium virens* consistently controlled the pathogens better than other micro-organisms tested and one isolate (Gl–21) was effective against both *P. ultimum* and *R. solani*.

Upon selection of *G. virens* as the micro-organism of choice, an appropriate formulation of the fungus for ease of preparation, application, and maximum efficacy was chosen. A suitable formulation, previously developed by the BPDL was based on alginate-wheat bran granules (prill)

263

(Fravel et al., 1985; Lewis and Papavizas, 1987). Three US Patents were issued for this formulation (US Patents No. 4668512; 4724147; and 4818530).

Formulations

The effective G. virens isolate formulated into alginate prill was thus protected by patents, and the technology was transferred through an exclusive licence to W. R. Grace & Co. whose research facilities are located near Beltsville in Columbia, Maryland. The close proximity of the two facilities was a major advantage in the development of the product since it enhanced interactions between scientists at the two locations. The cooperative venture was expedited by the Technology Transfer Act of 1986 passed by the US Congress to encourage and promote cooperative projects between US Government laboratories and private enterprise (Villet, 1992). As a result of this legislation, a Cooperative Research and Development Agreement (CRADA) was established.

The cooperative agreement involved improvement of the fermentation process, refinement of the formulated G. virens, improvement of the shelf life, and establishment of efficacy on several crop plants at several glasshouse locations (Lumsden et al., 1990). Achievement of these objectives was necessary before the Environmental Protection Agency (EPA) was approached for registration and eventual commercialization.

Registration

There are two primary regulatory agencies in the US that are concerned with regulation of micro-organisms used in agriculture. The Animal and Plant Health Inspection Service (APHIS) of the USDA reviews and regulates microbial control agents intended for importation, interstate shipment, or field release under the Federal Plant Pest Act (FPPA) (Coulson et al., 1991). This regulatory act is intended to protect US agriculture against non-indigenous micro-organisms and known plant pathogens that might pose a threat to food, fibre and timber production. Because of the world-wide distribution of G. virens (Domsch et al., 1980; Papavizas, 1985) and because it has not been reported to be a pathogenic organism (Farr et al., 1989), the concern that APHIS had in this fungus was minimal.

The process of registration of G. virens as a pesticide was exclusively the purview of the EPA (Charudattan and Browning, 1992). The EPA reviews and regulates micro-organisms used as pesticides if they are engineered, non-indigenous, or field tested on more than 10 acres (4.05 ha) of land or 1 acre surface of water (0.41 ha) (Betz et al., 1987). An experimental use permit (EUP) is required for these conditions. The primary legislation that charges the EPA with this responsibility is the Federal Insecticide, Fungicide, and Rodenticide Act (FIFRA).

Product evaluation

According to subdivision M of the EPA Pesticide Testing Guidelines (1989) (Federal Register, 1989), microbial agents for the control of plant pests are treated in many ways similarly to chemical pesticides, and companies that apply for registration must provide extensive information for approval to use microbial 'pesticide' products commercially. Product testing is set up in a tier system which recognizes the inherent risks and degrees of exposure associated with different uses of pesticides. In addition to production and taxonomic data, long- and short-term effects on a variety of organisms including plants, animals, and other non-target organisms may be necessary (Table 25.1). According to the regulations, studies may be required for effects that are toxicological (toxic), mutagenic (causing gene damage), carcinogenic (causing cancer), foetotoxic (toxicity to a foetus), teratogenic (causing birth defects), and oncogenic (causing tumours) depending on the envisioned use pattern (Table 25.2).

Detailed data requirements may include: (1) acute oral, dermal, respiratory, eye irritation, dermal irritation, dermal sensitization, and acute delayed neurotoxicity; (2) subchronic studies to include 90-day feeding studies, 21-day repeat dermal tests, 9-day dermal toxicity, 9-day inhalation, and 90-day neurotoxicity tests; (3) chronic and long-term feeding and oncogenicity studies; teratogenicity tests and reproduction studies; (4) mutagenicity studies to evaluate gene mutations, chromosomal aberrations and genotoxic effects;

Table 25.1. US Environmental Protection Agency toxocological test requirements for non-target organisms. Tier I environmental expression data[a]

| Type of data | Use pattern | | | | | | | | | |
| | Terrestrial | | Aquatic | | Greenhouse | | Forestry | | Domestic | |
	Food	Non-food	Food	Non-food	Food	Non-food	Food	Non-food	Food	Non-food
Acute avian	R[b]	R	R	R	CR	CR	R	R	CR	CR
Avian injection	R	R	R	R	CR	CR	R	R	CR	CR
Wild mammal	CR	CR	CR	CR	NR	NR	CR	CR	NR	NR
Fresh water fish	R	R	R	R	CR	CR	R	R	CR	CR
Fresh water invertebrate	R	R	R	R	CR	CR	R	CR	CR	CR
Non-target plant	R	R	R	R	NR	NR	R	CR	NR	NR
Non-target fish	R	R	R	R	CR	CR	R	R	NR	NR
Honey bee	R	R	R	R	CR	CR	R	R	NR	NR

[a]Exerpted from Federal Register. 1989. Data requirements for pesticide registration: final rule. 53: 15952–15999.
[b]R, required; CR, conditionally required; NR, not required.

Table 25.2. US Environmental Protection Agency acute toxicity test requirements. Tier I

| Type of data | Use pattern | | | | | | | | | |
| | Terrestrial | | Aquatic | | Greenhouse | | Forestry | | Domestic | |
	Food	Non-food	Food	Non-food	Food	Non-food	Food	Non-food	Food	Non-food
Acute oral	R[a]	R	R	R	R	R	R	R	R	R
Acute dermal	R	R	R	R	R	R	R	R	R	R
Acute inhalation	R	R	R	R	R	R	R	R	R	R
Intravenous, intracuticular and Intraperitoneal injection	R	R	R	R	R	R	R	R	R	R
Primary dermal	R	R	R	R	R	R	R	R	R	R
Primary eye	R	R	R	R	R	R	R	R	R	R
Hypersensitivity	CR	CR	CR	CR	CR	CR	CR	CR	CR	CR
Immune response	R	R	R	R	R	R	R	R	R	R
Tissue culture[b]	R	R	R	R	R	R	R	R	R	R

[a]R, required; CR, conditionally required
[b]Viruses only.

(5) general metabolism studies; (6) dermal penetration studies; and (7) domestic animal safety determinations.

These requirements were originally designed for the evaluation of synthetic chemicals and not for living organisms, but some latitude is recognized for the differences. The testing is done on a multi-tier system so that micro-organisms that pass certain requirements at the Tier I level need not be progressively tested at a more stringent, long-term testing Tier II level. In addition, the EPA is aware of a cost-benefit mode of assessment and the need to expedite introduction of safe alternative pest control strategies into agricultural production for minimization of the use of chemical pesticides. The EPA is thus willing and able to waive or minimize the impact of certain requirements of the legislated regulations when

Table 25.3. Toxicology test results for *Gliocladium virens* (GL-21) formulated as wheat bran-calcium alginate prill or as fresh fungal preparation

Study	Form of G. virens[a]	Results
Acute oral toxicity/pathogenicity	Alginate prill	No acute toxicity, pathogenicity or infections detected. *G. virens* expelled in faeces
Acute pulmonary toxicity/pathogenicity	Mycelium	No acute toxicity, pathogenicity or infections detected. Spores took 2–3 weeks to clear lungs
Acute intravenous toxicity/pathogenicity	Conidia	No apparent acute toxicity, pathogenicity or infections detected. Mycelium was trapped in lungs, liver and spleen and cleared after 14 days
Acute dermal/primal dermal	ND[b]	Waived due to product form
Primary eye	ND	Waived due to product form and requirement for users to wear protective eye covering

[a]Administered to test animals (rats) in physiological saline.
[b]ND, Not done.

appropriate. Each situation is judged individually. In the case of *G. virens* Gl–21, only the trials outlined in Table 25.3 were deemed necessary.

A system for evaluation of risks is essential for the protection of the world's environment and to assure the public that a product is safe (Hess, 1990). However, the regulations should reflect the potential risks in proportion to the benefits gained and facilitate the transfer of technology to industry for implementation. The following factors should be considered as possible warnings for potential concerns: (1) newly described or non-indigenous species of organisms should be quarantined before release; (2) micro-organisms that can grow at body temperature (37 °C) should be reviewed as possible concerns for human or animal health; (3) micro-organisms that may be ingested because of application to edible plant parts should be extensively reviewed; (4) micro-organisms that produce abundant air-borne spores that might be inhaled by workers handling a product should be reviewed as possible allergens, eye irritants, or opportunistic pathogens; and (5) micro-organisms that cause serious plant diseases that cannot be stringently controlled should be confined.

In view of these considerations, *Gliocladium virens* was evaluated and reviewed in an equitable but thorough manner for its safety as the active component of an agricultural product intended

for the control of damping-off diseases. Evaluations were conducted based on the following considerations, the results of which are summarized in Table 25.3.

Origin

G. virens, isolate Gl–21, was from the collection of the BPDL (Lumsden and Locke, 1989). The fungus was originally isolated by Dr Michael Dunn, a member of the laboratory, from sclerotia of *Sclerotinia minor* buried and recovered from a local Beltsville, Maryland soil. The isolate was originally designated as MTD-453, redesignated Gl–21, and is now deposited in the Northern Regional Research Laboratory (NRRL) collection, Peoria, Illinois as NRRL isolate No. 15948. Although the isolate of *G. virens* is indigenous to the USA, having been isolated from a Beltsville, Maryland soil, the species is widely distributed throughout the world (Domsch *et al.*, 1980; Farr *et al.*, 1989). Verification of the taxonomic identity of *G. virens* was done by the Systematic Botany and Mycology Laboratory of the USDA.

Biological characteristics

G. virens is a hyphomycete with no confirmed sexual stage. The possible sexual stage is *Hypocrea gelatinosa* (Domsch *et al.*, 1980). It proliferates as asexual conidia that are held in masses of moist spores, or it survives as vegetative segments of the mycelium, termed chlamydospores, usually embedded in organic matter. The spores are not air-borne and are dispersed only as spore suspensions in water or carried in soil or in organic debris.

G. virens is a common soil saprophyte and, as with many other soil-borne fungi, produces several antibiotic metabolites (Howell and Stipanovic, 1983; Taylor, 1986; Jones and Hancock, 1987) that are thought to enhance its soil competitiveness. The metabolite most likely to be associated with control of Pythium and Rhizoctonia damping-off is gliotoxin, an epipolythiopiperazine-3,6-dione antibiotic (Aluko and Hering, 1970; Roberts and Lumsden, 1990; Lumsden *et al.*, 1992). Gliotoxin has antibacterial, antifungal, antiviral, and antitumour activity. It also interferes with phagocytic cells and is immunosuppressive (Waring *et al.*, 1988). Since gliotoxin has moderate mammalian toxicity (50 mg/kg) (Taylor, 1986), its ingestion directly by an animal or human is of concern. Thorough evaluation of the formulated product has not shown the presence of the antibiotic in the alginate prill (unpublished results). Consequently, the prill would not be harmful if ingested. Moreover, gliotoxin is produced after incorporation of the alginate prill in soil, it remains active for a short period of time, and is inactivated by the soil microbiota (unpublished results).

Toxicological data

Conclusions regarding the safety of the formulation are supported by the demonstrated lack of toxicity in oral, dermal and pulmonary studies conducted on rats (Table 25.3). The toxicological data that were submitted for this formulation included results from an acute oral toxicity/pathogenicity study, an acute pulmonary toxicity/pathogenicity study and an acute intravenous toxicity/pathogenicity study. The studies were performed using rats. All studies were classified as acceptable. The EPA review of these studies indicated that *G. virens* is not toxic to, infective in, or pathogenic to rats by oral or pulmonary routes of exposure and not infective or pathogenic to rats by intravenous injection. Mycelium of the fungus was used for the injections and was found to be acutely toxic and lethal to some test animals due

to mechanical clogging of capillaries. However, these mortalities were not considered to be relevant in this case since injection is not a normal route of exposure and the product consists of chlamydospores rather than mycelium.

Acute dermal toxicity testing was not required for *G. virens* based on the fact that the end-use product consists of large pellets (1–2 mm diameter). The label will be required to state that gloves be worn when handling the product, and application will be by soil incorporation. A primary eye irritation study was not required since protective eye covering must be worn (Table 25.3).

Other toxicity testing was waived based upon submission of data which: (1) indicated that the fungus does not grow at or near the body temperature of mammals or birds; (2) demonstrated that toxins or antibiotics that were produced during the manufacture of the product were not considered to be of toxicological significance; (3) showed that personnel working with *Gliocladium virens* strain Gl–21 for several years showed no adverse toxicological effects attributable to that work; (4) provided the criteria used to determine the extent to which formulated preparations are free from contaminating micro-organisms; and (5) confirmed the exempt status of certain inert ingredients.

Ecological effects

Data requirement for the evaluation of adverse effects on birds, wild mammals, freshwater fish, freshwater aquatic invertebrates, estuarine and marine animals, non-target insects and honey bees (Table 25.1) were either not required or waived based on the nature of the product and its proposed use pattern. Since there will be minimal exposure to non-target organisms from the use of the product in glasshouse food and non-food crops, there is not expected to be a 'may affect' situation with regard to endangered species.

Data for environmental fate (Tier II) were not required for this fungus since there will be negligible environmental exposure associated with product use in glasshouses and the initial (Tier I) tests for environmental fate were waived or not required.

Tolerance assessment

An exemption was submitted from the requirement for a tolerance level for residues of *G. virens* Gl–21. Tolerance levels for chemical residues are normally required in or on all raw agricultural commodities when a fungicide is used. Since testing of *G. virens* showed no evidence of effects that would be of toxicological concern, an exemption was granted. In addition, the product will be used on seedlings of crop plants and will not be exposed to produce that are sold as food for humans or animal consumption.

Use patterns and formulations

Two formulations of *G. virens* (Gl–21) were approved. WRC-GL-21 is a manufacturing-use product (fungal biomass) for use in formulation of biocontrol products. WRC-AP-1 is an end-use formulated granular product containing calcium alginate, wheat bran and proprietary additives to prolong shelf-life. The granular material is mixed with soil or soil-less plant growing media at least 1 day prior to planting, or incorporated into the medium surface in plant beds prior to or at planting. The formulation is used at the rate of $1–1\frac{1}{2}$ lb/yd^3 (approx. 1 gm/l) of medium when mixed, or at the rate of $\frac{3}{4}–1$ ounce/square foot when applied to the bed surface.

Field testing: environmental use permit (EUP)

An EUP was not required by the EPA for *G. virens* because the isolate was indigenous to the USA and because field testing was confined to small glasshouse applications, well below the 10 acre (4.05 ha) minimum area for requirement of an EUP (Federal Register, 1989). Testing for efficacy proceeded in glasshouse locations in three states, (Lumsden *et al.*, 1990) and for horticultural acceptance and compatibility with crop plants at several locations across the USA.

Marketing assessment

Commercialization of *G. virens* or other biocontrol agents is dependent on marketing assessment and determination of market availability and profit margins. Several factors were considered for Gl–21: (1) products would be protected by patents issued to the USDA that pertained to the formulation and that were transferred by exclusive licence to industry; (2) the glasshouse bedding-plant production industry requires safe, reliable treatments for controlling damping-off diseases and non-chemical natural biocontrol systems are favoured; (3) a simple, inexpensive fermentation system is available for producing biomass of *G. virens* in large commercial scale fermenters; and (4) biological pest control agents are generally less damaging to the environment and cheaper to develop, register, and market than chemical control compounds.

Considering all of these factors, W. R. Grace & Co. concluded that production of a fungicide (Gliogard) that contained *G. virens* was a sound commercial venture for the agricultural market. Test marketing was carried out in 1992 in Florida and a few other states (Knauss, 1990). Larger scale supplies were delivered in 1993.

Benefits

The use of this product will be beneficial in the control of damping-off diseases of seedling plants, a major source of economic loss in the glasshouse production of ornamental and food-crop plants. The availability of a biological fungicide will provide a less toxic alternative to the currently registered chemical treatments. In addition, this achievement provides the impetus for future biocontrol technology, development, and application.

References

Aluko, M. O. and Hering, T. F. (1970). The mechanisms associated with the antagonistic relationship between *Corticium solani* and *Gliocladium virens*. *Transactions of the British Mycological Society* 55, 173–179.

Betz, F., Rispin, A. and Schneider, W. (1987). Biotechnology products related to agriculture. *Overview of Regulatory Decisions at the US Environmental Protection Agency.* ACS Symposium series 334, pp, 316–327. American Chemical Society, Washington, DC.

Charudattan, R. and Browning, H. W. (eds) (1992). Regulations and guidelines: critical issues in biological control. *Institute of Food and Agricultural Science,* University of Florida, Gainesville, Florida 32611.

Coulson, J. R., Soper, R. S. and Williams, D. W. (eds) (1991). Biological control quarantine: needs and procedures. *Proceedings of a Workshop Sponsored by USDA-ARS*. US Department of Agriculture, Agricultural Research Service, ARS-99. Nat. Tech. Info. Serv. Springfield, VA.

Domsch, K. H., Gams, W., Anderson, T. (1980). *Compendium of Soil Fungi*, vol. 1. Academic Press, London.

Farr, D. F., Bills, G. F., Chamuris, G. P. and Rossman, A. Y. (1989). *Fungi on Plants and Plant Products in the United States*. APS Press, The American Phytopathological Society, St Paul, MN.

Federal Register (1989). Data requirements for pesticide registration; final rule 53, 15952–15999.

Fravel, D. R., Marois, J. J., Lumsden, R. D. and Connick, W. J. Jr. (1985). Encapsulation of potential biocontrol agents in an alginate-clay matrix. *Phytopathology* 75, 774–777.

Hess, C. E. (1990). International Conference on Risk Assessment and Agricultural Biotechnology. In *Risk Assessment in Agricultural Biotechnology: Proceedings of the International Conference*, (ed. J. J. Marois and G. Bruening), pp. 1–2. University of California Publ. No. 1928 Oakland, CA.

Howell, C. R. and Stipanovic, R. D. (1983). Gliovirin, a new antibiotic from *Gliocladium virens*, and its role in the biological control of *Pythium ultimum*. *Canadian Journal of Microbiology* 29, 321–324.

Jones, R. W. and Hancock, J. G. (1987). Conversion of viridin to viridiol by viridin-producing fungi. *Canadian Journal of Microbiology* 33, 963–966.

Knauss, J. F. (1992). *Gliocladium virens*, a new microbial for control of *Pythium* and *Rhizoctonia*. *Florida Foliage* 18, 6–7.

Lewis, J. A. and Papavizas, G. C. (1987). Application of *Trichoderma* and *Gliocladium* in alginate pellets for control of Rhizoctonia damping-off. *Plant Pathology* 36, 438–446.

Lumsden, R. D. and Lewis, J. A. (1989). Selection, production, formulation and commercial use of plant disease biocontrol fungi, problems and

progress. In *Biotechnology of Fungi for Improving Plant Growth* (ed. J. M. Whipps and R. D. Lumsden), pp. 171–190. Cambridge University Press, Cambridge.

Lumsden, R. D. and Locke, J. C. (1989). Biological control of damping-off caused by *Pythium ultimum* and *Rhizoctonia solani* with *Gliocladium virens* in soilless mix. *Phytopathology* 79, 361–366.

Lumsden, R. D., Locke, J. C., Lewis, J. A., Johnston, S. A., Peterson, J. L. and Ristaino, J. B. (1990). Evaluation of *Gliocladium virens* for biocontrol of Pythium and Rhizoctonia damping-off of bedding plants at four greenhouse locations. *Biological and Cultural Control Tests* 5, 90.

Lumsden, R. D., Locke, J. C. and Walter, J. F. (1991). Approval of *Gliocladium virens* by the U.S. Environmental Protection Agency for biological control of Pythium and Rhizoctonia damping-off. *Petria* 1, 138.

Lumsden, R. D., Locke, J. C., Adkins, S. T., Walter, J. F. and Ridout, C. J. (1992). Isolation and localization of the antibiotic gliotoxin produced by *Gliocladium virens* from alginate prill in soil and soilless media. *Phytopathology* 82, 230–235.

Papavizas, G. C. (1985). *Trichoderma* and *Gliocladium*: biology, ecology and potential for biocontrol. *Annual Review of Phytopathology* 23, 23–54.

Roberts, D. P. and Lumsden, R. D. (1990). Effect of extracellular metabolites from *Gliocladium virens* on germination of sporangia and mycelial growth of *Pythium ultimum*. *Phytopathology* 80, 461–465.

Taylor, A. (1986). Some aspects of the chemistry and biology of the genus *Hypocrea* and its anamorphs, *Trichoderma* and *Gliocladium*. *Proceedings of the Nova Scotia Institute of Science* 36, 27–58.

Villet, R. (1992). A product orientation for agriculture. *Agricultural Research* 40, 2.

Waring, P., Eichner, R. D., and Mulbacher, A. (1988). The chemistry and biology of the immunomodulating agent gliotoxin and related epipolythiodioxopiperizines. *Medical Research Review* 8, 499–524.

26

Economics of classical biological control: a research perspective

J. M. Cullen and Max J. Whitten

Introduction

The Industries Assistance Commission of Australia, in their review of the economics of agricultural research using the CSIRO Division of Entomology as an example, concluded that the majority of projects were 'dry holes' in that they yielded negligible economic return, but that it only needed one or two successes to pay for the entire operation of a large research agency over a decade or more (Marsden *et al.*, 1980). Biological control was a major area considered in that review and these two principal conclusions are as true for this field as they are for the general case. Less than half of biological control projects produce substantial success, but those that do are the really big winners.

We would like to review briefly the history of application of economics to biological control in Australia and then consider the benefits and costs in a more general sense. There is nothing unique about biological control from an economic viewpoint, though it does have some special properties that assist the process and increase benefits, and others that require particular consideration.

The examples that we propose to cover concern classical (inoculative) or inundative biological control where the natural enemies have been selected from naturally occurring species or biotypes. We do not cover the release of genetically improved natural enemies, whether modified by conventional methods or by genetic engineering. Many of the concepts we canvas are equally relevant to the 'new technologies' in terms of assessing the costs of research, the likelihood of

success and the ensuing economic benefits. Even the risks are not dissimilar in nature. Later, we discuss the risk of untoward or unexpected damage, for instance where the intended benefits are outweighed by some unintended damage to non-target organisms or the environment and point out they are of low probability in biological control, although possible. We would expect a similar situation to prevail with the 'new technologies' so long as sensible and definable precautions are taken. Indeed, sometimes the modifications to an engineered organism are so precise that a higher level of confidence might be justified.

The Australian experience

In 1975 and 1976, at the request of the Australian government, the Industries Assistance Commission, a publicly funded statutory authority, investigated and reported on the financing of rural research in Australia. As part of that exercise, the research of the CSIRO Division of Entomology was used as an example to evaluate the benefits arising from research over the 1960–75 period against the total cost of the Division's operations over that same period (Marsden *et al.*, 1980). Not all projects and benefits could be evaluated, the total benefits were based finally on only 13 projects, four of them being 'classical' biological control projects in a fairly broad sense; the control of white wax scale, *Gascardia destructor*, on citrus trees by introduced parasites, the control of orchard mites by insecticide resistant predaceous mites, the control of sirex wood wasp, *Sirex noctilio*, by introduced parasites and a nematode, and the control of skeleton weed, *Chondrilla juncea*, principally by an introduced rust fungus, *Puccinia chondrillina*. The last project in particular, by its demonstration of enormous benefits, focused attention on biological control of weeds and the ability to carry out quite effective economic analyses.

At a discount rate of 10%, the estimated accumulated benefits from 1960 to 2000 (expressed in 1976 Australian dollars) for these four projects were 1.5, 14.4, 12.8 and 261.2 million dollars, respectively. The major causes of variation were in the severity of the problem for the industry, the level of control and the size of the industry. There was initially some resistance to accepting the figures for the skeleton weed project, but the field data and analyses proved quite robust and there was a general realization that economic returns from biological control could be very high. The skeleton weed analysis was subsequently repeated in the light of later field data (Cullen, 1985) while the sirex project had an interesting sequel in that recently, the system for the establishment of nematodes in areas newly invaded by sirex broke down. Sirex is still spreading and can cause considerable losses until such time as the nematode, the principal control agent, establishes itself at adequate levels in the infested area. Nematodes are, therefore, normally inoculated into the area ahead of and at the time of initial invasion. When this did not happen, particularly when a drought predisposed millions of trees to attack, there were considerable losses, with the potential for the industry to face further losses of thousands of millions of dollars (over the 30 year cycle of pine trees). A rapid re-establishment of the proven technology, avoided this loss, in effect a repeat performance of solving the problem, with higher pay-offs, for a small additional cost.

Amid some reawakened interest in the economics of agricultural research and biological control in particular, a study of control of red scale on citrus by the parasite *Aphytis melinus* in two of the major irrigated citrus areas of Australia also showed substantial benefits (Anonymous, 1977), though in this case, the research was in effect the implementation of results from elsewhere, which has some bearing on the interpretation of benefit (see later).

During the 1980s biological control in Australia was again in the public spotlight and again the targets were weeds. The controversy over whether Paterson's curse, *Echium plantagineum*, and blackberry, *Rubus fruticosus*, should actually be controlled led to the need to justify these projects economically before their implementation could be approved. There would be tangible losses to some sections of the community if biological control was successful and therefore the concern was over the overall balance of benefit resulting from control. In both these cases, reduction through biological control would cause losses to beekeepers, and in the case of Paterson's Curse, to some sectors of the grazing industry. In each case, the benefits of control substantially outweighed the losses. In the case of blackberry the estimated benefits were $40 million per annum against $0.7 million loss and it was not necessary to analyse this in any depth (Field and Bruzzese, 1984). For Paterson's curse a major public enquiry was necessary, the conclusion being that control would yield $30 million per annum potential benefit for a $2 million per annum loss (IAC, 1985). Both these cases are examples of *ex ante* analyses identifying the potential benefits of biological control projects and could be considered useful examples of priority setting procedures, though at a substantial cost in the case of Paterson's curse. The existence of appropriate legislation that permitted orderly debate before a decision to proceed was taken, has proved an important element in government policies assisting the orderly implementation of biological control projects.

More recently, the Australian Centre for International Agricultural Research (ACIAR) has been concerned to monitor the impact of the projects it sponsors and has commissioned economic analyses of some of these projects. These include two in biological control; the use of entomopathogenic nematodes to control lepidopterous pests of apples and shade trees in China (Doeleman, 1990), and the control of salvinia in Sri Lanka by the weevil *Cyrtobagous salviniae* (Doeleman, 1989).

Strictly speaking, the nematode project is one of augmented/inundative control rather than classical, in that it involves continuous production of nematodes for repeat applications. This throws a slightly different light on the economics in that the recurring expense needs to be taken into account, but nonetheless the benefits were estimated as being substantial e.g. for a medium rate of implementation and a discount rate of 10%, the estimated return over 25 years was calculated as $97 million giving a benefit/cost ratio of 69 and an internal rate of return of 80%.

The salvinia evaluation is particularly interesting in that it was the first of these analyses to consider benefits in terms other than dollars and cents. For aid projects, returns expressed in terms of the monetary currency of the donor country are often not appropriate. 'Hours worked' is a currency common to donor and recipient, independent of exchange rates, and often more relevant to the lifestyle of the recipient country. For a medium estimate of the cost of salvinia without biological control and a discount rate of 10%, the estimated benefit was 62 million hours for a return ratio of 1675.

In each case considered above, the returns, whether actual and projected in the case of *ex post* analyses, or potential in the case of *ex ante* analyses, have been assessed in a standard economic manner and have been demonstrated to be enormous. There is considerable reason to suppose that this is a general feature of biological control (Tisdell, 1990), though very few projects worldwide have been subjected to rigorous analysis.

However, biological control has other benefits than straight dollars and cents, as implied by the approach taken in the Sri Lanka analysis. Though the projects considered to date in Australia have been accounted in strictly economic terms, there has been substantial consideration, but no quantitative costing, of a whole set of 'externalities', e.g. reduced hazards to users and the public from reduced chemical usage, lack of undesirable residues and decreased disruption of the environment and effects on non-target species. In effect, the benefits are captured by the community at large and commercialization is neither possible nor appropriate, making them difficult to demonstrate by standard economic methodology. Increasingly, research and development managers and investors assume that research must create intellectual property which, in turn, generates revenue by royalties, licensing fees or a saleable product. This resulting income serves to justify the original investment. The private enterprise paradigm ignores the important cases where public benefits arise out of publically funded research and development, but which are not attractive to, or taken

up by private enterprise because adequate benefits cannot be captured by the private investor. Doeleman (1989) develops this argument convincingly in his salvinia study. It is important that a means is found to quantify the community benefit in order to demonstrate the value of biological control.

Conway (1987) suggested that 'sustainability' in the sense of income to the farmer should be taken into account in economic analyses, along with stability of income and equitable distribution of income. Biological control also tends to produce better results in the latter area, the benefits not being solely available to those who can afford the control, but to all producers and to the public in general. There is no doubt that in the current state of agriculture, sustainability of production systems should be taken into account. If, as Tisdell and Auld (1989) suggest, there is no satisfactory methodology for accommodating the approach of Conway, compared with standard social cost–benefit analysis, then this would seem to be an area where present economics need to match current needs. There seems little doubt that if the methodology could be established, the gains from biological control would appear even higher.

The costs

The emphasis in the discussion so far has been on benefit derivable from biological control. Does it have any particular costs? Economically, probably the biggest cost is the high risk of failure. While the discipline of biological control seeks to decrease the probability of failure, there will always be projects for which classical biological control does not produce a satisfactory solution, even after prolonged effort. If high financial return is the concern then the projects attempted should be those with the biggest potential pay off, in which case the *ex ante* analyses, as in the case of Paterson's curse and blackberry, are useful exercises in their estimation of the enormous potential benefits to be obtained from success in these programmes. Predicting the probability of that success or the level of success in a programme, is still difficult however, though it is clear that lower levels of success can still have substantial benefit. Also, we need to bear in mind that several programmes are commenced on grounds other than purely potential economic pay off. There may be no other satisfactory approach, even to a small problem and the possibility of biological control should not be dismissed, particularly in view of the additional benefits from externalities.

While it is a basic tenet of classical biological control that an introduced biological control agent is sufficiently host specific not to cause any damage to non-target species, there can be an economic cost associated with the risk of damage to non-target species. At one end of the spectrum is the risk of environmental or economic disaster, i.e. gross unpredicted damage through inadequate screening or changes in host-specificity. Such risks have been, and are being, continually decreased through tightened (and more realistic) screening, while significant changes in specificity are at this stage unknown in biological control. At the other end of the spectrum, a more common scenario is the tolerance of some non-target damage if the economic effect is still heavily out-weighed by the benefits of control of the target species. Many countries, including Australia, have a mechanism to take this into account. For economically important non-targets, the effects are as quantifiable as the overall benefits, though it is often not considered worthwhile to cost them in detail e.g. the possibility of damage to the herbs, borage and comfrey from the moth *Dialectica scalariella* imported for control of Paterson's curse. Where the non-target is part of the indigenous flora and fauna, there have been no attempts so far to evaluate this potential cost. As with sustainability, there probably needs to be a better framework for considering such factors in the equation, though a major problem at present is again a lack of ability to predict the extent of possible damage to non-targets, particularly in natural habitats.

There will always remain the risk that the perception of justifiable non-target damage will change, either by the development of new economically important crops susceptible to a previously introduced biological control agent, or by the realization of the greater significance of a susceptible member of the native flora or fauna. It is difficult to do more than make an assessment on the basis of the best information available at the time, so this risk needs to be taken into account as it cannot ever be completely removed. In the case of the introduction of the blackberry rust, *Phragmidium violaceum*, into Australia, it was minimized by the inclusion of all likely future parents of *Rubus* cultivars in the testing programme.

The fear of the risk of damage to non-target species can also put at risk unnecessarily the successful prosecution of other programmes. For example, in recent years the Australian Conservation Foundation, concerned over one proposed introduction, publicly called for a general moratorium on biological control. Its spokesperson cited the Queensland cane toad (an introduction carried out in the 1930s without any consideration of the consequences and where the evidence at the time indicated that the toad was unlikely to be effective against, and certainly not specific to, its target, a group of sugar cane grubs) and the introduction of rabbits into Australia (clearly nothing to do with biological control). Wiser counsel prevailed when the public debate demonstrated the speciousness of these arguments and the considerable environmental cost that would result from suspending biological control programmes, especially for exotic weeds threatening Australian natural ecosystems. Indeed the Foundation has now assumed a more constructive role, championing the cause of conventional biological control. We would hope that this outcome will provide a useful model for resolving the ongoing debate on genetically engineered arthropod and microbial natural enemies where the issues are similar, albeit less widely understood and emotions are more in evidence.

The third main area of economic loss particularly relevant to biological control has already been considered in relation to Paterson's curse and blackberry, i.e. when the target itself is of benefit to a sector of the community which will suffer loss by its control. The intense consideration noted for these two cases was precisely to quantify this issue and it was possible to derive estimates to set against the benefits. *Ex ante* analyses obviously have a special role to play in such situations.

Necessary information for economic evaluation

Many biological control projects have not been evaluated in a sufficiently rigorous manner or even in a preliminary way, due to lack of relevant data.

This is unfortunate for a field capable of demonstrating significant returns on investment in scientific research. Much can be gained before or during a programme if attention is paid to this aspect.

Some measure of abundance of the problem is required, that can be monitored, and on the one hand, be related to the causative agents of biological control and on the other, to the economic loss caused. It is the latter that is often the most difficult relation to establish and *ex ante* analyses are useful in drawing attention to any shortcomings.

A necessary adjustment to this basic information is knowledge of significant seasonal and geographic variation, not only in the seriousness of the problem but also in the economic effect of the benefit, e.g. the market price of commodities, extension and contraction of agricultural enterprises.

The standard measure of benefit from biological control in Australian studies has been improvement over the next best alternative control system. While direct comparison with a current method of control (if available) is straightforward, projections into the future can cause considerable problems due to the uncertainty of other research that might have been applied to the problem. An increased discounting of benefits is a common method of dealing with this possibility.

A common feature of biological control programmes around the world is implementation of research results obtained elsewhere, i.e. the redistribution of agents already known to be effective. The economic benefits can be very substantial for a very small outlay, but there is a note of caution sounded by some economists (Marsden *et al.*, 1980), who point out that the next best alternative in such situations may well be distribution naturally or by other agencies. To some extent, the benefits from a programme of distribution of the blackberry rust in Australia have been compromised by the illegal introduction of a partially effective strain in 1984, but as this was almost certainly also a result of the original research, the combined benefits would still be attributable to the original programme.

Further, perhaps it should be considered a difficult-to-quantify, but nevertheless significant spin-off benefit of biological control that partici-

pation in such an international field allows the possibility of technology transfer from other countries where the costs of the research have already been covered, thereby allowing the flow on of benefits for little cost, i.e. in exchange for the cost of developing a control programme that might be applicable in another country, there is the possibility of benefits at little cost from programmes also developed elsewhere.

Discounting of benefits can be used to take into account the other major factor that needs to be considered in projecting future benefits, the possibility of replacement of the target by another problem species. The specificity of classical biological control, one of its main benefits, renders it susceptible to this scenario. The later analysis of skeleton weed (Cullen, 1985) took into account the increase in abundance of forms of the weed immune to the strain of rust imported, and the salvinia analysis (Doeleman, 1989) considered the possibility of replacement by water hyacinth, *Eichhornia crassipes*. Estimated benefits still remained high in both cases.

The above factors have particular relevance for biological control. Other factors that have considerable bearing on the size of benefits in most analyses are adoption rate and size of the industry affected. Adoption rate equates with the dispersal and build up of the biological control agent to equilibrium with the target throughout its range. The more rapidly it does so, the higher the benefits, indicating that initial rapid rearing and distribution is economically beneficial. The size of the industry affected is obviously always going to affect the scale of benefits.

Biological control in natural habitats

Biological control of pest problems of natural habitats was not included in the original brief for this review, but is an increasingly important area of biological control activity. Particularly for weeds invasive in conservation areas, a recent Australian report (Humphries *et al.*, 1991) favours biological control as often being the only approach possible. Moreover, recent work in New Zealand (Greer and Sheppard, 1990) has demonstrated the feasibility of giving this an economic basis. Old man's beard, *Clematis vitalba*, is not an agricultural problem, but is a major threat to much of

New Zealand's native bushland. Using contingent valuation and a survey of a large representative sample of the population, it was possible to conclude that New Zealand valued control of this weed at $44 million per annum, even at the most conservative interpretation of the analysis. Biological control, therefore, again has the potential to yield enormous returns on investment, this time in terms of protection of a natural resource, the New Zealand environment. It would seem that there is considerable scope for adaptation and refinement of this technique to cover not only many similar situations, but also to help assess some of the intangibles of sustainability and environmental acceptability that are characteristics of biological control in general and even to assist in judging the value of possible risks to non-target native species.

One problem that does not yet seem to have been adequately addressed, however, is that of discounting in relation to long-term preservation of natural resources. A rate of 10% is often used for economic projects, but it does not take too much imagination to realize that there is something inappropriate in a method that suggests that whatever value we put on a natural resource, e.g. the environment, now, it will be worth only a fraction of that in 25 years time.

Conclusion

Since the results of the skeleton weed analysis first became known, they have been used to highlight the actual and potential returns from research, not only in biological control, but in biological research in general. In a political climate that has seen the erosion of scientific research budgets, apparently based on the impression that research is an expensive luxury rather than an investment in a high return area, the demonstration of hard economic benefit plays a vital role in convincing the non-scientist that scientific research pays off. While there have been, and always will be, failures in biological control, the benefits of success outweigh them by an order of magnitude and are often such as to command widespread attention.

It must be admitted that biological control also has a few natural advantages that make it a good vehicle for demonstration of benefits. It is a very visible process, where the problem being addressed is generally obvious, as is the nature of the solution. A biological control agent is a tangible entity and when it is successful, the result is also quite apparent. The fact that the whole process is readily comprehensible, essentially a human factor, also makes it a good subject for presentation. Backing it up with a good cost–benefit analysis gives it enormous strength and an even broader appeal.

Quite apart from helping to get the message across in understandable terms, economic analysis of biological control projects is clearly useful in the justification of research funding by government or by external agencies, and also in giving some priority to projects beforehand. From the preceding discussion, the principal barriers seem to be:

1. a lack of relevant data on the nature of losses and the impact of biological control on them.
2. lack of predictive ability in estimating likely success against the target or risk against non-targets.
3. a lack of a clear economic methodology to deal with sustainability issues and preservation of long-term natural resources.

The first two are clearly the responsibility of the biologist and underline the crucial importance of good follow-up and evaluation studies. The third point is the responsibility of economists and may benefit from the current debate on natural resource economics. Improvements in these areas will allow more reliable analyses and almost certainly demonstrate even bigger benefits from biological control. These will continue to emphasize its key role in protection of agriculture and the environment for the foreseeable future.

References

Anonymous (1977). Report on Irrigated Horticultural Research in the Sunraysia-Riverland Region. Report to Standing Committee on Agriculture 1977. Canberra.

Conway, G. R. (1987). The properties of agroecosystems. *Agricultural Systems* **24**, 95–117.

Cullen, J. M. (1985). Bringing the cost benefit analysis of biological control of *Chondrilla juncea* up to date. *Proceedings of the VIth International Symposium on Biological Control of Weeds*, 19–25 August 1984, Vanconver, Canada, (ed. E. S. Delfosse), pp. 145–152. Agriculture Canada, Ottawa.

Doeleman, J. A. (1989). Biological control of *Salvinia molesta* in Sri Lanka: an assessment of costs and benefits. *ACIAR Technical Reports No. 12*, Canberra.

Doeleman, J. A. (1990). Benefits and costs of entomopathogenic nematodes: two biological control applications in China. *ACIAR Economic Assessment Series* No. 4.

Field, R. P. and Bruzzese, E. (1984). Biological control of blackberry. A case for consideration by Australian Agricultural Council, Standing Committee on Agriculture. Keith Turnbull Research Institute unpublished report 1984/2, September 1984.

Greer, G. and Sheppard, R. L. (1990). An economic evaluation of the benefits of research into biological control of *Clematis vitalba*. *Agribusiness and Economics Research Unit, Lincoln University, Canterbury, New Zealand, Research Report* No. 203.

Humphries, S. E., Groves, R. H. and Mitchell, D. S. (1991). Plant invasions of Australian ecosystems: a status review and management directions. *Report to Australian National Parks and Wildlife Service, Endangered Species Program, Project* No. 58, June 1991.

IAC (1985). *Industries Assistance Commission Report on Biological Control of Echium species (Including Paterson's curse/Salvation Jane)*, 30 September 1985, Australian Government Publishing Service, Canberra.

Marsden, J. S., Martin, G. E., Parham, D. J., Ridsdill Smith, T. J. and Johnston, B. G. (1980). *Returns on Australian Agricultural Research: The Joint Industries Assistance Commission – CSIRO Benefit-Cost Study of the CSIRO Division of Entomology*. CSIRO, Melbourne.

Tisdell, C. A. (1990). Economic impact of biological control of weeds and insects. In *Critical Issues in Biological Control* (ed. M. Mackauer, L. E. Ehler and J. Roland), pp. 301–16. Intercept Ltd; Andover, Hants.

Tisdell, C. A. and Auld, B. A. (1989). Evaluation of biological control projects. *Proceedings of VII International Symposium Biological Control Weeds*, 6–11 March 1988, Rome, Italy (ed. E. S. Delfosse), pp. 93–100. Instituto Sperimentale per la Patologia Vegetale MAF, Rome.

27

Economics of biocontrol agents: an industrial view

Timo Törmälä

Introduction

The use of chemical pesticides has resulted in a considerable increase in crop yields throughout the world. There is, however, increasing public pressure against the use of chemical pesticides due to their adverse effects on human health and the environment. The very concept of using chemical pesticides has been challenged by the green movement, which is constantly gaining momentum. World-wide, annual agrochemical sales amount to approximately $US 25 billion (Powell & Jutsum, 1993). Biocontrol agents and biopesticides currently contribute only about 0.5% to the total (Meneley, 1990; Anonymous, 1992). Products based on *Bacillus thuringiensis* account for 90–95% of biopesticide sales (Feitelson *et al.*, 1992). *Bacillus thuringiensis* is a biologically produced pesticide, but the active ingredient is not a living organism.

The economics of pesticide use can be considered at three levels: the producer/distributor, the user and society. A pesticide must eventually bring profit for the producer and it must give ample economic return to the user. From society's point of view, the economic benefits from use of a pesticide must clearly offset the adverse effects (Pimentel *et al.*, 1991).

There are scant historical data on the economics of biopesticides other than for *Bacillus thuringiensis*-based products. Most of the data are proprietary. In this chapter, I discuss the economic aspects of biopesticides from the viewpoint of a company commercializing biopesticides (see Cullen and Whitten, Chapter 26 for a discussion of the economics of biocontrol from a research perspective). The scope is limited to microbial biopesticides. Special emphasis is given to the risks involved in various steps of the development and commercialization process.

The biopesticide industry

The biopesticide business can be divided into two main categories: small biopesticide/biotech companies and large multinational (chemical) corporations. The objectives, strengths and strategies of the two groups are somewhat different. Neither, however, functions in isolation from the other. There are numerous research, distribution, financing and ownership arrangements between the multinational companies and the small companies.

There are several reasons why multinational companies are interested in the development of biopesticides. By being involved in the emerging biopesticide business they ensure their participation in any significant breakthroughs in the field. In addition to this largely defensive rationale, multinational companies see biopesticides as possible niche products, which can be introduced into the market at relatively low cost. A biopesticide product line is also good for the image of a chemical company. The advantages that a large established company has include its Research and Development (R&D) resources and experience in the chemical pesticide field. These types of companies also have the required financial resources and established distribution outlets. There may, however, exist internal conflicts of interest between production of chemical and biopesticide products. The economic potential of some biopesticide products may be too small to arouse the interest of a large company. The culture of the multinational companies is chemical-orientated, which may result in difficulties in the development of biological pesticides. Abbott, Ciba, W. R. Grace, Kemira, Novo and Sandoz are examples of the major corporations active in the biopesticide market.

The small biopesticide companies have been founded on the basis of product ideas. In the initial phase the company raises money from private investors. It can also sell future marketing rights to larger companies in exchange for funding R&D programmes and other costs. If the company develops smoothly and creates some confidence, it can go public and raise substantial funds on the stock market. The amount of cash that can be raised in the initial public offering depends largely on the expectations for the company's product line and the consequent future value of the company. Ecogen, EcoScience and Mycogen are examples of small biotech companies that produce and market biopesticides. They have all been successful in raising funds on the stock markets. The biopesticide companies concentrate on one business idea, which may give them an advantage over larger companies pursuing several business ideas simultaneously. The culture in the biotech companies favours innovation and the management is normally much more flexible than in larger corporations. The biotech companies may lack the necessary experience in the field. This has been, however, often been compensated for by recruiting experienced executives and other key personnel from the chemical industry. Funding is often a problem for the small biotech companies. The key issue is how to raise sufficient capital to fund research programmes during the long period when the company is making a loss. It is likely that more than ten years will elapse before the product sales begin to create positive cash flow in a new biotech company. All the existing small biopesticide companies still make considerable losses, mainly due to heavy R&D costs.

Economic aspects of biopesticide development

General

The commercialization of biopesticides is a relatively complex process, which consists of numerous sequential and overlapping activities. The following tasks are invariably involved in the process (Meneley, 1990; Woodhead et al., 1990): (1) discovery; (2) market analysis; (3) confirmation of efficacy; (4) patenting; (5) greenhouse and field tests; (6) process development; (7) formulation; (8) scale-up; (9) tests to establish optimal application rates and compatibility studies with chemical pesticides; (10) registration studies and application; (11) fine-tuning of the production process and cost optimization; (12) construction/contracting of the production facilities; (13) assignment of distributors; and (14) development of marketing and pricing policy.

Discovery

There are two main processes that can lead to discovery of an organism which has potential as a biopesticide agent. An observation from basic research may lead to a new product. A more systematic approach identifies a problem and then a variety of organisms are screened for their potency.

Most of the initial leads in the biocontrol field have come from universities and research institutes. Some of this research has been financed by interested companies. The company and the university or discoverer usually agree on compensation in the case of the microbe being commercialized. The most frequent arrangement is that the company pays a royalty, which is normally in the range 1% to 5% of the net sales. This kind of discovery–research is more flexible and less costly for the companies than total concentration on in-house research. The economic risks for the industrial partner in this phase are limited to the funds allocated to the research teams in the universities or research institutes. Mycostop biofungicide is an example of a product that is based on an 'accidental' discovery at a university (Lahdenperä, 1991).

Market analysis

The analysis of the potential market size for a candidate biopesticide should be conducted as early as possible. The development of a biopesticide requires substantial resources and the market size must be large enough to justify the costs. The potential market is determined for each crop. The actual market size is much smaller than the potential one because the severity of pest problems varies geographically and also the biocontrol agent may be influenced to a large extent by the environment. The value of the crop per unit area, and consequently the value of the yield increase that results from application of the biopesticide, may be so low that growers cannot afford the cost of using the biopesticide. This further reduces the actual market size. Many of the new biopesticides will be targeted in the short-term at high value cash crops.

Competition with chemical pesticides is a major issue determining the actual market penetration of a biopesticide. Prediction of the competition is difficult, but new niches are opening for biopesticides as effective and cheap chemicals are scrutinized and banned due to their adverse effects. The re-analysis and re-registration process is currently in progress e.g. in the US (see Klingauf, Chapter 28, for registration requirements in Europe). A requirement for pure 'biologically' grown food by consumers may also create substantial new market openings for biopesticides in the developed countries.

Confirmation of efficacy

If a biocontrol agent is sold as a registered pesticide is must be effective to be a commercial success. The efficacy is confirmed during the initial phase in the laboratory and in small-scale tests in the greenhouse and field. The traditional pesticide industry, which is accustomed to screening thousands of compounds in the quest for a new active ingredient, has had very high requirements for efficacy. Such efficacy is seldom observed with potential biocontrol agents, whose mode of action is frequently based on several mechanisms, which do not even collectively result in complete control. Often biopesticides work better when they are used in a preventive manner before the problem becomes too severe. It is not reasonable to expect that biopesticides should perform as consistently as chemicals or under as wide a range of circumstances. From industry's point of view it is not yet clear how much less

efficient a biopesticide can be than a chemical pesticide and still successfully compete in the market.

Patenting

Protection of intellectual property rights is essential when a company develops a new product. Patents can be obtained for biological processes, methods and organisms. They have been applied for and granted in great numbers. An alternative to patenting is to retain the innovation as a trade secret within the company. In biopesticides development this approach is compromised by the fact that the active ingredient (micro-organism) can be isolated from the product. The general strategy in the field has been to file for extensive patent protection. This is expensive as the total cost of a patent with adequate geographical coverage ranges from $US 50 000 to $US 100 000. For example in 1988–91 for *Bt* technology alone, 39 patents were issued in the US (Feitelson *et al.*, 1992). The strength of the patents is not yet clear and many more court-cases and out-of-court settlements can be expected in the near future. Genetic engineering will add to the complexity of the patent issue.

If the viability of a small biopesticide company is based on a sole product or technological idea, the protection of intellectual property rights is naturally crucial for the future of the venture. The outcome of patenting may take years and it is always possible that other competitiors may have filed overlapping patents. In this situation the entire development is jeopardized unless the two parties can find a mutually acceptable solution to the problem.

Greenhouse and field tests

Large scale greenhouse and field trials serve several purposes for a biopesticide company. They primarily provide final proof of whether the product will work reliably in the 'real world'. They also provide information about the range of crops and environments in which the biopesticide performs satisfactorily.

These types of trials are also required to demonstrate the efficacy of the product, to the registration authorities. Large-scale trials are necessary prior to product introduction to convince distributors and customers, as well as the experts working in independent agricultural extension organizations, of the benefits of the product.

The first question to be answered in field trials is simple but essential: does the product work in practice? The next phases of trials should then address the more detailed questions of fine-tuning the application rates and timing of application. International field tests can be relatively inexpensive for the company, if they are run by the future distributor or a research institute, but even in those cases, to be successful, they require intensive management.

There is a fundamental difference between basic research and the approach of industry, which has resulted in many disappointments. Many ideas for new products have worked nicely in the laboratory and in small-scale field trials but have not made their way on to the market because they have not met the expectations of industry. This is disappointing for both parties. I estimate that far less than 5% of the initial leads developed in the universities will ever result in a commercially viable biopesticide, because for one reason or an other they fail to provide unequivocal results in field trials. The failure may also depend entirely on economics: efficacy may be sufficient but the cost of use is too high.

Process development and formulation

The development of the production process and formulation is begun in the initial phases of the commercialization process. The production process and formulation should in principle be ready by the time greenhouse and field trials begin. This is important for the registration procedure.

Bacillus thuringiensis differs fundamentally from the other biopesticides because the active ingredient is not a living organism. With other biopesticides, where the active ingredient is a live micro-organism, the production process differs from conventional fermentation processes, e.g. production of antibiotics, where the viability of cells after harvesting is not an issue. Thus, the vast experience many pharmaceutical companies have, cannot directly be applied to biopesticide production. The basic problem the industry has encountered is how to harvest, dry and formulate the micro-organism in such a manner that the shelf-

life is sufficiently long. In this respect the biopesticides have a competitive disadvantage compared with chemicals. Most products have a guaranteed shelf-life of 3 to 6 months when stored at low temperature (Lahdenperä, 1991). The industry is actively working on new formulations and packaging methods. Development of rigid quality control systems is crucial, because product recalls and adverse publicity can ruin the whole project.

There is a lot of room for innovative solutions in biopesticide development. One example of a novel approach to production and formulation is the *Bacillus thuringiensis* product line of Mycogen. The gene responsible for production of *Bt* toxin has been transferred into a species of *Pseudomonas*. After the fermentation, *Pseudomonas* cells are killed and stabilized and the *Bt* toxin remains encapsulated within the cells. This gives the product better persistence in the natural environment and thus improves the economic competitiveness of the biopesticide (Feitelson *et al.*, 1992).

Scale-up and production

The initial process research can be done in fermenters ranging in volume from 10 to 100 litres, while the actual production tanks may have a volume of 10 000 litres or more. The process cannot normally be simulated so reliably that a direct leap to a production scale process would be feasible. In experimental scale-up development, the process is adjusted step-wise to a larger scale (Rosen, 1985). The final production cost analysis is performed latest point during the scale-up work. The necessary background data required include the definition of unit operations of the process, raw materials, energy balances, etc. The equipment needed for the scale-up is generally so expensive that a biopesticide company may choose to contract the scale-up research, or ally with a company that has the necessary fermentation and downstream facilities.

For the actual production, pharmaceutical and enzyme-producing companies (e.g. Abbott and Novo) have the necessary fermentation capacity in-house. Newcomers have two options: to contract the production or construct the production plant. Investment in production capacity is risky if the company has only one or few products. The capital cost is high (millions of dollars) and if the product(s) are not selling as expected, the capacity will not be fully utilized and high production costs will ensue.

There is much fermentation capacity available in Europe and in the US. Most of the investment cost has already depreciated. Thus, the cost of custom fermentation can be competitive. In a custom fermentation arrangement the biopesticide company has to share the technology and some of the profit with the company doing the fermentation. Several agreements of this kind have been realized in the biopesticide field.

IPM

Due to their specificity, biopesticides cannot solve all of the pest problems a grower might have. Thus, for in the foreseeable future biopesticides will have to be applied together with chemicals. This has two implications for a company that develops biopesticides. It has to perform extensive studies to determine which chemicals can be safely used with the biopesticide, and the biopesticide has to be formulated so that it can be applied with existing equipment and using existing methods. This can be a painstakingly slow and costly process.

Registration

A registration process is required in most countries for biopesticides, but the systems vary between countries. In Denmark, for example, at present no registration is needed for biopesticides, while in other countries the procedure is parallel to that required for chemicals. In the US a specific registration system for biopesticides has been established. The toxicity package costs roughly $US 200000–$US 500000 and the process can be completed in 2 years if no problems arise during the 'Tier I' tests. The registration procedure is step-wise and allows a case by case approach (Meneley, 1990). The costs are relatively low compared with chemical pesticides, which require a $US 7–10 million data package. The total registration costs are relatively high also for biopesticides, since the products have to be registered in each country where they will be marketed. The costs are highest and time lags longest for the pioneering products in each country. In Finland, the registration of Mycostop bio-

fungicide took 7 years. The were no regulations for biopesticides and both the authorities and the company were naturally inexperienced with the procedures. In future, the regulatory system will be smoother for biopesticides, which will reduce the risk of delays and extra cost for the biopesticide industry.

The registration of genetically engineered biopesticides will be more complex and expensive than that for biopesticides based on naturally occurring microbes.

Distribution, marketing and pricing

There are several options for the distribution of biopesticides. An agrochemical company can naturally use its own outlets. A company without its own distribution organization can either assign other companies to distribute it products or build-up its own network. Biopesticides will be mostly niche products in the near future and it is unlikely that the sales volume can justify dedicated worldwide distribution networks. When choosing the distributor from the existing companies the basic selection is between the big multinationals and the smaller specialized companies. The big distributing companies have experience in the market and a sizeable sales forces, but their commitment to small niche products remains to be proved. On the other hand, niche products are more important for smaller companies and they may be more dynamic and versatile in the distribution of biopesticides, which require more attention and user education than do chemicals.

The industry has to base its marketing primarily on the economic benefits that the biopesticide brings to the farmer or grower. The secondary arguments – safety for humans and environment – will however constantly gain in importance.

Pricing of biopesticides is an extremely complex issue for industry. The returns to the producer have to cover: (1) the variable, fixed and capital (depreciation, interest) cost of the production; (2) the overhead costs associated with the product (R&D, marketing, general administration); and (3) the profit margin. The margin has to be high enough so that during the life time of the product it can cover the development costs, a share of the cost of failed products, a portion of the costs of development of new products and finally a profit for the company. This means that the gross margin has to be high to make the product economically feasible. On the other hand, the market limits the amount of margin the company can collect. The limiting mechanisms include the price of competing products, the value of the crop and the prospective gain accruing from using the product. A factor the company has to consider is that different markets can afford different prices.

Bacillus thuringiensis is the only biopesticide sold in significant amounts. In its prime market, the US it is sold for approximately $US 25–30 per kg. The price of biopesticides based on living microorganisms, and currently sold in low volumes is higher, but price comparisons are difficult, because the formulations, microbe contents (cfu) and application rates vary greatly. Of course, for the user the cost per unit area is more meaningful than cost per kilogramme, since the application rates vary considerably.

Conclusions

At present the introduction of a chemical pesticide into the market may cost $US 40–80 million and take 7–12 years (Meneley, 1990; Woodhead *et al.*, 1990). Such development cost requires substantial turnover (hundred of millions of $US at least). It is difficult to give general figures for biopesticides, but the overall development cost may be anything between $US 1 and 10 million. The costs are much lower than those for chemicals, but higher than anticipated some years ago. The R&D inputs of the large multinational companies to the biocontrol field are not published, but the R&D expenditures of three small biopesticide companies (Ecogen, EcoScience and Mycogen) ranged in 1990 between $US 2 and 10 million (Hodgson, 1992).

The biopesticide business has been disappointing for industry (e.g. Patel, 1991). *Bacillus thuringiensis* has been the only significant development so far despite huge efforts both in the public and private sector. The *Bacillus thuringiensis* market has grown steadily by about 25% annually and is expected to increase further and by 2000 capture 19% of the pesticide market (Anonymous, 1992). Other commercial microbe-based biopesticides are insignificant and probably all are still lossmarking. One advantage claimed for biopesticides is the lack of pest resistance. There are, however,

indications that resistance may at least develop against *Bacillus thuringiensis* toxins (Gibbons, 1992). If this proves to be a widespread phenomenon, it will add to costs of product development.

Despite the slower than expected developments, industry will continue to invest in biopesticides, because their time will inevitably come. The growth is most likely to be achieved through exploitation of numerous niches in the markets. The forecasts for future market share of biopesticides vary considerably, but it is realistic to assume that the total world-wide sales will reach $US 1–2 billion by 2010 (Meneley, 1991). Economically, this opportunity, in the long-term, is worth pursuing.

References

Anonymous (1992). IB Market forecasts. *Industrial Bioprocessing* Jan., 4–5.

Feitelson, J. S., Payne, J. and Kim, L. (1992). *Bacillus thuringiensis:* insects and beyond. *Bio/Technology* 10, 271–275.

Hodgson, J. (1992). Biotechnology: feeding the world. *Bio/Technology* 10, 47–51.

Gibbons, A. (1992). Moths take the field against biopesticides. *Science* 254, 646.

Lahdenperä, M. L. (1991). Streptomyces – a new tool for controlling plant diseases. *Agro-Industry High-Tech* 2, 25–28.

Meneley, J. C. (1990). A shining star in the future of agricultural industry. In *Biotechnology: Science, Education and Commercialization* (ed. I. K. Vasil), pp. 129–150. Elsevier, New York.

Patel, M. (1991). Confident Sandoz stays happy to grow alone. *European Chemical News* 9 September, 21–22.

Pimentel, D., McLaughlin, L., Zepp, A., Lakitan, B., Karaus, T., Kleinman, P., Vancini, F., Roach, W. J., Graap, E., Keeton, W. S. and Selig, G. (1991). Environmental and economic effects of reducing pesticide use. *BioScience* 41, 402–409.

Powell, K. A. & Jutsum, A. R. (1993). Technical and commercial aspects of biopesticides. *Pesticide Science* 37, 315–321.

Rosen, C. G. (1985). Biotechnology commercialization and scaling-up. *Chemical Economy and Engineering Review* July/August, 5–11.

Woodhead, S. H., O'Leary, A. L., O'Leary, D. J. and Rabatin, S. C. (1990). Discovery, development, and registration of a biocontrol agent from an industrial perspective. *Canadian Journal of Plant Pathology* 12, 328–331.

28

Registration requirements of biological control agents in Germany and in the European Union

Fred A. J. Klingauf

Introduction

The number of plant protection products and active ingredients has been considerably reduced in Germany during the 5 years since the new Plant Protection Act (Gesetz zum Schutz der Kulturpflanzen) of September 1986 came into force. The number of registered plant protection products now amounts to 851 formulations with 216 active ingredients (18 August 1993, Fig. 28.1). The decline is a result of the consistent application of more stringent requirements for product registration.

In the coming years, the number of authorizations is expected to decrease further. Authorizations made before the Plant Protection Act of 1986 have expired and today's requirements for re-registration are tougher. Only about 25% of the plant protection products that were registered in West Germany in 1986, have already become subject to the new Plant Protection Act. In the new Bundesländer (Federal States) in eastern Germany, old registrations are valid until the end of 1995 with the exception of some products banned in Germany and some whose use is not acceptable. It is assumed that about 90% of them do not comply with the requirements for re-registration.

Although the range of plant protection products has decreased in Germany, there was no marked effect on their usage until 1989. In the late 1980s consumption of active ingredients in West Ger-

many amounted to about 36 000 tonnes p.a., but dropped after 1989. About 33 000 tonnes were sprayed in West Germany in 1990. In 1992 33 570 tonnes were sold in united East and West Germany, although the acreage nearly doubled after unification (Table 28.1).

Public perception

The more rigorous rules of authorization and the decline in available plant protection products must be seen in connection with the low public reputation of chemical plant protection. An enquiry by von Alvensleben (1989) in Germany revealed a predominantly negative attitude towards

- chemical plant protection,
- vegetable cultivation without soil,
- artificial fertilizers.

Of the interviewees 81% were of the opinion that chemical plant protection products and fertilizers have more disadvantages than advantages, only 7% held the opposite opinion. In contrast, only 51% expressed a negative attitude towards nuclear power. In other European countries the assessment of chemical plant protection shows the same negative tendency. People are particularly critical in countries where only a small proportion of the population is employed in agriculture.

Over-production of many agricultural products,

283

the rising expenses of the European Union (EU) in agriculture, and the poor public reputation of modern agricultural technologies will also increase the requirements on chemical plant protection products in future.

Authorization of biocontrol agents – general aspects

In the German Plant Protection Act plant protection products are defined as substances, which are, among other purposes, intended to protect plants against organisms or non-parasitic impairment, or to protect plant products against harmful organisms. There is no German legal definition of the term 'substance', but chemicals are obviously classified as substances. The term 'substance' should represent the widest range of substances possible.

Can biocontrol agents be referred to as 'substances'?

- Sometimes naturally occurring substances, be they extracted or synthesized, of microbial, plant or animal origin and exerting pesticidal or behaviour-mediating effects, are regarded as biological agents. They are, however, classified according to German regulations as 'substances' and during authorization procedures are examined in the group of chemicals. The Council Directive that concerns placing plant protection products on the market (91/414/EEC) also defines 'natural pesticides', e.g. rotenone and extracts from the neem tree as chemicals: substances are defined as 'Chemical elements and their compounds, as they occur naturally or by manufacture,

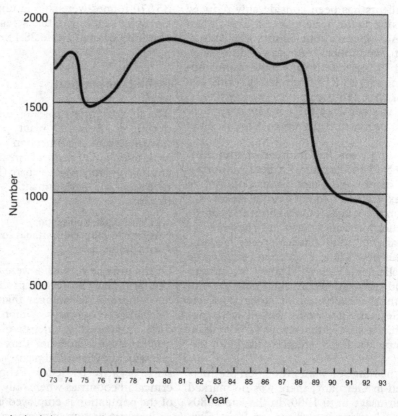

Fig. 28.1   Number of authorized plant protection products in Germany.

Table 28.1. Active ingredients sold in Germany (1987–1990 West Germany; 1991–1992 united Germany)

Year	Active ingredients (tonnes)
1987	36367
1988	36774
1989	34625
1990	33146
1991	36944
1992	33570

including any impurity inevitably resulting from the manufacturing process' (Bode, 1992).

- Micro-organisms including viruses are by definition not chemical compounds. The Council Directive mentioned above defines plant protection products as 'active substances and preparations containing one or more active substances ...' and defines active substances as 'substances or micro-organisms including viruses, having general or specific action against harmful organisms or plants, parts of plants or plant products'. The Council Directive comprises detailed lists of data requirements for the authorization of plant protection products for chemical substances or chemical preparations, and different requirements for micro-organisms and viruses or preparations of them. The German authorization procedure also requires different data for chemicals and micro-organisms including viruses (Bode, 1992).
- There are no generally adhered to regulations for authorization of beneficial insects and other macro-organisms for biological control. In some Member States introduction and release of non-indigenous beneficials is strictly controlled for reasons of nature conservation and is subject to an authorization procedure. Harmonization within the EU is under discussion, and there are efforts to introduce the 'FAO Code of Conduct for the Import and Release of Biological Control Agents' into the EU.

Authorization of plant protection products including preparations of micro-organisms and viruses in Germany

Legal basis

The legal basis of the authorization of plant protection products in Germany seems to be of general interest. First, German requirements are the most rigorous in Europe and, secondly, the EEC Directive on Authorization (91/414 EEC) is similar to German provisions.

The Plant Protection Act of September 1986 (Gesetz zum Schutz der Kulturpflanzen, 1986) provides the legal basis for the examination and authorization of plant protection products in the Federal Republic of Germany. Products include those used for the control of weeds, pests and diseases as well as growth regulators and additives. Biopesticides are also included as are baculoviruses, beneficial fungi and other micro-organisms, but not macro-organisms such as arthropods.

Further important provisions are stipulated in the Regulatory Ordinance on Plant Protection Products and Plant Protection Equipment of 28 July 1987. The Ordinance for Bans of Plant Protection Products (Regulatory Ordinance on the Prohibition of the Use of Plant Protection Products) of 27 July 1988, amended 3 August 1993, completely bans the use of 44 plant protection products. Sixty substances are prohibited from use in drinking water protection areas and medicinal spring areas.

Purpose of the Plant Protection Act

The purpose of the Plant Protection Act is:

1. The protection of plants, particularly crop plants, against harmful organisms and against non-parasitic impairments.
2. The protection of plant products against harmful organisms.
3. The avoidance of dangers which may result from the use of plant protection products or other plant protection measures, in particular where the health of man and animals and the natural balance are concerned.

The balance of nature is defined by the components soil, water, air, species of wild animals and plants as well as the interactions between them.

Requirements for authorization

According to Article 15 of the Plant Protection Act the Federal Biological Research Centre for Agriculture and Forestry (Biologische Bundesanstalt für Land- und Forstwirtschaft, German abbreviation BBA) grants an authorization if the application fulfils the requirements. The examination of the plant protection products must show that:

1. The plant protection product is sufficiently effective in the light of scientific knowledge and technique.
2. The precautions necessary for the protection of human and animal health in dealing with dangerous materials do not require otherwise.
3. The plant protection product, when used for its intended purpose, and in the correct manner, or as a result of such use,
   (a) does not have any harmful effects on human and animal health or on groundwater, and
   (b) does not have any other effects, particularly with regard to the natural balance, which are not justifiable in the light of the present state of scientific knowledge.

What are 'other effects'? The interpretation of this concept has been defined in a sentence from the Senate of the Federal Administrative Court (Bundesverwaltungsgericht) of 10 November 1988: 'Other effects' (according to Article 15(1), point 3(b) of the Plant Protection Act) are all those effects which cannot be excluded with a probability next to security. For the decision as to whether effects of a plant protection product are 'not justifiable' in the light of scientific knowledge, the likelihood of the occurrence of the effects, the significance of the disadvantage of the effects, the possibility of replacing the product and the disadvantage for not using the product are to be mutually balanced. When deciding on 'not justifiable' concerning these other effects the authority has no scope for its assessment. Thus, an authorization can only be granted if other effects, especially those that affect the balance of nature, almost certainly can be ruled out.

Authorization procedure

According to Article 15, the Federal Biological Research Centre decides on the compliance of the requirements as to health in consent with the Federal Institute for Health Protection of Consumers and Veterinary Medicine (Bundesinstitut für gesundheitlichen Verbraucherschutz und Veterinärmedizin) and as to the avoidance of harm through the contamination of water and air as well as wastes of plant protection products, in consent with the Federal Environmental Office (Umweltbundesamt). The consent of both federal offices guarantees a thorough examination of the effects of plant protection products (Fig. 28.2).

The authorization of plant protection products

has been made much more stringent with this new legislation. While efficiency, residues in food, toxicity to human beings and analyses had been given principal attention until the Plant Protection Act came into force, now the aspects that concern the balance of nature are of equal significance. Not only the probable contamination of ground and surface water, but also the effects of plant protection products on the aquatic biocoenosis, on beneficial arthropods and their fate in the air, and many other problems, are of importance.

Many standard guidelines exist on how to conduct the different tests. Thus, the tests have to be carried out in a standard manner to generate results that can be compared. Criteria for reaching decisions have been formulated for every sector and flow charts have been drawn up for examination and evaluation. They comprise trigger values necessary to request further analyses and in individual cases threshold values for decisions on the authorization (Biologische Bundesanstalt für Land- und Forstwirtschaft, 1993).

The application is presented to an expert advisory committee, which has 25 members appointed by the Federal Ministry of Food, Agriculture and Forestry. It consists of competent experts of the Plant Protection Service of the Federal States (Länder), universities and other research institutes. The Expert Committee acts as a consultant. The BBA decides on the clearance of a product after it has consulted this committee.

When clearance is granted (by certificate), the BBA may stipulate directions on how to apply the pesticide, which have to be printed on the label, and which also indicates the imposition of a fine in the case of violation. Directions will be stipulated e.g. for a substance that has the tendency to leach, and if there is the danger that the pesticide or its metabolites may contaminate the groundwater.

Harmonization of regulations in the EU

EEC directives in different fields of crop protection

Directives issued by the EU are of great importance for crop protection in Europe (Petzold, 1991). The field of plant inspection has been largely harmonized within the EU (Directive 77/93 EEC of 31 January 1977). In view of the single market, adjustments will have to be made as

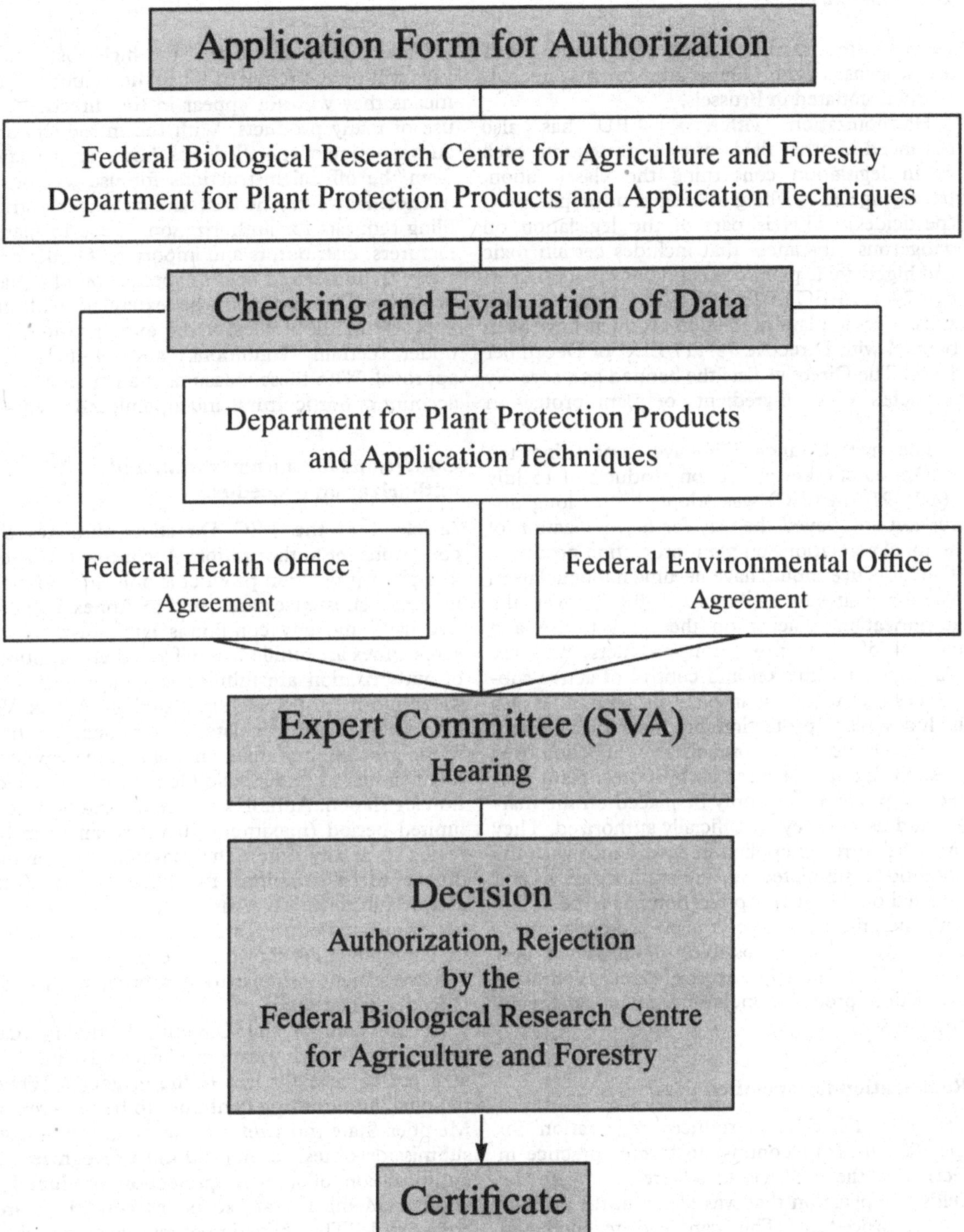

Fig. 28.2 Course of the authorization procedure for plant protection products in Germany. *FIHPCVM = Federal Institute for Health Protection of Consumers and Veterinary Medicine.

checks at the internal borders of the community are dispensed with. These adjustments are currently negotiated in Brussels.

Harmonization within the EU has also advanced in other fields of crop protection, such as in legislation concerning the classification, packaging and labelling of dangerous preparations (pesticides). This is part of the legislation on dangerous substances that includes certain toxic and highly toxic plant protection products (Directive 78/631/EEC of 29 July 1978). Certain active substances in plant protection products have been banned with Directive 79/117/EEC of December 1978. The Directive lists the banned and severely restricted active ingredients of plant protection products.

The new Council Directive concerning the marketing of plant protection products of 15 July 1991 (91/414/EEC) was adopted after long discussion and paved the way for harmonization of national regulations on plant protection products. The Directive should have become national law in Member States by 19 August, 1993. It covers the authorization, placing on the market, use and control of plant protection products, and the placing on the market and control of active substances meant for use in plant protection. It also includes plant protection products that contain genetically engineered organisms or are composed of such. Except for research purposes, plant protection products may only be placed on the market and used if they are officially authorized. They must be correctly applied in accordance with the conditions stipulated in the authorization and detailed on the label. Application has to be in line with the principles of good plant protection practice and, whenever possible, of integrated pest control. The Council Directive concerns all plant protection products including micro-organisms and viruses.

Registration for specified uses

The new Directive introduces registration for specified uses. In contrast to former practice in Germany, the user has to adhere strictly to the field of application that was tested in the authorization procedure. This can lead to increased problems for use in minor crops as producers of plant protection products will concentrate requests for authorization on potentially profitable i.e. major uses. Because of the high cost, minor uses will not be reported for authorization, which means they will not appear in the directions for use of many products. With the introduction of authorization of specified uses, however, deviation from the official instructions for use will be an offence. However, the Directive no longer restricts filing requests for authorization of use to manufacturers, distributors and importers. On the contrary, an authorized field of application of a plant protection product may be extended with the approval of the holder of the authorization, but under certain conditions also without his approval. With these measures the Directive takes account of horticultural and agricultural needs.

Conditions for authorization and authorization procedure

Article 4 of the EEC Directive stipulates the conditions for authorization. According to Article 4, a plant protection product is only authorized if its active substances are listed in Annex I ('positive list') and any conditions laid down in the same annex are fulfilled, and if all other conditions of authorization are fulfilled in compliance with the uniform principles stipulated in Annex VI. The authorization stipulates requirements relating to the protection of man, animal and the environment similar as under valid German plant protection legislation. Authorization is also granted for a limited period (maximum 10 years) and can be reviewed at any time. The requirements for the dossier to be submitted for the inclusion of an active substance in Annex I, part B microorganisms and viruses, are comprehensively listed in Table 28.2. As shown in the list, an extensive risk assessment – similar to that for chemicals – is considered necessary.

At the core of the Directive is the mutual recognition of authorizations (Articles 10 and 11) as a prerequisite for free traffic of goods. While national authorization continues to be required, a Member State must refrain from requesting new submission of test results and must recognize the authorization of a plant protection product by another Member State so far as conditions are comparable. The obligation of mutual recognition of authorizations has been restricted to those plant protection products the active substances of which are listed in Annex I ('positive list'). The inclusion

Table 28.2. Important requirements for the dossier to be submitted for the inclusion of an active substance in Annex I, part B, micro-organisms and viruses, Council Directive of 15 July 1991 (91/414/EEC)

Identity of the organism

Biological properties of the organism

Further information on the organism: function, effects on harmful organism, field of use envisaged etc.

Analytical methods

Toxicological, pathogenicity and infectivity studies including oral dose, percutaneous dose, inhalation dose, intraperitoneal dose, skin and eye irritation, skin sensitization, short- and long-term toxicity (90 days exposure), carcinogenicity, mutagenicity, teratogenicity, metabolic studies, neurotoxicity, immunotoxicity, pathogenicity, toxic effects on livestock and pets, medical data

Residues in or on treated products, food and feed

Fate and behaviour in the environment

Ecotoxicological studies
  Birds
  Fish
  Daphnia magna
  Algal growth
  Important parasites and predators of target species
  Honey-bees
  Earthworms
  Other non-target organisms to be at risk
  Contamination on adjacent non-target crops, wild plants,
    soil and water
  Effects on other flora and fauna

In cases where toxins are produced, further data are required as outlined under Part A, chemical substances

Table 28.3. Commercially available micro-organisms and viruses in Germany (authorization necessary)

Bacillus thuringiensis
  var. *kurstaki*
  var. *tenebrionis*

Metarhizium anisopliae

Codling moth granulosis virus (CpGV)

of active substances in the positive list, or their rejection, is decided on a majority basis by the Standing Committee on Plant Health.

A total of 700 active substances are currently used in plant protection products in the EU. The Commission has worked out a programme to reevaluate these active substances as soon as possible. With participation by all Member States, 90 active substances are to be evaluated per year, 12 of them in Germany. If the programme is carried out on time, all old active substances would be reevaluated within about ten years.

Consequences of stringent authorization requirements

With the EU Directive largely corresponding to current German regulations, authorization of plant protection products will probably be tightened throughout the EU. The consequences could be derived from the development in Germany since the new Plant Protection Act came into force in 1986. As the requirements for authorization have increased the number of plant protection products registered decreased dramatically.

The resulting tendency is that:

- A smaller number of pesticides is being applied more frequently on the same area
- Danger of the build-up of resistance or reduced product efficacy increased
- Risk of ground and drinking water contamination may become higher
- Adequate strategies for minor uses are increasingly unavailable, which may hinder crop rotation with consequences for the environment
- The industry stops or at least reduces research for modern products especially for selective pesticides which have, of course, only a small market and cannot cover the costs of research and development.

It is difficult to say if biological plant protection products, having a selective effect and an accordingly smaller market, have increasingly less chance of being developed by industry. In fact, the number of biological agents did not increase considerably during a promising stage of development approximately 10 years ago (Table 28.3). Also natural substances, e.g. neem-extracts, did not make real breakthrough on the European market despite the substantial efforts of scientific investigation.

Conclusions

In principle the European Community demands high safety standards before permission is granted to use a biological control agent. As increased safety requirements are requested, they will have immediate consequences on agriculture, the environment and the consumer. The ideal biological control agent, which is economically feasible, has a selective effect and which after doing its task merely fades into the environment, does not yet exist.

The avoidance of danger in the domain of plant protection products must be considered successfully. Now it is necessary to improve the quality of the testing standards step-by-step, according to up-to-date scientific knowledge, and develop them in such a way that also agents, such as those based on new biological investigations, would have a chance. Up to 20 species of beneficial predators and parasites are commercialized in Germany and other European countries, mostly for the control of pest insects and mites in protected crops. There are only a few microbial pesticides available. Despite the numerous reports on the successful experimental use of microorganisms, the progress in practical utilization is very slow; partly due to the extensive authorization requirements. Since micro-organisms are effective also against some plant diseases and even weeds, which represent the greatest number of damaging agents in temperate climates, the renunciation of those biological agents cuts back many chances of biocontrol.

I consider the development of regulations towards the Good Agricultural Practice (GAP) in plant protection as an important task for the future. This allows the description of the application conditions of all plant protection measures –

chemical pesticides and alternatives. Hence, this provides a better estimation of real risks than is currently possible by isolated examination of pesticides within an authorization procedure more or less independent of its position in integrated control. The more practical view, together with post authorization monitoring under field conditions, gives biological agents a better chance.

References

Alvensleben, von, R. (1989). Die Beurteilung moderner Agrartechnologien durch den Verbraucher. *Agrarwissenschaftliche Fakultät, Universität Kiel*. Heft 71.
Biologische Bundesanstalt für Land- und Forstwirtschaft (ed.) (1993). Criteria for Assessment of Plant Protection Products in the Registration Procedure. *Mitteilungen aus der Biologischen Bundesanstalt für Land- und Forstwirtschaft, Berlin-Dahlem*, 285. Paul Parey Verlag, Berlin/Hamburg.
Bode, E. (1992). Naturally occurring substances of plant origin used as plant protection products: authorization requirements. In *Practice Oriented Results on Use and Production of Neem Ingredients* (ed. H. Kleeberg), pp. 123–127. Trifolio-M GmbH.
Council Directive of 15 July, 1991, Concerning the Placing of Plant Protection Products on the Market (91/414/EEC). *Official Journal of the European Communities* No. L 230.
Gesetz zum Schutz der Kulturpflanzen (Pflanzenschutzgesetz – PflschG) – Plant Protection Act, BGBl. I, 1505 ff., 15 September 1986.
Petzold, R. (1991). EG-Harmonisierung der Zulassung von Pflanzenschutzmitteln. *Gesunde Pflanzen* 43, 350–354.

Index